Geophysical References, Volume 2
T. Norman Crook, Editor

A Practical Introduction to Borehole Geophysics

An Overview of Wireline Well Logging Principles for Geophysicists

J. Labo

Edited by Samuel H. Mentemeier and Charles A. Cleneay

Society of Exploration Geophysicists
POST OFFICE BOX 702740 / TULSA, OKLAHOMA 74170-2740

ISBN 0-931830-47-8 (Series 2)
ISBN 0-931830-39-7 (Volume 2)

Library of Congress Catalog Number: 8760425

Published 1987

Printed in the United States of America

Table of Contents

INTRODUCTION

Geophysicists are intensive users of well log information. The log analyst is interested primarily in restricted parts of the log — the potentially productive formations. Only a part of the remainder is used for log quality checks and parameter selection. Geologists are interested primarily in well-to-well correlations for structural and stratigraphic details for subsurface work. For the geophysicist the entire log, including productive and nonproductive intervals, is critical in the tie between geophysical data and "hard" well information. And yet geophysicists often need more information concerning the use of logs and their pitfalls.

Borehole geophysics has an aura of magic if a wireline logging operation and a drilling rig are not familiar. An explanation of how logs are obtained, an area not usually covered in well log interpretation courses, eliminates some of this mystery. The introduction to borehole geophysics presented here emphasizes hardware, operational aspects, key geophysical measurements along with their pitfalls, and an overview of well log interpretation principles. This introduction gives an explanation of what is seen at the wellsite while the interpretation chapters aid in understanding how logs are used for formation evaluation, their most immediate purpose. This overview will help in understanding how each piece of a logging course fits together. By understanding well-logging principles, an explorationist will have a better knowledge of geophysical well logging than is provided by an interpretation course alone and will develop a better background from which to make log quality judgements.

A comment is needed on the units of measurements presented in this book. The units in the log examples shown do not conform to SI metric standards adopted by the Society of Exploration Geophysicists. Figure 8-29 (on page 170) was taken from a European publication and shows the common mixing of metric and English units, as depth is measured in meters and borehole diameter (caliper) is measured in inches. Most acoustic logs are still recorded in $\mu s/ft$. The prime exception is that modern Canadian logs are completely metric. Since this is a practical text the examples are presented as they really occur. Most charts presented in this book, however, contain metric references. Appendix F contains conversion of well logging units to SI metric units.

Chapter 1

BOREHOLE GEOPHYSICS TECHNIQUES AND APPLICATION

Borehole geophysics is the primary technique used in initial well evaluation. When management must decide whether to complete or abandon a well, several types of information are available: drilling measurements, hydrocarbon detecting mud log, sample cutting log, drill stem tests, and wireline geophysical logs. Mud logs and sample work indicate the presence of hydrocarbons within a borehole interval, but do not determine whether the show is potentially commercial. Drill stem tests can indicate formation producibility but, if an interval is incorrectly chosen, the data can be misleading.

Geophysical well logs economically provide the best detail on relative formation changes. Logs can accurately define the intervals of potential hydrocarbon bearing formations that need further testing and can indicate their commercial nature. Logs also point out prospective intervals not originally considered in picking the prospect location. Many wells have been converted from "dusters" to "keepers" by careful examination of the *entire* logged section.

Ideally a single recorded measurement would indicate the presence of hydrocarbons. If this measurement were scaled in barrels per acre-foot, evaluating a well would be simple. Unfortunately, a single direct measurement for hydrocarbons has not yet been developed. An indirect approach used relates measurements that can be obtained to a hydrocarbon indicator.

Borehole geophysics began with Conrad Schlumberger's idea of using surface measurement techniques to detect mineral bodies. Because of the metallic nature of many sought after minerals, electrical methods were employed. Early work used electric current and mapped the resultant equipotential surfaces (Figure 1-1 a), the concept being that metallic, and thus conductive, bodies would disturb the equipotential surface distribution as compared with the smooth patterns over a homogeneous subsurface. Schlumberger developed techniques and equipment to make accurate, precise measurements that demonstrated the practicality of the method. However, interpretation of the equipotential methods results was difficult — especially if it was necessary to move the current electrodes to cover

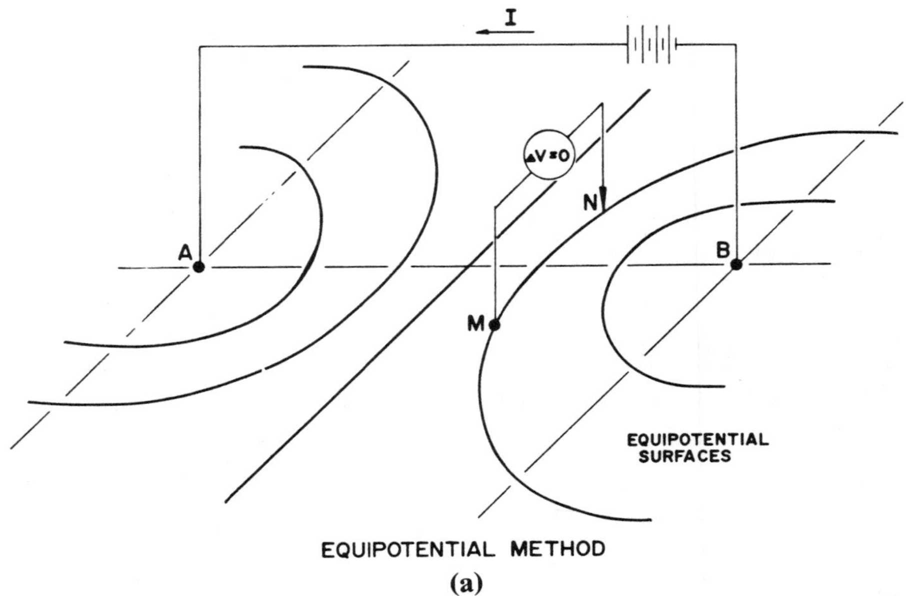

EQUIPOTENTIAL METHOD

(a)

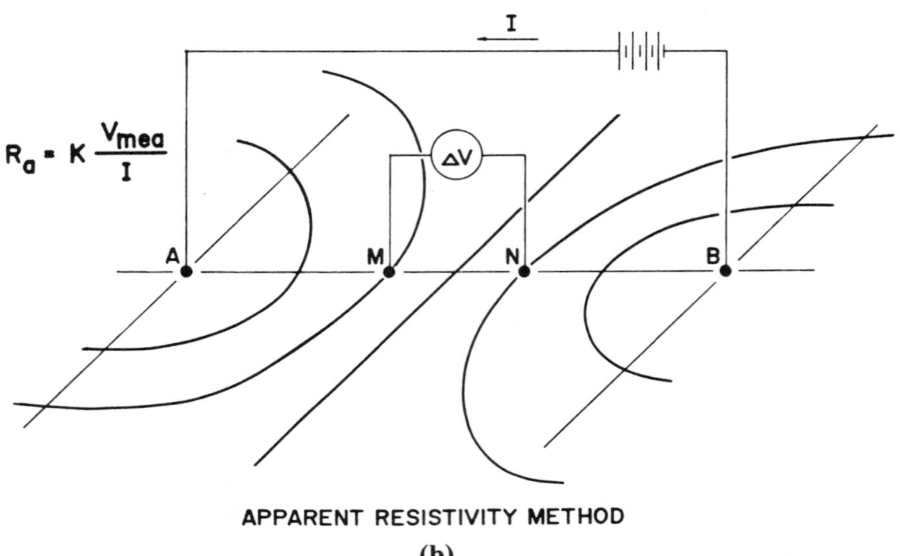

APPARENT RESISTIVITY METHOD

(b)

FIG. 1-1. Surface resistivity measurement system where a resistivity profile is made by moving the array along the surface and making readings at each point. A resistivity sounding is made by expanding the electrode array spacing and taking readings at each stage.

a larger area. The method was dramatically refined with the concept of apparent resistivity measurement (Figure 1-1 b). Here was a measurement that could be linked to subsurface geology and made interpretation of the results easier. The concept of apparent resistivity also made movement of the measurement system to cover any size area immaterial. Finally younger brother Marcel Schlumberger joined Conrad in this work.

Using the apparent resistivity method, a resistivity sounding can be made deeper into the earth by expanding the electrode spacing around a point. Alternately, a lateral resistivity profile can be made by keeping the electrode spacing constant and moving the device, as a unit, along a line. In 1927 this later approach was applied to the borehole environment at Pechelbronn, France (Figure 1-2). ''Electrical coring'' was a station-by-station plot of a single formation resistivity measurement (Figure 1-3) which was used to obtain some indication of well-to-well structure by correlating the resistivity curve with curves from nearby wells (Figure 1-4). This is still one major use of well logs. As other types of measurements were developed, logs were used to evaluate individual intervals, first qualitatively and much later quantitatively. Well logging has now grown to a bewildering set of measurements used in seemingly infinite combinations.

Borehole measurements are obtained by lowering one or more tools into a wellbore and monitoring a variety of parameters at a surface located

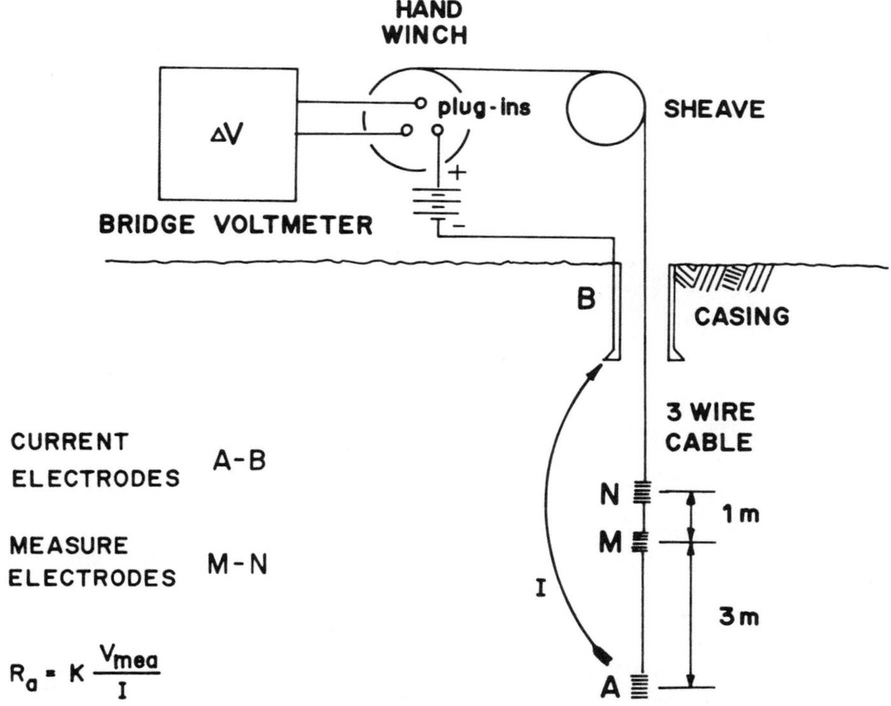

Fig. 1-2. An inverted AMN lateral configuration of the 1927 Pechelbronn logging setup. The surface equipment was unplugged each time the cable was moved to a new location in the borehole.

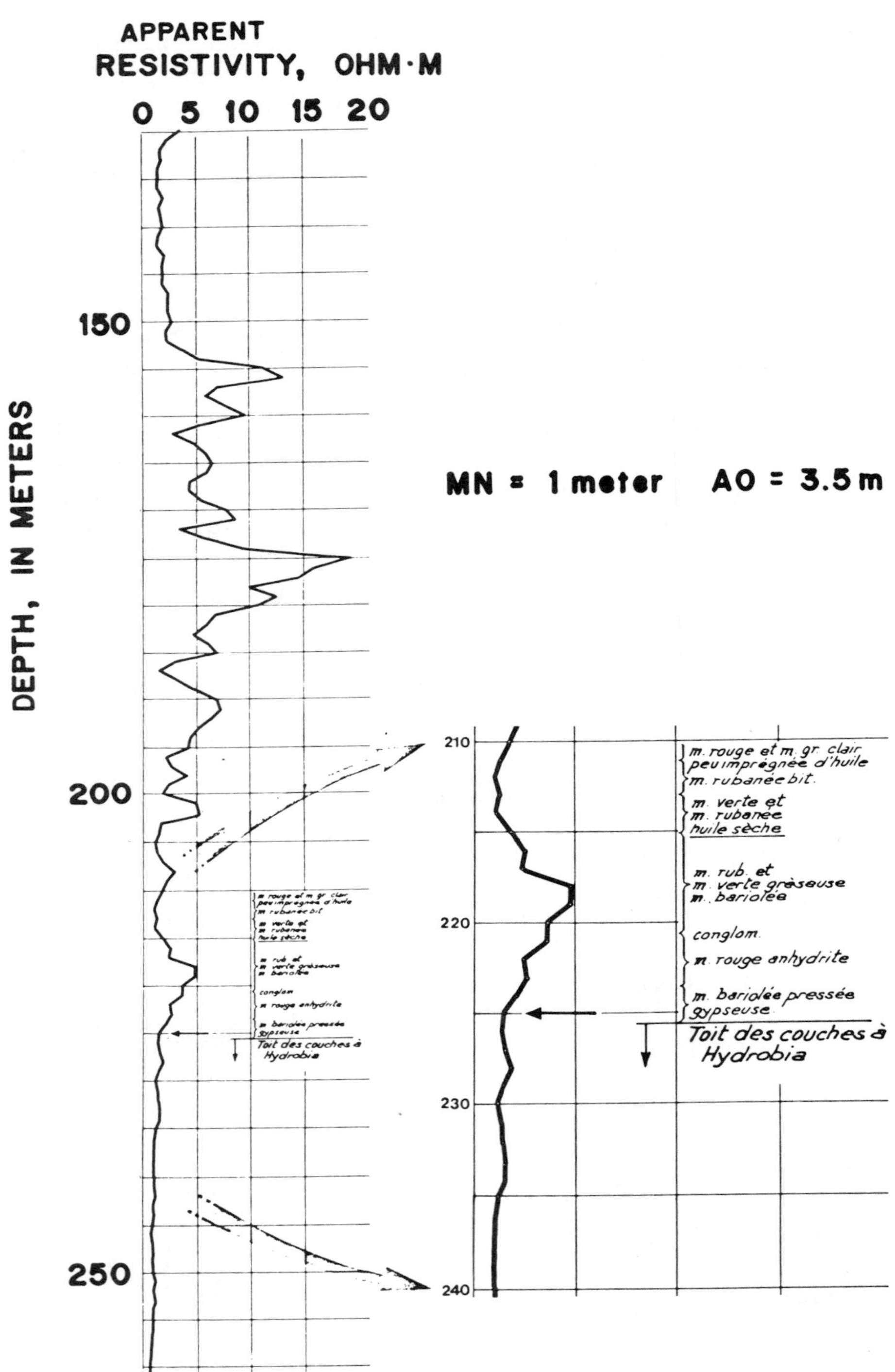

FIG. 1-3. The first subsurface electrical log (September 1927) showing a station-by-station hand plot of a single resistivity spacing measurement. The character changes of the log allowed for well-to-well depth correlations with later logged wells and yielded the subsurface structure of the field.

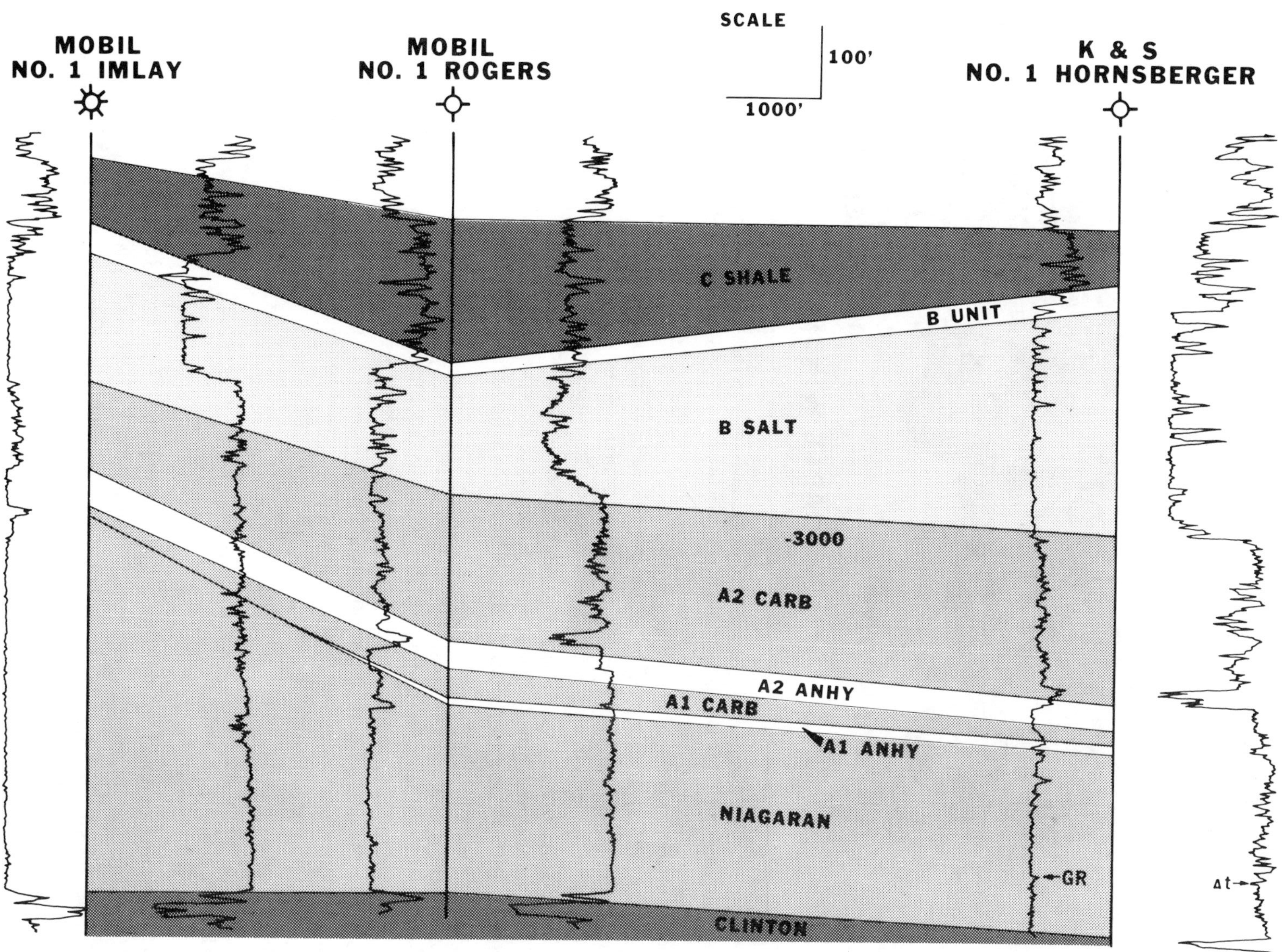

FIG. 1-4. Determining well-to-well structure using well logs.

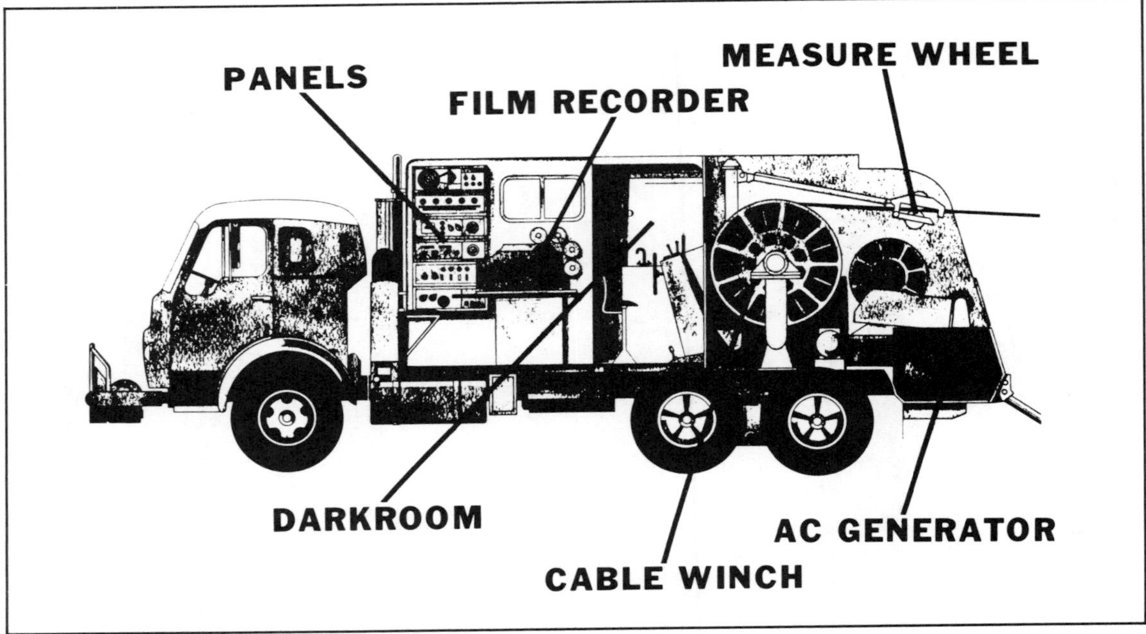

FIG. 1-5. Cut-away diagram of a circa 1970 Schlumberger logging truck containing an analog recording system. This combination truck was used for open-hole and completion services.

truck (Figure 1-5) or at a skid mounted recording unit. Because all log measurements are apparent readings, the recorded values often need correction before use in computations. Some borehole and adjacent formation effects are automatically accounted for, or the tool used was designed to minimize their effects. Service company interpretation chart books are full of corrections which cannot be applied, however, until after the entire log is available.

Open-hole log measurements can be divided into three types: resistivity, porosity, and permeability or correlation. Porosity is obtained through acoustic and radiation (both gamma-ray and neutron sources) measurements. The natural gamma ray, spontaneous potential (SP), and caliper measurements aid in determining permeable intervals and in checking log-to-log depth control.

THE MICHIGAN BASIN

Many examples used here are from the Michigan Basin where the prime drilling objective for 16 years has been Niagaran reefs. These reefs were formed in a ring about a shallow Silurian aged sea (Figure 1-6), and subsequently were buried, now draped, under a carbonate-evaporite sequence (Figure 1-7).

Productive Niagaran reefs were drilled as early as the nineteenth century in Southwestern Ontario in the extensive barrier reef complex. Drilling

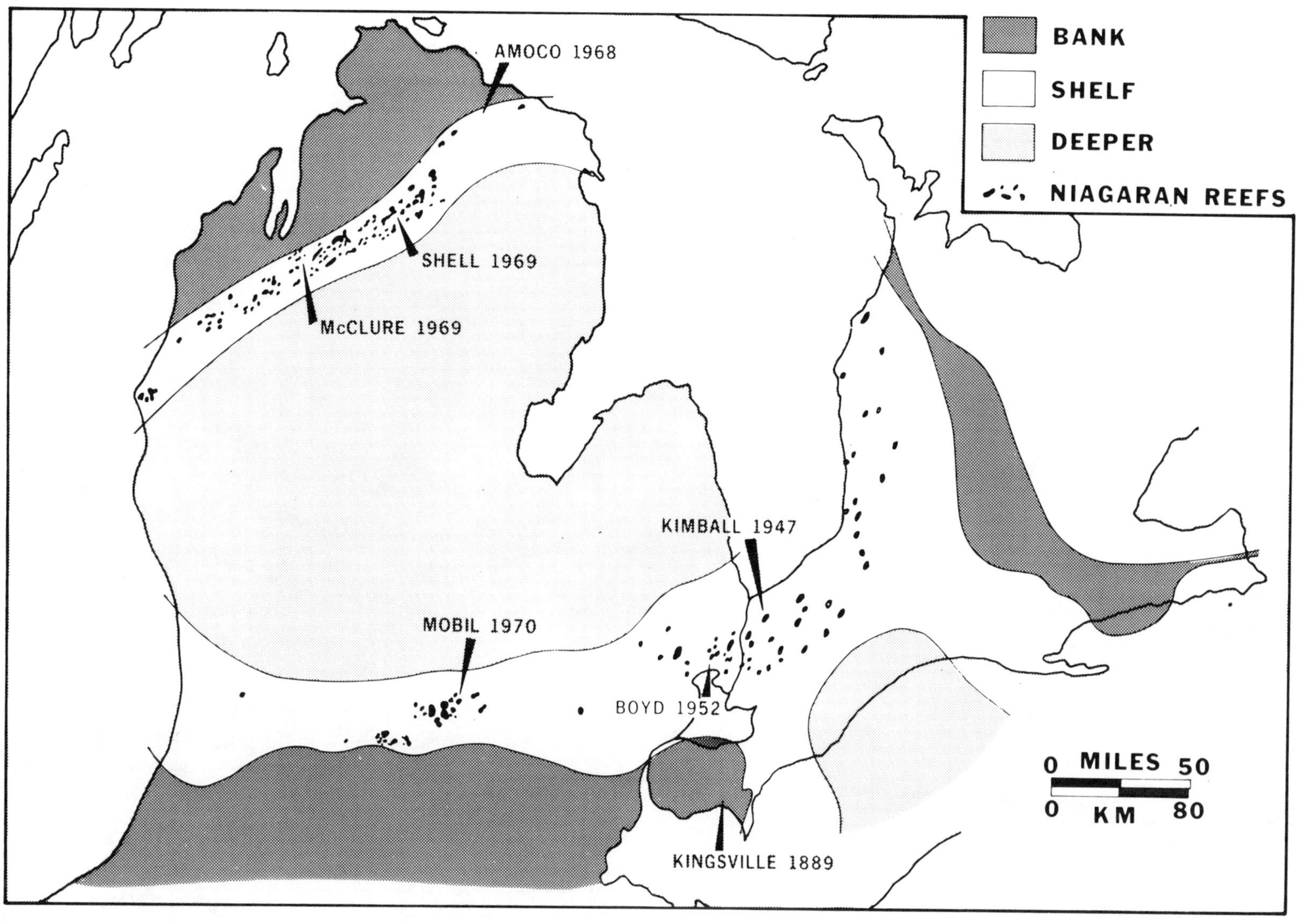

FIG. 1-6. Subsurface geologically deduced reef depositional shelf trend (Mantek, 1973).

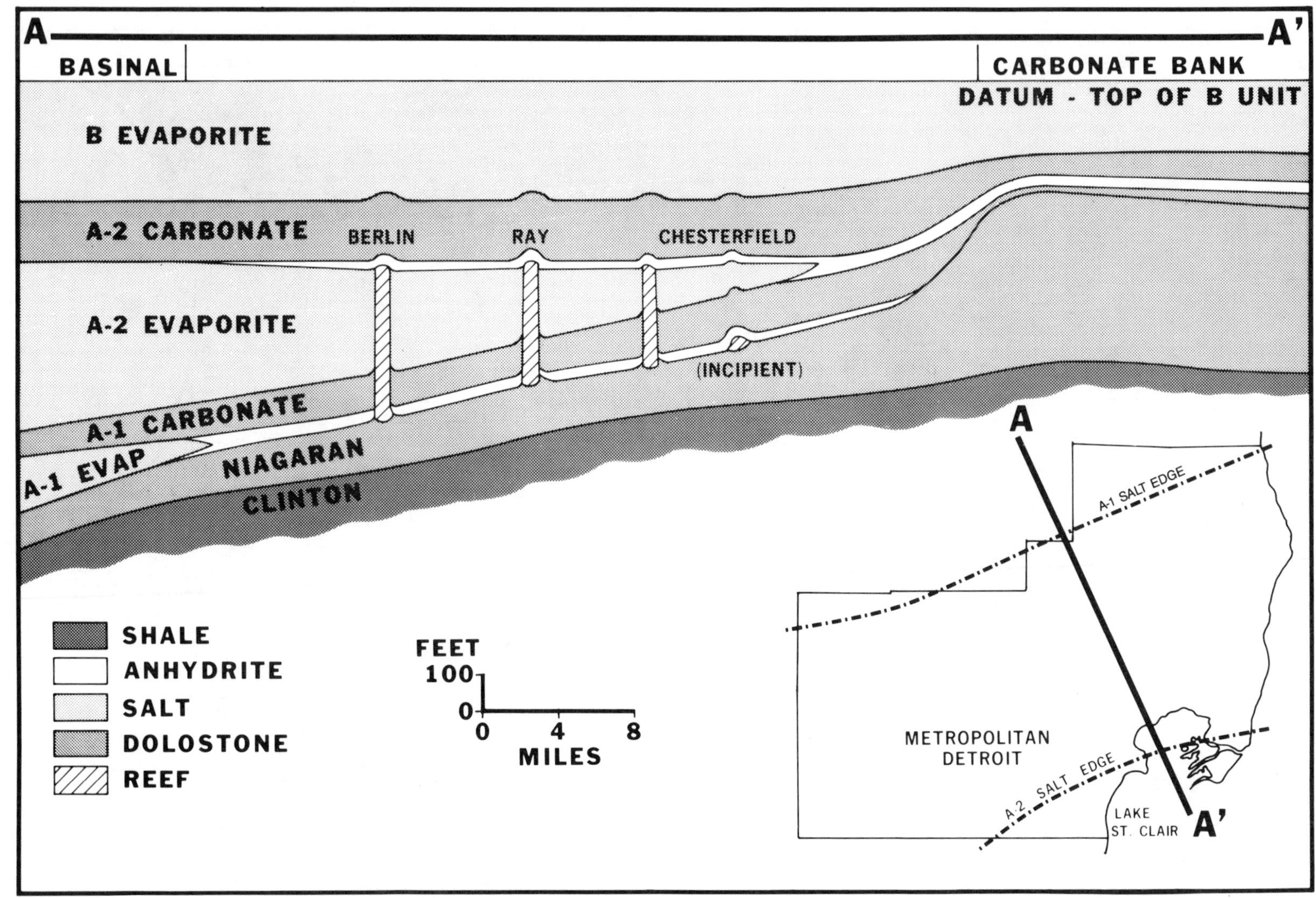

FIG. 1-7. Niagaran shelf cross-section in the southeast of the Michigan basin (Mantek, 1973).

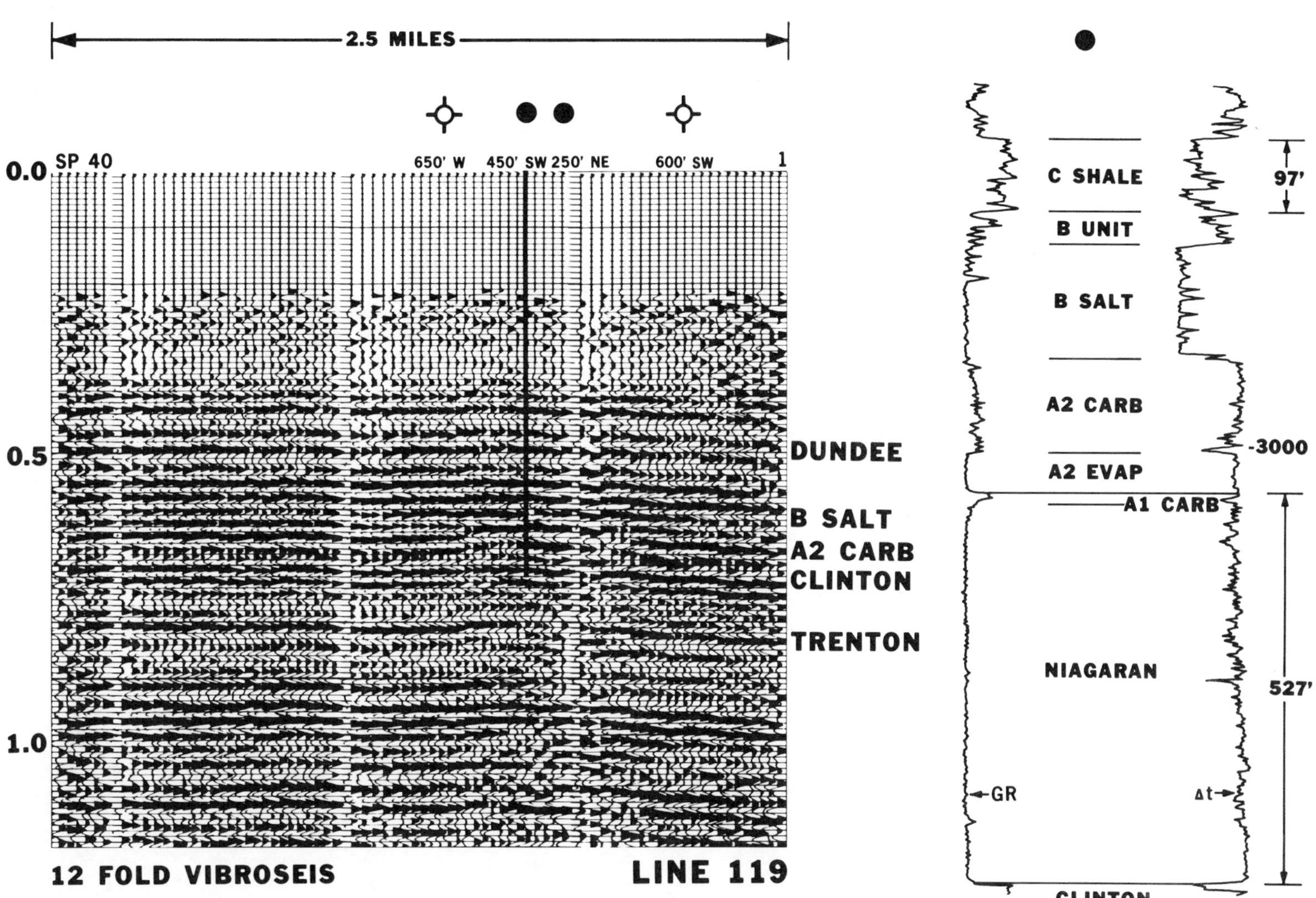

FIG. 1-8. Seismic section over a Niagaran reef showing the reef expressed in the thickening between the A2 Carbonate and Clinton reflectors.

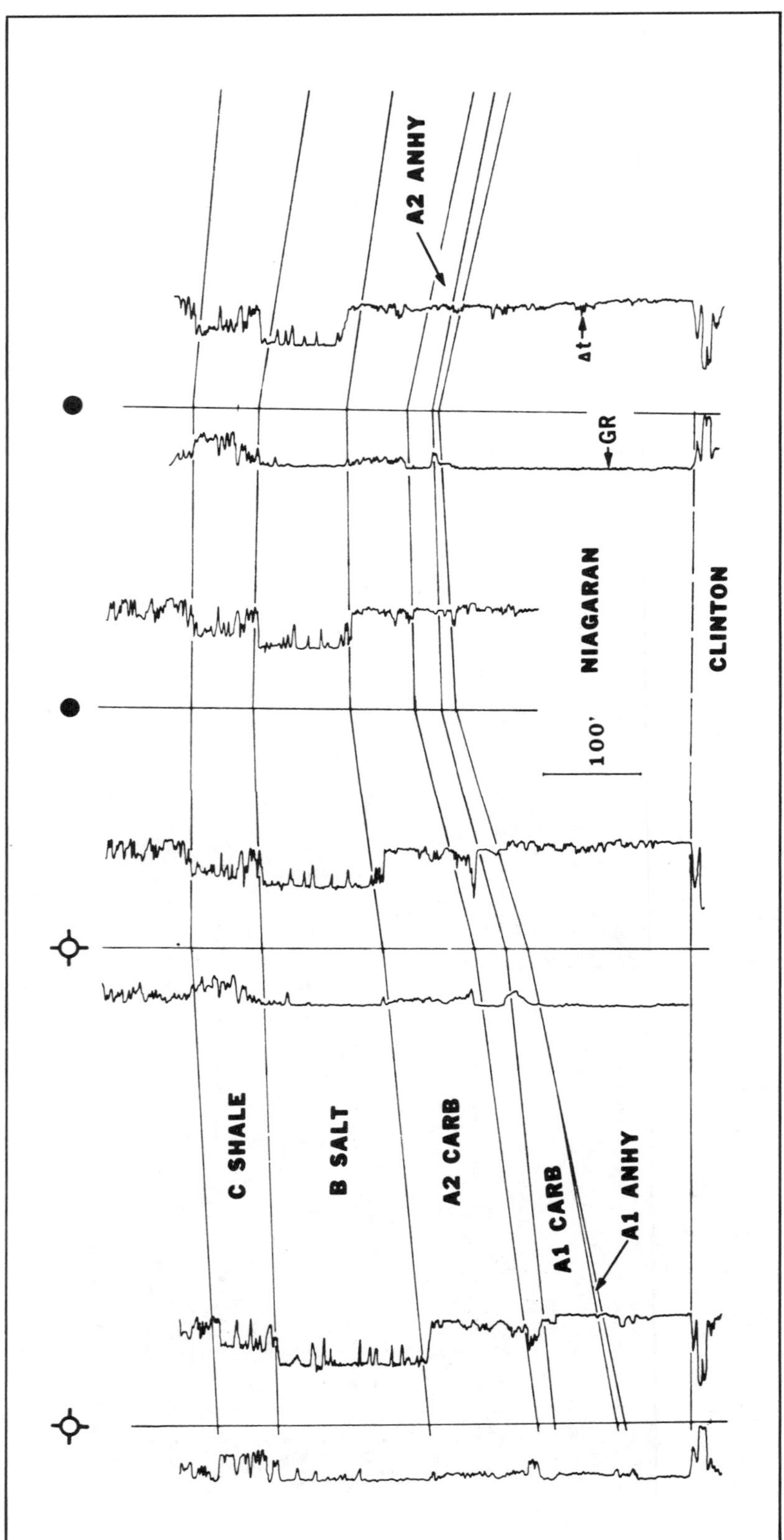

FIG. 1-9. Log derived cross-section showing Niagaran reef geometry.

for smaller, basinward patch reefs began in the 1930s with moderate success, but an accurate surface method for locating these productive features was needed.

In the late 1940s the surface gravimeter was successfully applied to the problem because it could detect the density contrast between the dense carbonate reef and the less dense surrounding salts. Using this method the play was expanded in the 1950s to southeastern Michigan. However, the highly variable, near-surface glacial drift produced the same reef-like gravity expression and prevented expansion of the play west of Detroit. Single-fold seismic was unsuccessful because of the variable glacial drift. Geophysical technology would have to make another quantum jump before the play could extend into the known geologic trend of the reefs.

In 1968 Pan American drilled the No. 1 Draysey in northern lower Michigan based on interpretation of common depth point (CDP) seismic. The No. 1 Draysey well penetrated a hydrocarbon bearing, back-barrier reef. While the well was not a commercial success, it did prove the validity of CDP seismic and began a play that continues with great momentum 16 years later. This discovery was followed in 1969 by the commercial success of Shell and the Michigan independent McClure Oil Company in the northern trend, and in 1970 by Mobil Oil Corporation in the southern reef trend. They all used CDP seismic (Figure 1-8) to determine drilling locations.

The Silurian Niagaran reefs, in the south central Michigan trend, range from 20 to 200 acres. The regional, off-reef Niagaran is 140 ft to 200 ft thick (Figure 1-9), with a 10 ft thick A1 evaporite between the Niagaran and A1 carbonate. At the reef crest the A1 evaporite is missing and the A1 carbonate and Niagaran thickness is typically 500+ ft. The reef flanks are relatively steep sided with reef draping exhibited in the A1 carbonate, A2 carbonate, and B unit. Infilling of the interreefal areas was completed by C shale time. Over the reef the B salt is thin and off-reef it is thick.

REFERENCE

Mantek, W. 1973, Niagaran Pinnacle Reefs in Michigan; 12th Ann. Conf. of Ontario Petro. Inst.

Chapter 2

LOGGING SERVICE COMPANIES

Because it is not economical for each energy company to conduct their own wireline logging, as the equipment would be idle most of the time, wireline service companies provide these services. The major open-hole service companies are Birdwell, Dresser-Atlas, Gearhart (formerly GO Wireline), Schlumberger, and Welex. Although Schlumberger serves a major share of the market (Figure 2-1), an understanding of the techniques and differences of all the major service companies is necessary. Each company uses slightly different terms and trade names for similar logging services (Table 2-1). The finer details can, however, become very significant during log analysis or log editing for geophysical uses.

All five major companies also handle cased-hole logging and completion services. Because it is easier to enter the cased-hole market, many medium and smaller companies, including one and two truck operations, compete for this market. NL McCullough is the largest predominately cased-hole service company (Figure 2-1).

BIRDWELL

When James Bird returned to Bradford, Pennsylvania in 1946 from the U.S. Navy, to start James M. Bird Surveys, his purpose was to conduct electrical surveys in the water flood projects of the old Pennsylvania oil fields. Single-point resistivity and normal resistivity measurements were the prime services offered.

In 1949 Bird Surveys was the first company to introduce radioactive tracers to fluids in the borehole. In 1951 they expanded electrical measurement services by introducing their multicurve electrical survey. Bird Surveys

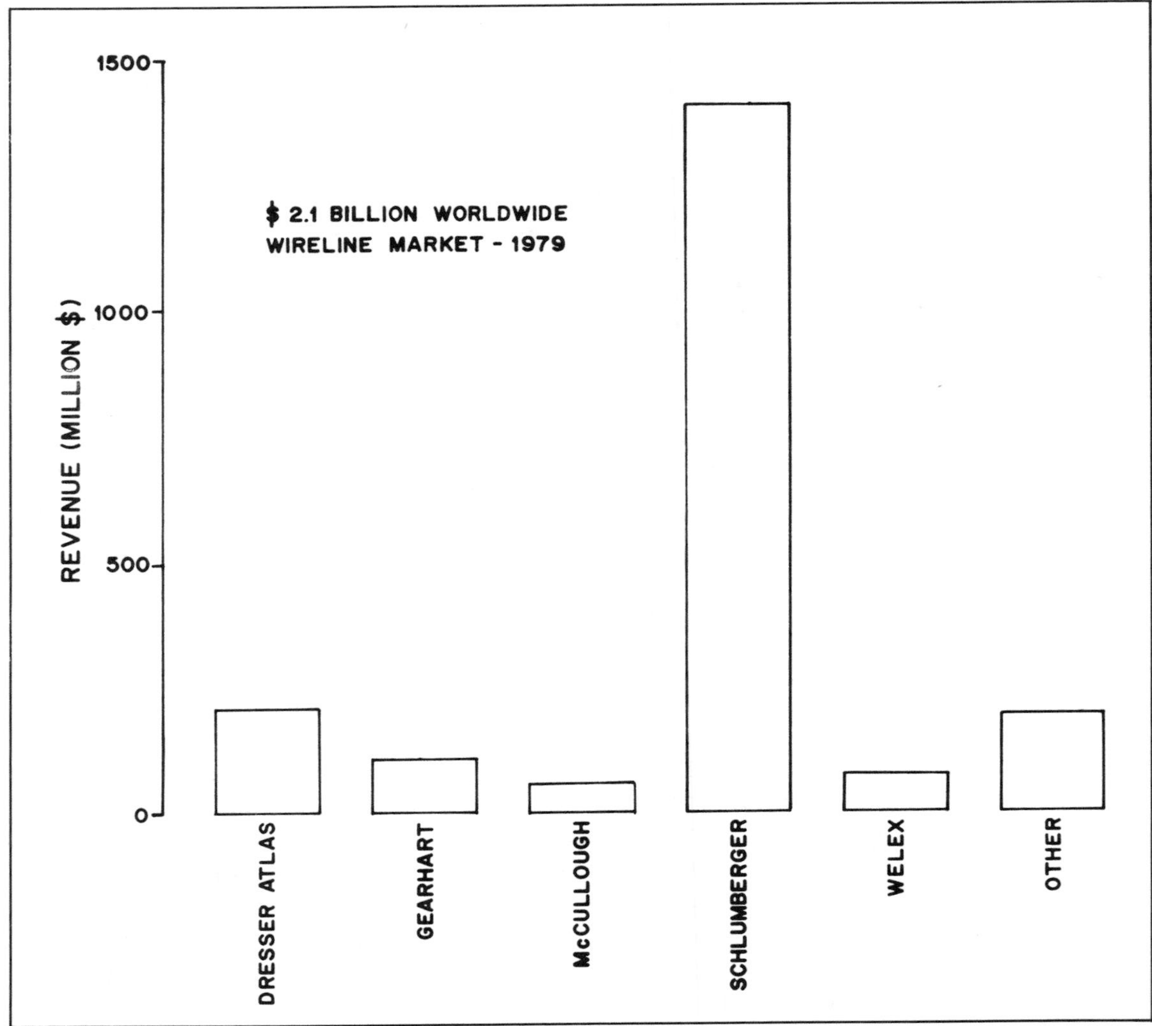

FIG. 2-1. The wireline service company market competition. Source: Howard, Weil, Labouisse, Friedrichs.

began cased-hole jet perforating in 1954, with the introduction of the gamma-perforator tool. This tool allowed detecting and correlating the formation behind the casing with the casing collars, and then setting perforating depths with reference to the casing collars — all in a single trip into the cased borehole.

In 1956 Birdwell, Inc. was formed with the merging of James M. Bird Surveys, Birdwell of Illinois, and Birdwell Surveys of Oklahoma. From headquarters established in Bradford, Pa., they served Kansas, Oklahoma, Illinois, Michigan, Ohio, West Virginia, New York, and Pennsylvania. In 1959 Seismograph Service Corporation (SSC) bought Birdwell and moved the headquarters to SSC's Tulsa offices.

Seismograph Service Corporation began well logging research in 1937 and, along with an industry research group that would become Well

Surveys, Inc., developed a method to measure gamma rays in a borehole. In 1939 SSC developed the Bairoid Mud Log and in 1953 they made available the first commercial continuous velocity log, under a patent license from Magnolia Petroleum. This device provided detailed correlation of the well-to-surface seismic data. In 1959 existing SSC wireline operations were merged with the newly acquired Birdwell Division of SSC.

In 1960 the three-dimensional (3-D) velocity log was introduced. This was the first commercial service to display the acoustic wave train on a reproducible film (Figure 2-2). The Birdwell Division introduced the first dry borehole geophone in 1962, which allowed recording of velocity surveys in air- or gas-drilled boreholes. Birdwell then began vertical seismic profiling (VSP) on a commercial scale in 1977 in the United States market. In December 1982 borehole seismic operations were transferred to the Velocity and Special Projects Division of SSC and in 1984 the Birdwell Division was sold to Dresser Industries.

DRESSER ATLAS

Dresser Atlas Division of Dresser Industries was formed by the merger of Elgin Corporation, Lane-Wells Company, Pan Geo Atlas Corporation, and later Birdwell; to become the second largest wireline service company.

The Lane-Wells Company, started in California in 1932 by Walt Wells and Bill Lane, patented and began commercial gun perforating well completion services. They were restricted to completion services, because Schlumberger held the patent rights to electric logging in the United States. In 1937 a patent suit was settled which granted Lane-Wells a license to offer electric logging and Schlumberger a license to offer perforating services.

Lane-Wells began marketing electric log surveys in 1939. The company expanded into international operations in 1946 with the acquisition of Petroleum Technical Services in Venezuela. In 1954 they acquired Well Survey, Inc. of Tulsa which had marketed the first commercial gamma-gamma density tool in 1950. The tool had been developed in conjunction with McCullough and California Research Company.

In 1958 Lane-Wells, purchased by Dresser Industries, became the Lane-Wells Division, and actively began to compete for the open-hole logging market. Elgin Corporation was acquired and merged into the new division, Lane Wells was moved from Los Angeles to Houston in 1958, and Well Surveys (which had become the research group of Lane-Wells) was moved from Tulsa to Houston in 1960. With acquisition of Pan Geo Atlas Corporation in 1968 the Dresser Atlas Division was established.

Elgin Corporation started in Dallas in 1953 as an open-hole logging company with two logging trucks. By 1958 when they were acquired by the Lane-Wells Division their operations had expanded to support thirty-five logging trucks.

Pan Geo Atlas Corporation was begun in 1946 as Atlas Research Corporation by former Schlumberger engineers Paul Charrin and Jacques

Table 2-1. A comparison of terminology for open-hole logging tools. See Appendix A for logging tool abbreviations.

Logging tools	Birdwell	Dresser Atlas	Gearhart	Lane-Wells
Resistivity	Electric Log	Electrolog	Electric Log	Electrolog
Focused conductivity	Induction Electric Log	Induction Electrolog	Induction Electric Log	Induction Electrolog
Dual focused conductivity	Dual Induction Focused Log	Dual Induction Focused Log	Dual Induction Laterolog	
Focused resistivity	Guard Log	Laterolog	Guard Log	Focused Log
Dual focused resistivity		Dual Laterolog	Dual Laterolog	
Micro-resistivity	Micro-Contact Caliper	Minilog	Micro-Electrical Log	Minilog
Micro-focused resistivity		Microlaterolog Proximity Log	Microlaterolog	Minifocused Log
Dielectric		Dielectric Log	Dielectric Constant Log	
Two receiver acoustic	Continuous Velocity Log	Acoustilog	Sonic Log	Acoustilog
Compensated acoustic	Borehole Compensated Acoustic	BHC Acoustilog	Borehole Compensated Sonic	Dual Spaced Acoustilog
Acoustic amplitude		Acoustic Parameter Log	Sonic Formation Amplitude	Acoustic Amplitude
Acoustic wavetrain — wiggle trace		Signature Log	Signature Display	
— variable density	3D Velocity Log	Variable Density Log	Seismic Spectrum	
Density	Formation Density Log	Densilog	Density Log	Densilog
Compensated density	Borehole Compensated Density	Compensated Densilog	Compensated Density Log	
Photoelectron/compensated density		Z-Densilog		
Thermal neutron (n-γ, n-t)	Neutron Log	Neutron Log		Neutron Log
Epithermal neutron (n-e)	Epithermal Neutron	Epithermal Sidewall Neutron	Sidewall Neutron Log	Neutron Log
Dual detector neutron	Borehole Compensated Neutron	Compensated Neutron Log	Compensated Neutron Log	
Pulsed neutron capture		Neutron Lifetime Log		
Wellsite computer interpretation	Stratalog	Prolog	Laserlog	

Table 2-1, cont.

Logging tools	PGAC	Schlumberger	Welex
Resistivity	Electrical Log	Electrical Log	Electric Log
Focused conductivity	Induction Electrical Log	Induction Electrical Log	Induction Electric Log
Dual focused conductivity		Dual Induction Laterolog Dual Induction Spherically Focused Log	Dual Induction Guard Log
Focused resistivity	Laterolog	Laterolog	Guard Log
Dual focused resistivity		Dual Laterolog	Dual Guard Log
Micro-resistivity	Micro Survey	Microlog	Contact Caliper Log
Micro-focused resistivity	Microlaterolog	Microlaterolog Proximity Log	FoRxo Log
Dielectric		Electromagnetic Propagation Log	
Two receiver acoustic	Acoustic Log	Sonic Log	Acoustic Velocity Log
Compensated acoustic	Borehole Compensated Acoustic Log	Borehole Compensated Sonic Log	Compensated Acoustic Velocity Log
Acoustic amplitude	Acoustic Parameter Log	Sonic Amplitude	Acoustic Amplitude (Frac Finder)
Acoustic wavetrain — wiggle trace — variable density		Waveforms Variable Density Log	 Micro-Seismogram
Density	Gamma-Gamma Density Log	Formation Density Log	Density Log
Compensated density	Compensated Gamma-Gamma Density	Compensated Formation Density Log	Compensated Density Log
Photoelectron/compensated density		Litho-Density Log	Spectral Density Log
Thermal neutron (n-γ, n-t)	Neutron Log	Neutron Log	Neutron Log
Epithermal neutron (n-e)	Neutron Log	Sidewall Neutron Porosity	Sidewall Neutron Log
Dual detector neutron		Compensated Neutron Log	Dual Spaced Neutron Log
Pulsed neutron capture		Thermal Decay Time	Thermal Multigate Decay
Wellsite computer interpretation		Cyberlook	Computer Analyzed Log

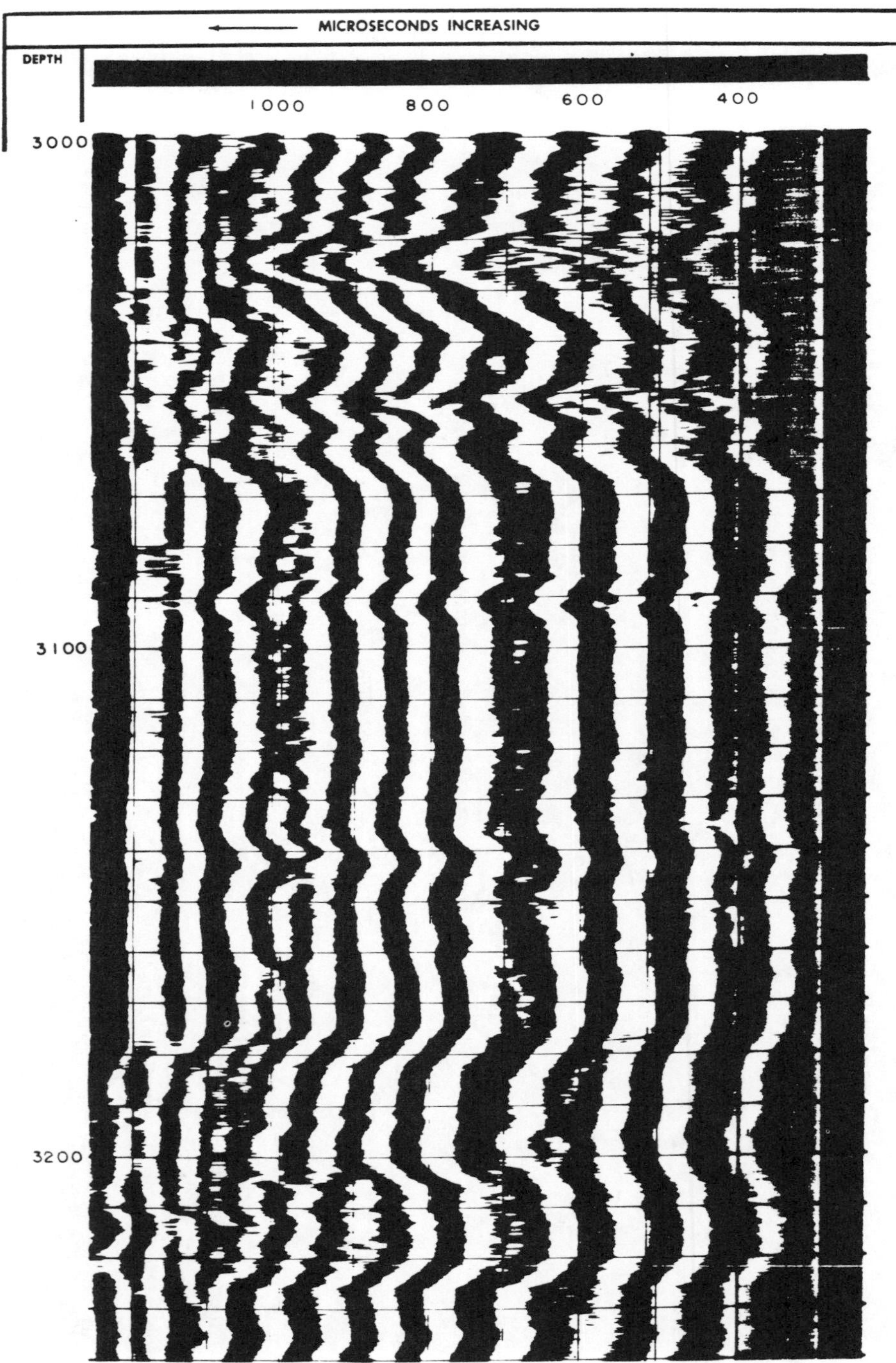

FIG. 2-2. Birdwell 3-D acoustic log, the first full wave train display for the acoustic logging tool.

Castel who began commercial perforating service in 1947. They acquired the assets of Perforating Guns, Inc. in 1948 and adopted the name Perforating Guns Atlas Corporation in 1949. At that time twenty-three completion units were in operation and radioactive cased-hole logging had been introduced. They began open-hole electric logging operations in 1954 and overseas operation in Venezuela in 1955.

The company name was changed to Pan Geo Atlas Corporation in 1959 before the Acoustic Logging Division of Empire Geophysical, Inc., of Fort Worth was acquired in 1960 to support their open-hole market operations. They merged with the Lane-Wells Division in 1968.

GEARHART

Gearhart was begun in 1955 in Fort Worth as GO Oil Well Services, Inc. and was a partnership of Marvin Gearhart (previously security engineering director for Dresser Industries and chief logging engineer for Welex) and Harrold Owen (previously chief explosive design engineer for Welex). Their operations were centered in the perforating field and they sought entry into the jet perforating business. However, patents for jet perforating were tightly held by a group of seven companies headed by Byron Jackson Company. Unable to obtain a license from the pool, they designed their own shaped charge perforators, but were eventually sued by one of the pool members. GO counter filed two suits and, while waiting settlement, decided to get out of wireline operations and become a wireline equipment manufacturer and supplier for companies that would enter the wireline service market once their suit was won.

The suit was won in 1962 and GO Industries began supplying equipment to hundreds of small logging companies including six of their own spinoffs. At the same time the company diversified greatly by acquiring a seismic computing company, a powerline installation equipment company, a conductor cable equipment engineering company, and a manufacturer of earth boring equipment.

In 1969, after deciding that they would eventually reach a limit in manufacturing but that wireline operations allowed for substantial continued growth, GO International, Inc., a subsidiary of Gearhart-Owen Industries, Inc., was created by acquiring the spinoff companies, GO Inc., GO Jet Services, GO Services Inc., GO Western Inc., Well Perforators Inc., and Jet Technical Services Inc. and became the fifth largest wireline service company. While the company became very competitive in cased-hole services, it did not make a substantial impact on the open-hole market.

In 1977 two changes occurred: (1) the partnership split with Gearhart retaining GO International, Inc., and the wireline related business, and Owen spinning off Pengo and the nonwireline related business, and (2) the field introduction of the first digital recording and operating wireline systems.

The latter was aimed at first improving log quality by reducing human error in the calibration and operation of the logging tools. The system was primarily developed by Max Moseley and Jack Burgen. Moseley and Burgen had founded Electronic Instruments, Inc. (which had been bought by GO in the early 1970s and was eventually sold to Teledyne) but returned to GO and began to develop their ideas of applying computers to improving logs.

The logging company became first GO Wireline and then Gearhart after the split. Gearhart is now strongly competitive in both the cased-hole and open-hole markets, especially in wells in the 10 to 12,000 ft range. They operate primarily in North and South America with operations outside the United States often being joint ventures with national interests.

SCHLUMBERGER

Schlumberger Well Services, Inc., originated from the initial tests of electrical exploration methods for the detection of metallic minerals conducted by Conrad Schlumberger in 1919. These initial tests mapped induced equipotential surfaces, and applied geologic interpretation to the resultant surfaces. The Schlumberger Brothers introduced the concept of apparent resistivity, which greatly reduced the survey presentation and interpretation problems. In 1926 The Societe de Prospection Electrique (SPE) was formed to exploit and promote these surface techniques commercially. The methods were first applied to the borehole for the exploration of hydrocarbons in 1927.

In 1928 Royal Dutch/Shell tested the technique and contracted for a crew in Venezuela to do downhole surveys with surface prospecting on a secondary basis. In 1929 Schlumberger conducted tests for Shell in California. Because of equipment troubles and difficulty in correlating the widely spaced wells, the contract was terminated after a few jobs. The crew moved to Oklahoma to work for a Gulf Oil affiliate. In 1930 Schlumberger operated for six months on the Gulf coast under exclusive U.S. contract to Humble, Gulf, and Shell. However, again because of difficulties, the contract was not renewed.

Schlumberger expanded overseas into Russia in 1929. Geologic conditions in Russia were very favorable for the technique. Some correlation between the resistivity measurements and hydrocarbon production from the wells was noted. After 1931, the effort was split with Compangnie Generale de Geophysique operating the surface prospect methods and, Societe de Prospection Electrique taking over all measurements and operations in boreholes.

In 1932 they reentered the U.S. market with another contract with Shell Oil Company for one crew in California and one crew in the Gulf coast area. By 1933 there were, outside of Russia, eight logging crews in France, Morocco, Rumania, the United States, Venezuela, and Trinidad. By 1934 there were four crews in the Gulf coast, three crews in California, and one crew in Oklahoma. Because growth of the business in the United States now required better legal organization, Schlumberger Well Surveying

Corporation, with headquarters in Houston, was created with S. de P.E. providing technical support. At this time the principle service was still the electrical survey with spontaneous potential.

The 40 logging crews operating in 1935 increased by 1939 to 140 logging crews, half in the United States, and the others spread out in 20 countries, with major concentration in Venezuela and the Dutch East Indies.

Settlement of the United States patent lawsuits in 1937, allowed Schlumberger to receive a license from Lane-Wells to perforate in the United States while Lane-Wells received a license to perform electrical logging services. Because Schlumberger was already conducting perforating services overseas, they were able to offer commercial perforating in 1938.

With the start of World War II the Paris headquarters were cut off from the rest of the organization. During the war Paris could not conduct field research and development, as had been anticipated, so much technical information and equipment was transferred to Houston and Houston began to operate independently. Operations contracted mainly to the United States and South America.

After the war a simultaneous administrative and technical reorganization began. In 1946 a sequential electric log survey (ES) was developed, which allowed three resistivity measurements and the SP to be recorded with a single pass over the borehole. The new ES caused the replacement of the four rubber-insulated steel conductor and textile-braid covered cable with a six, later seven, conductor steel armored cable.

Originally introduced in 1948, the induction log was created to measure resistivity in boreholes drilled with oil-based muds, an environment where the ES would not work. However, by 1957 the induction log virtually replaced the ES because it gave better resistivity measurements in a wider range of borehole conditions and geologic provinces.

WELEX

Welex, a Halliburton Company, was formed by the merger of Halliburton's well logging division and Welex Jet Services. Halliburton created their Electrical Wireline Services (EWS) Division in 1936 to conduct wireline resistivity and temperature services. The new division, under F. T. Robidoux from Humble Oil, obtained a license from Humble Oil to run resistivity logs to offer competition to Schlumberger which was then the only wireline service company.

In 1937 their first experimental logging unit began operating out of the Houston EWS headquarters. Commercial operations began in 1938 using a photographic technique to present their SP and single point resistivity log and the first commercial wireline caliper survey.

Their first markets were in south Texas and south Louisiana. By 1940 over twenty logging units were operating, with expansion into Kansas, Oklahoma, and Illinois markets. Expansion continued to foreign markets, first to Venezuela and then to Alaska. In the early 1940s the old resistivity and SP tool was replaced by a three-resistivity measurement tool offering greater depths of investigation. The second generation Halliburton electrical survey sonde introduced in the late 1940s was unique in that it contained downhole electronics that FM-telemetried the two lateral, one normal, and SP measurements up the single conductor cable. In 1950 they began offering the Guard Log to measure resistivity in salt mud drilling conditions.

In the early 1950s, with Lane-Wells offering both logging and perforating services and Schlumberger still holding a major share in both markets, Halliburton began looking for expansion into the cased hole market.

Well Explosives, Inc. was incorporated under Texas law in 1945 to manufacture perforating shaped charges. However perforating service companies were still using the bullet perforating technique introduced in 1940 and were not interested in the new untested technique. This caused creation of Welex Jet Services, whose name was derived from Well Explosives, Inc., as a shaped charge marketing method.

Shaped charge concepts were refined by R. W. Wood of John Hopkins University. Hollow cone shaped explosive charges when detonated create a high velocity gas jet out of the cone open end. By lining the cone interior with a thin metal liner the jet became powerful enough to cut a clean hole through steel casing, cement sheath, and formation. Shaped charges were developed and patented for oil field use by a group of Fort Worth engineers, including Henry Mohaupt and R. H. McLemore. Byron Jackson Company in Los Angeles was named under the patent as primary licensor, with Well Explosives as one of the patent pool manufacturing members.

In 1947 Welex began operating by establishing their first location in Wichita Falls, followed by locations in Odessa and Ardmore. By 1957 they had grown to over 30 locations with 400 employees. Jet Research Center was established in Arlington, Texas as a separate company to continue research and manufacture of perforating charges for Welex and other licensees. Welex corporate headquarters remained in Fort Worth, where trucks and equipment were still manufactured.

Halliburton expanded their wireline market share in 1957 by acquiring cased hole oriented Welex Jet Services. The two units were combined under the name Welex, Inc., with headquarters in Fort Worth. Manufacturing units were combined with Halliburton facilities in Houston. In 1961 Welex, Inc. was made a division of Halliburton and the division headquarters were moved to Houston. In the mid-1960s they introduced their Micro-Seismogram display of the acoustic tool wave train.

REFERENCES FOR GENERAL READING

Allaud, L. A., and Martin, M. H., 1977, Schlumberger, the history of a technique: John Wiley and Sons Inc.

Johnson, H. M., 1962, A history of well logging: Geophysics, **27**, 507-527.

Schlumberger, A. G., 1983, The Schlumberger adventure: Arco Publ. Co. Inc.

Steinert-Threlkeld, T., 1979, Gearhart-Owen Industries: The mouse that soared: Fort Worth Star-Telegram, 2 and 9 December.

Welex, Experience runs deep: Houston, Welex, a Halliburton Company.

———1965, Two decades/'45-'65: *The Welex Log*, summer, 1-4.

———1981, Atlas' role in Dresser's 100-year history: A.O.S.G. Recorder, **6** (1) 8-9.

Chapter 3

LOGGING HARDWARE

Until the last few years well-logging equipment was completely analog, a continuous representation of the downhole measured parameters, with an individual panel and downhole tool for each different logging service (Figure 3-1). The service companies have just undergone a transition to a universal, digital (noncontinuous or sampled) surface recording system along with computer aided calibration and wellsite data processing, while retaining the downhole analog equipment (Figure 3-2). Analog downhole tools are being replaced with tools that can take full advantage of digital techniques.

Computer assisted calibration and running of the logs eliminates many of the myriad of small details which could affect an otherwise good log. It also allows wellsite playback of the logs with better scale selection, proper depths, etc. without using valuable rig time to rerun the log.

Another current trend is to combine as many measurements as possible on one tool string and thereby cut down the rig time needed for logging. However, log files mostly contain older, individually run analog recordings.

DOWNHOLE MEASURING TOOL

The downhole logging tool is made up of at least two parts, the sonde and the cartridge. The sonde contains the detectors (resistivity electrodes, radiation detectors, acoustic transducers, etc.) that actually make the formation measurements. The cartridge contains the power supply and electronic circuitry for the sonde detectors. The cartridge also contains circuitry for any signal processing necessary before sending the measurements through the cable conductors to the surface equipment.

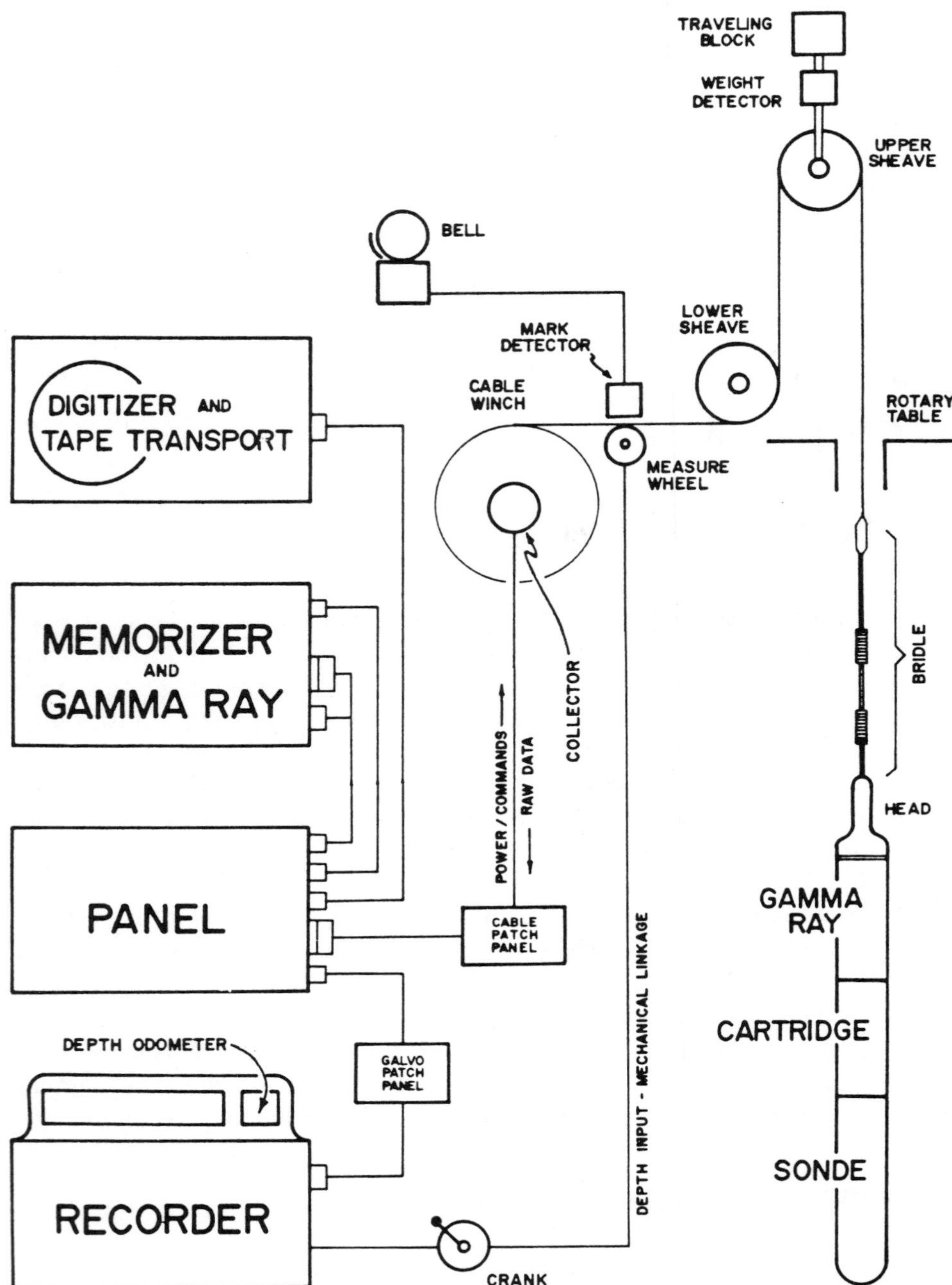

FIG. 3-1. The complete analog logging system where a different surface panel is used for each downhole tool.

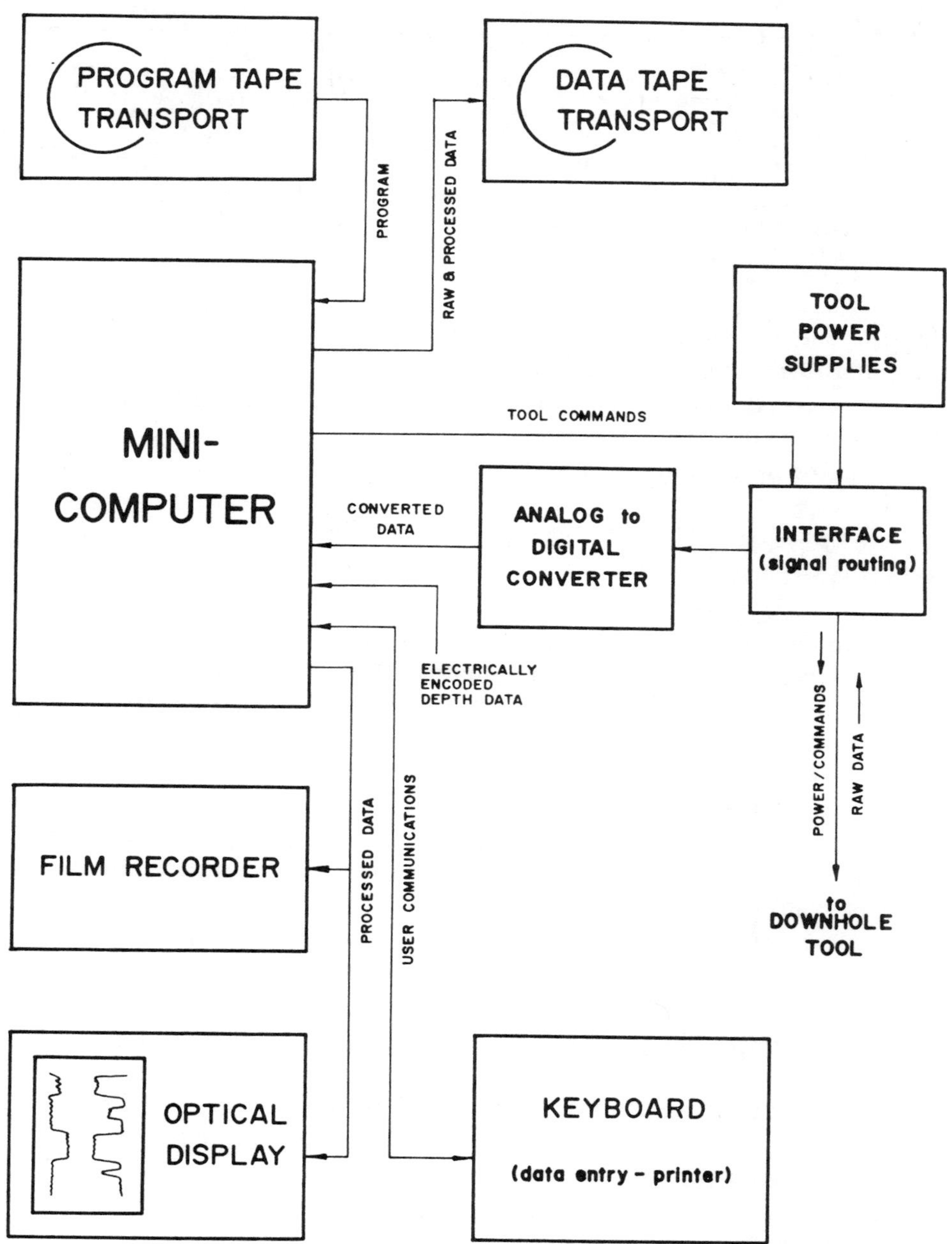

FIG. 3-2. The universal digital surface system with a different computer program loaded into the minicomputer for each downhole tool.

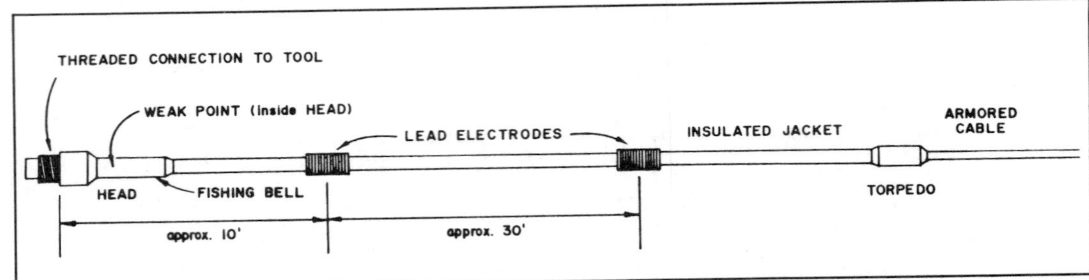

FIG. 3-3. When the bridle and cable are pulled free from the head the fishing bell is designed to fit exactly inside a drill pipe mounted fishing overshot. The torpedo mechanically and electrically connects the bridle and cable.

Because the borehole is narrow and the electronic circuits must be protected from high borehole temperature and pressure, the parts used are just barely convenient in length and weight for two operators to handle. An auxiliary natural gamma-ray (GR) measurement can be recorded with most logging tools by adding a self-contained unit between the head and cartridge.

HEAD AND BRIDLE

The downhole tool is either connected to the cable through a head assembly or a head and bridle combination (Figure 3-3). The head has four functions: (1) mechanical and electrical connections between the tool and cable, (2) electrical connections to bridle electrodes and cable wires, (3) a controlled "weak point" between the tool and cable, and (4) a precision sized fishing surface. Because a number of different tools are run at each well, the required mechanical and electrical threaded connection must be easy to make. The bridle electrically isolates the sonde from the conductive cable armor while providing distant current returns (electrodes) for resistivity logging. Nonresistivity tools do not need electrodes but can use either the head and bridle or a bridleless head.

The weak point and fishing bell are used when a tool becomes stuck and cannot be freed by pulling on the cable. The controlled weak point allows the cable and bridle to separate from the head if a specific tension, above the normal logging tension but below the breaking strength of the cable, is exceeded. A fishing overshot, attached to the bottom of the drill string, is designed to draw the fishing bell surface into a grapple thereby securing the head and tool. The force of the drill string then can be used to free the stuck tool.

CABLE

There are two types of logging cable: (1) multiconductor used primarily for open-hole logging, and (2) monoconductor used for completion services (perforating, setting plugs and packers, etc.). The multiconductor normally contains seven insulated conductors protected by two counter-wrapped layers of conductive steel armor. Two wires are used to send power to the cartridge, with the remaining five wires used for switching, measuring current returns, recording signals coming uphole, etc. It is not uncommon for each wire to serve several functions. A few companies use the monocable for open-hole logging; sending power down the single conductor and using multiplexed telemetry to both send commands downhole and to send measured signals to the surface.

PATCH PANEL

The patch panel is the surface termination of the logging cable before connection to the panel. It allows convenient access to the cable wires and grounds for checking continuity, insulation, and quality of the raw signals.

ANALOG PANEL

The analog panel accepts measurement signals from the sonde and converts them to galvanometer currents for the recorder. The panel also contains scale selectors, calibration controls, correction inputs, and panel test inputs (Figures 3-12 and 10-10).

The new universal surface equipment converts the analog measurements to digital form before entering them in the computer. Calibration, scale selection, memorization, and signal quality testing are done by the computer.

RECORDER

The galvanometer currents from the panel are recorded against depth on either photographic film or a paper strip chart. Depth is input through a mechanical linkage system turned by the movement of the cable past the measure wheel (Figure 3-1). The recorder can produce two films simultaneously: a 1 or 2 inch per 100 ft of borehole (1:1 200 or 1:600 scale) for well-to-well correlations; and a 5 inch per 100 ft of borehole (1:240 scale) for detailed formation evaluation. Depth numbers are added at 100 ft intervals with 10 ft and 2 ft depth lines indicated on the 2 inch and 5 inch films, respectively. Metric logs come in 1:1 000, 1:500, 1:200, and 1:20 scale and are marked at 50 m depths to distinguish them from United States logs.

The three primary field grid formats are linear, logarithmic, and split (Figure 3-4). All three formats have one single linear track to the left of the depth track and two tracks to the right. Track I is used for correlation measurements to compare relative log-to-log depths. The predominate correlation measurements, spontaneous potential (SP) and natural gamma ray (GR), are also relative shale content indicators. Measurements usually

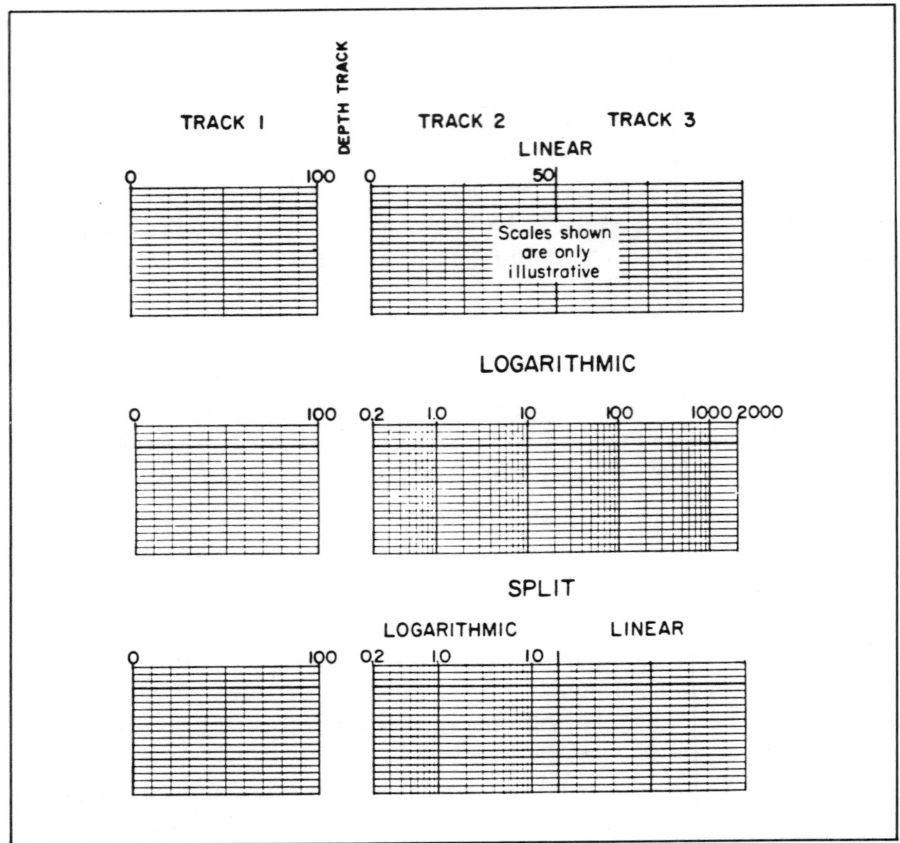

FIG. 3-4. The three common log grid formats: linear, logarithmic, and split.

read shalier to the right and cleaner, or less shaley, to the left depending on the scale selected. A caliper of borehole diameter is also recorded in Track I along with an indication of the logging speed; either a mark or a left edge grid line interruption every 60 s (Figure 3-5, minute marks). Tracks II and III are for presentation of the primary tool measurements, with porosity increasing and resistivity decreasing to the left.

Resistivity scales. — The older electrical surveys and induction logs use a linear resistivity presentation in Track II, with a "×10" backup curve for higher resistivity readings (Figure 3-5). Induction conductivity is presented in Track III and can be used to determine accurate resistivity values in low formation resistivities. The newer dual induction and dual laterolog use a logarithmic format which handles a large formation resistivity range better (Figure 3-6). The logarithmic resistivity scale maintains a constant degree of needed accuracy, as the difference between 1 and 10 Ω•m is of the same importance as between 100 and 1 000 Ω•m for formation evaluation (Figure 5-1). Thus the "error" in using logarithmic grid readings is constant, while that for the linear ×10 format increases as the resistivity becomes larger. The conventional logarithmic grid is four cycles, scaled from 0.2 to 2 000 Ω•m, with a single

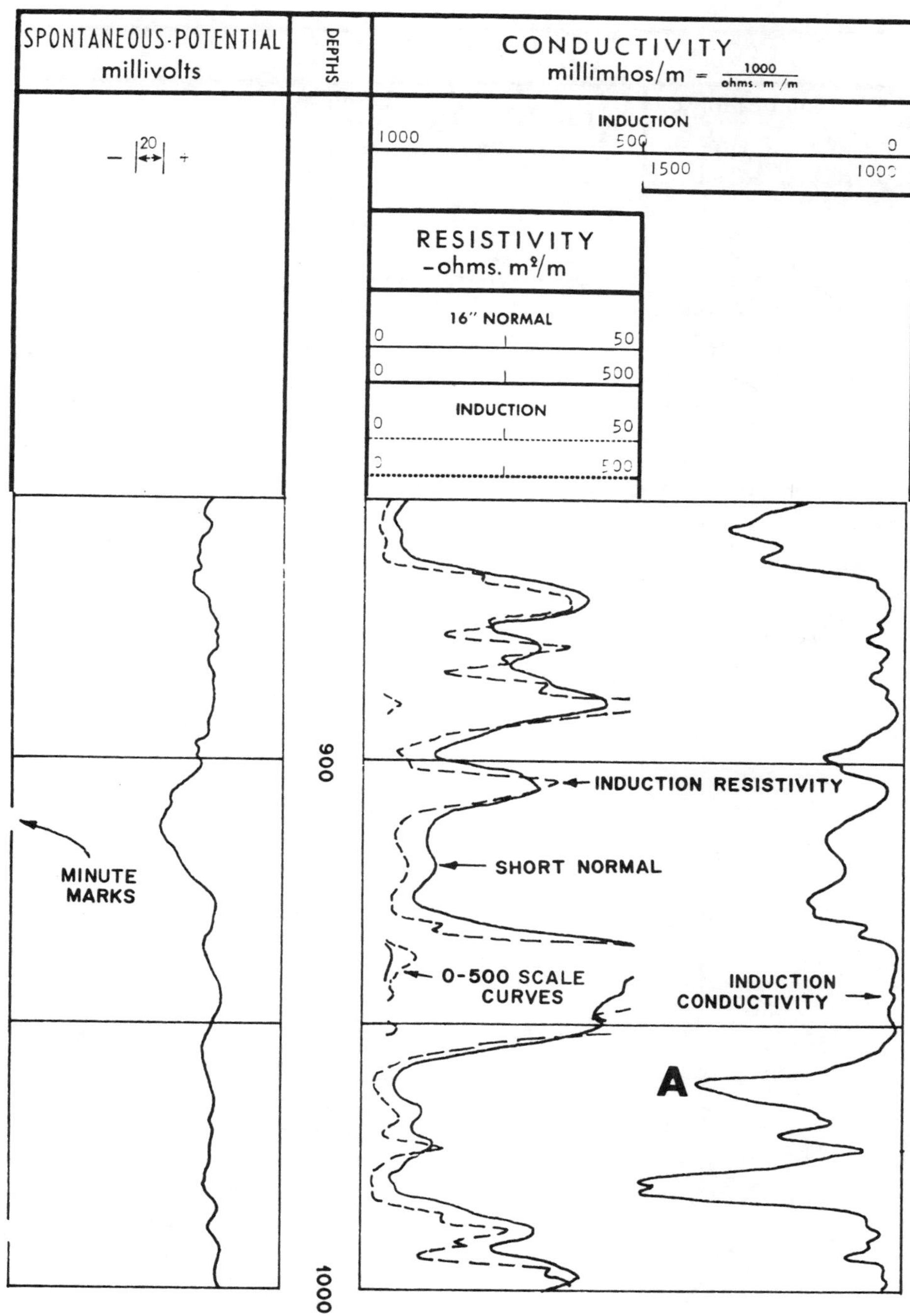

FIG. 3-5. The linear resistivity format: an induction electrical log (IES) with linear resistivity and an ×10 backup curve, i.e., a 0-50 Ω•m primary scale and a 0-500 Ω•m backup, all in Track II. The deep induction is presented in resistivity units in Track II and conductivity units in Track III. The induction conductivity curve accurately determines low-resistivity values, i.e., at A the conductivity is 2.7 Ω•m (1 000/375 mmhos/m).

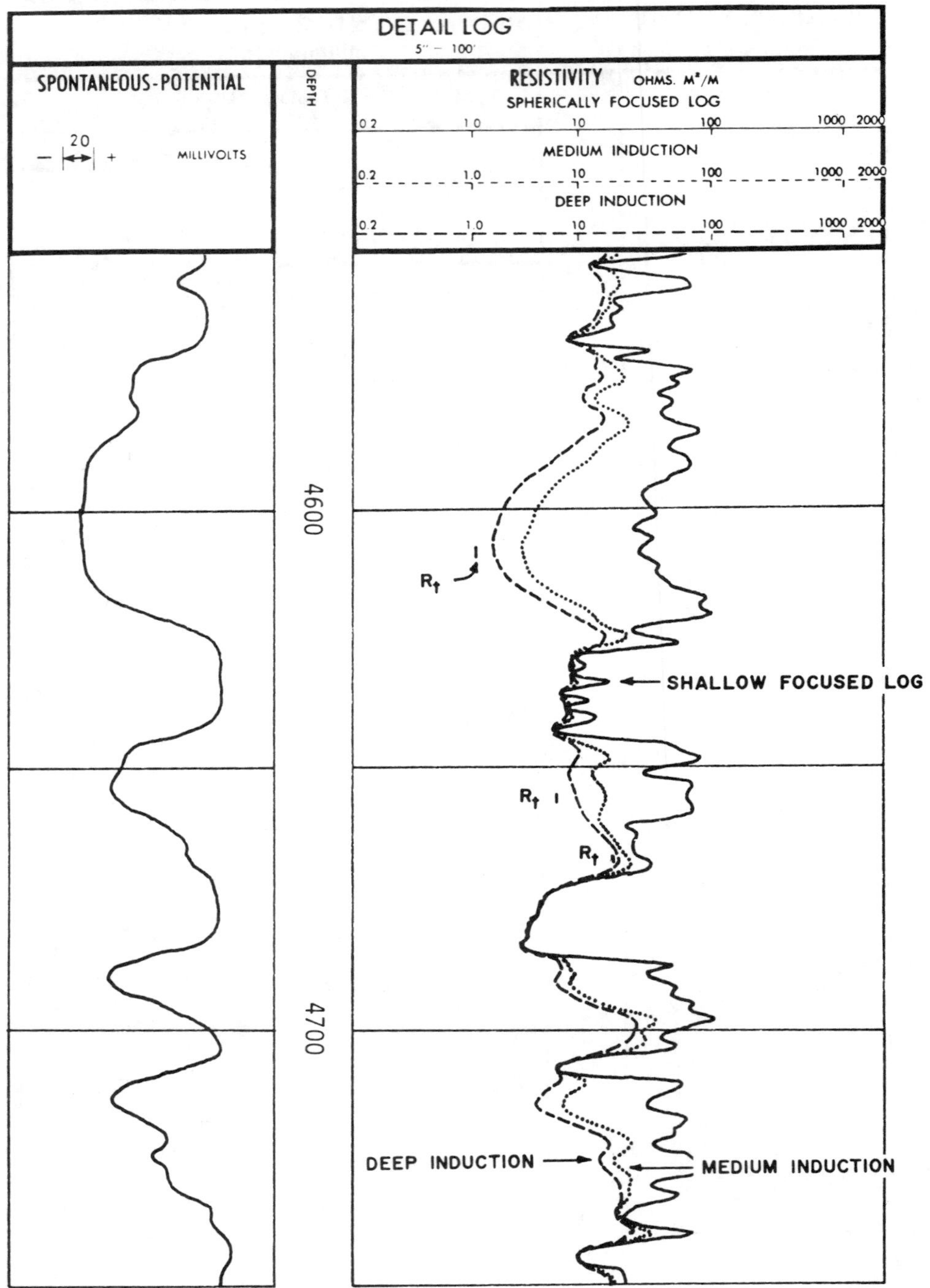

FIG. 3-6. The logarithmic resistivity format showing a dual induction log with a four-cycle grid and one curve for each resistivity device. The logarithmic format allows a large resistivity range presentation with only one curve. The R_t values at 4 608, 4 656, and 4 658 ft are determined by crossplotting the three resistivity measurements at each depth, Figure 5-8.

curve for each resistivity device. Because the conductivity-seeking induction device looses accuracy above 200 Ω•m, less than 5m S/m or m m hos/m, these curves are often electronically stopped at 2 000 Ω•m. Deep reading laterologs can measure high resistivities accurately, and are allowed to "go off the page" with a backup curve coming on from the left edge in a 2 000 to 2 000 000 Ω•m scale (Figure 5-15). Where there is a limited resistivity range, such as in the Gulf Coast tertiary sands, the two-cycle logarithmic portion of the split format grid scaled 0.2 to 20 Ω•m is adequate and also leaves Track III available for presenting a porosity measurement thereby allowing a simultaneous resistivity and porosity presentation such as the induction-density-neutron (Figure 3-7) and the induction-sonic (Figure 6-2) combination tools.

Logarithmic scales have many advantages for formation evaluation. However, in areas having a number of linear scale logs, the tendency is to continue with the older presentation to allow ease of comparison for geologic correlations. A compromise is often made with a linear resistivity 2-inch correlation log and a logarithmic 5-inch detail presentation (Figure 3-8).

A fourth resistivity presentation is the hybrid scale (Figure 3-9). This is an older compromise scale designed to handle a large dynamic resistivity range with a single curve and is found mainly in older laterologs and microlaterologs.

Porosity scales. — Porosity measurements are presented in Tracks II and III with increasing porosity to the left, or toward the depth track. The bulk density (ρ_a) and acoustic interval transit time (Δt) measurements are converted to a scaled porosity curve [assuming a matrix-lithology (Table 3-1)], using the traditional Wyllie time-average equations

$$\phi_D = (\rho_{ma} - \rho_a)(\rho_{ma} - \rho_f) \quad (3\text{-}1)$$

and

$$\phi_{SV} = (\Delta t_a - \Delta t_{ma})(\Delta t_f - \Delta t_{ma}) \quad (3\text{-}2)$$

Table 3-1. The "common" parameters used to transform acoustic traveltime and bulk density measurements into porosity. The exact parameters used are recorded in the heading (Figure 3-10, Item A). A porosity curve is only valid if the matrix lithology and pore fluid assumptions are correct for the formation being evaluated, i.e., if the formation is sandstone, then obviously a curve scaled in limestone porosity units is presenting an incorrect porosity.

	Matrix/Lithology				
	V_{ma}		Δt_{ma}		ρ_{ma}
	ft/s	m/s	μs/ft	μs/m	g/cm³
Quartz Sandstone	18 000	5 486	55.5	182	2.65
Well cemented, Limy Sandstone	19 500	5 944	51.3	168	2.68
Limestone	21 000	6 400	47.6	156	2.71
Dolomite	23 000	7 010	43.5	143	2.87
	Pore Fluids				
	V_f		Δt_f		ρ_f
Fresh Water	5 300	1 615	189	619	1.00
Salt Water	5 550	1 692	180	591	1.10
Saturated Salt Water					1.20

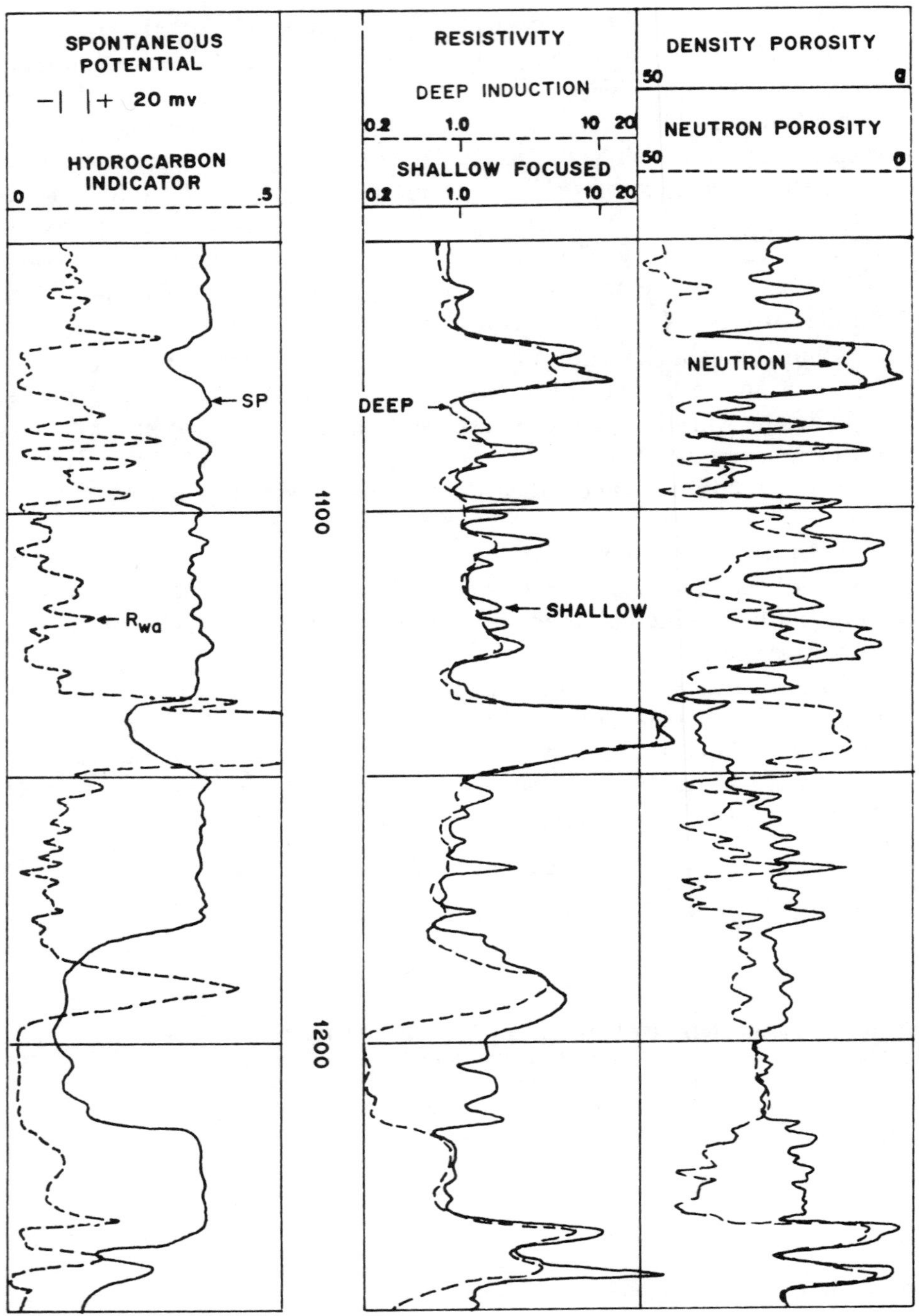

FIG. 3-7. The split format using a simultaneous induction-density-neutron combination tool shows limited, two-cycle logarithmic resistivity in Track II and linear porosity measurements in Track III. A linear resistivity only format, shown in Figure 3-5, is used for the 2 inch per 100 ft correlation log.

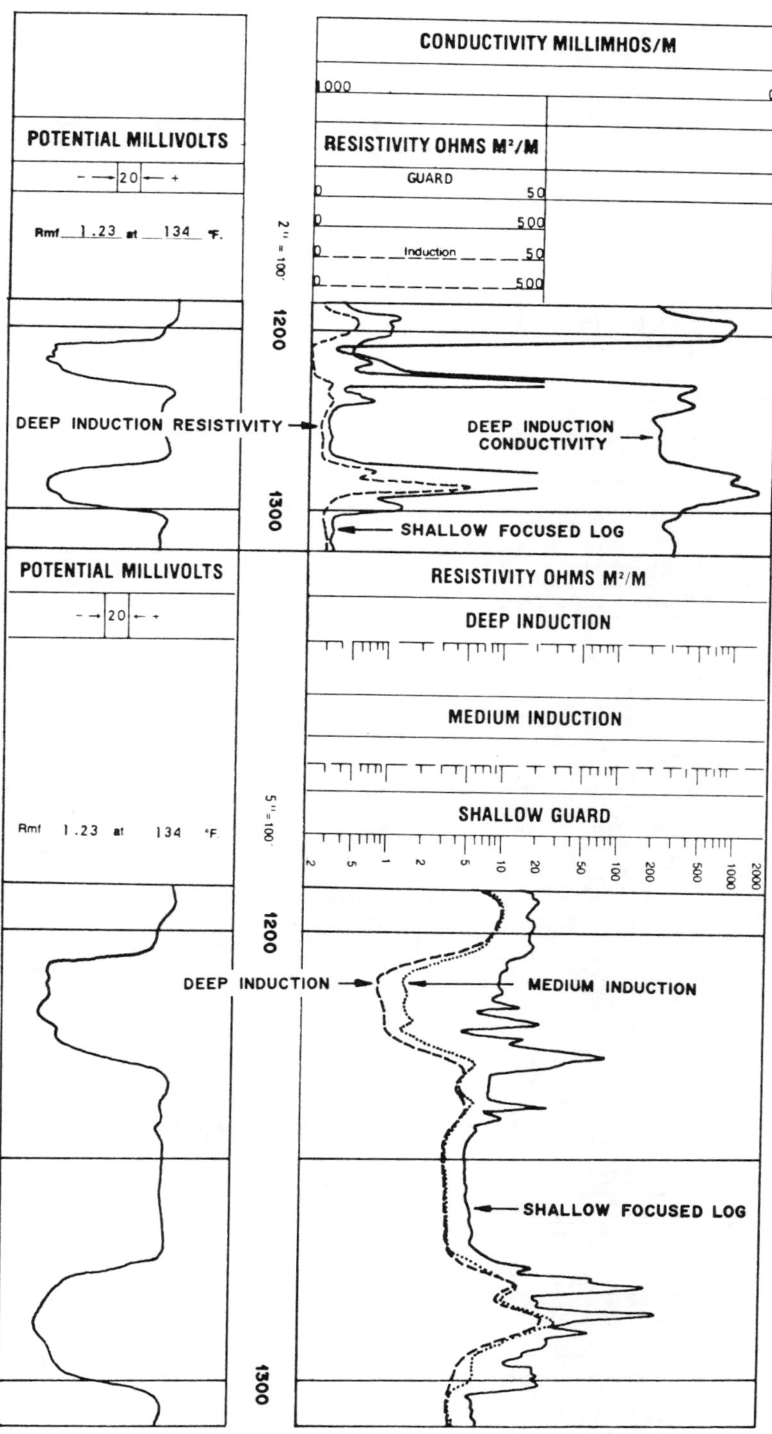

FIG. 3-8. A log-linear resistivity format using the dual induction log with a linear 2-inch correlation log (top) and a logarithmic 5-inch detail log (bottom) presentation of the same measurements.

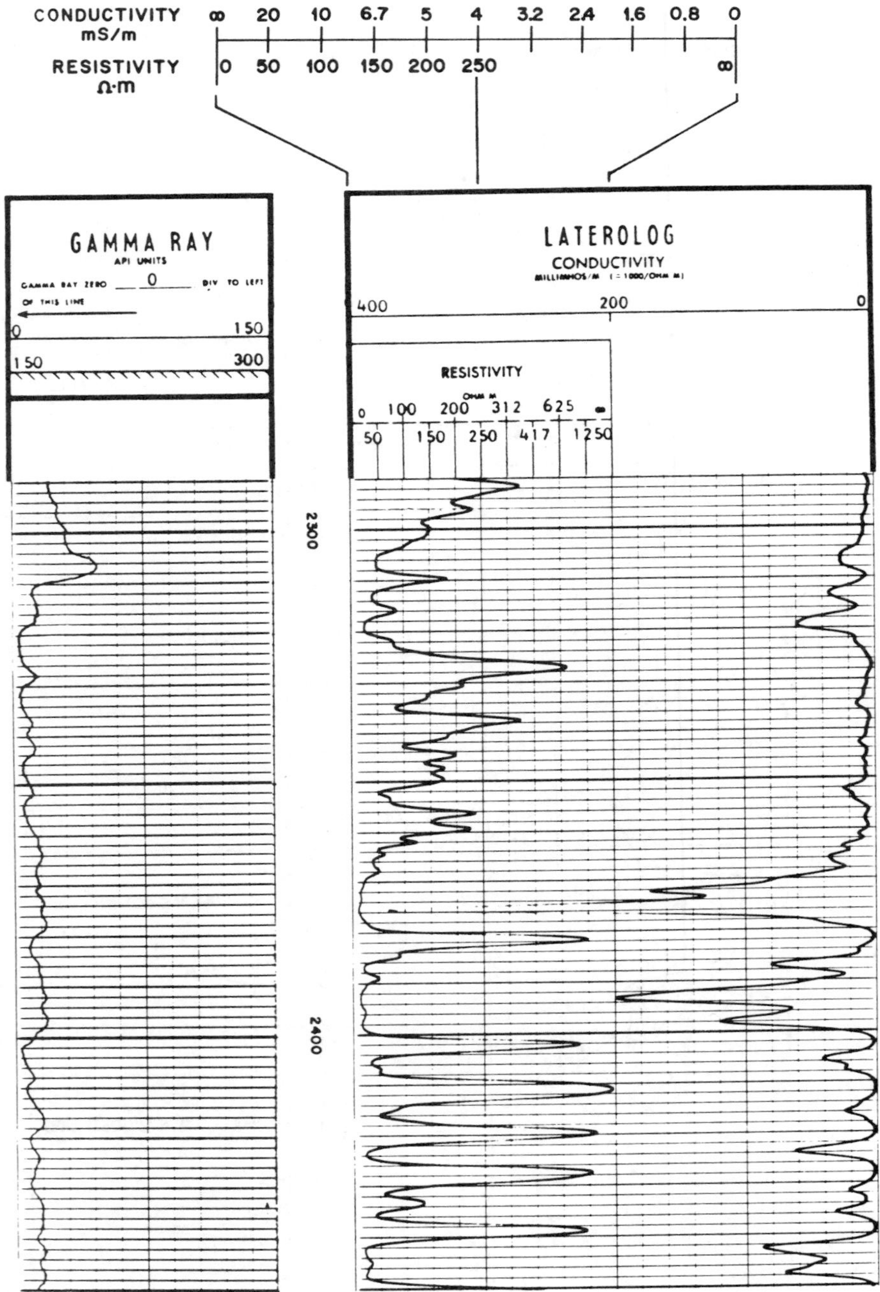

FIG. 3-9. Hybrid resistivity format with the resistivity curve in Track II linear in resistivity in the left-half of the track and linear in conductivity in the right-half thereby yielding a compressed higher-resistivity scale. Again, low resistivities can be determined accurately using the conductivity curve.

and using the subscripts *a* for apparent log measurement, *ma* for matrix, and *f* for pore fluid. All except the log measurements are assumptions for an analog porosity computer in the panel. These assumptions should be indicated on the log heading (Figure 3-10, item A).

When a scaled porosity curve is presented, the three most common formats are (1) with zero porosity in the middle of Track III for left to right scales of either 45, 15, −15, or (2) 30, 10, −10, and (3) zero porosity at the right of Track III in a 60, 30, 0 porosity unit (PU) scale (Figure 3-11).

MEMORIZER PANEL

The recorder odometer is normally set to the depth of the primary measurement of a tool. Most tools contain several measuring devices. When logging up, the devices on the upper part of the tool enter a formation before those on the bottom of the tool, causing the upper readings to need depth correction for presentation at the proper relation with the primary device readings (Figure 3-12). If a paper strip chart recorder is used, the length along the paper of higher measurements pen is adjusted for the proper scaled distance above the primary recording pen. When using a film recorder, the measurements must be stored electronically and delayed in the memorizer panel until they can be put on the film at the same time (correct depth) as the lower primary measurement.

A properly memorized log (Figure 3-13) should show curves deflecting at the same depth. There may be deflections in the primary measurement that are not reflected in auxiliary measurements, such as gamma ray and SP, but an auxiliary measurement deflection is usually expressed in the prime curve.

To record total depth (TD) of the borehole accurately, several feet of slack cable are run into the hole after the tool is sitting on the bottom, thereby driving the recorder odometer reading below the hole bottom. On resistivity and acoustic tools, straight line recordings indicate when the tool is stopped. Tool first readings (FR) are indicated by movement from these constant readings (Figure 3-12, item D). The radiation measurements, GR, neutron, and density, have statistical variations and thus it is not obvious when the tool lifts off bottom. Pickup for these tools is indicated by movement of the caliper (Figure 3-14, item B). Log TD is the depth of the FR plus the distance from the prime measurement point to the bottom of the tool.

The measurement detectors on the upper part of the tool do not record data as low as the prime measurement when the tool is sitting on bottom. Because this is commonly the gamma ray, it can indicate misleading readings for the memorized distance above the primary measurement FR. The depths of the first valid readings for important measurements are shown, usually by FR marks in the depth track (Figure 3-14, item A).

Incorrect interpretation of false readings before pickup can be a problem especially with the combination density-neutron tool. Figure 3-15 shows

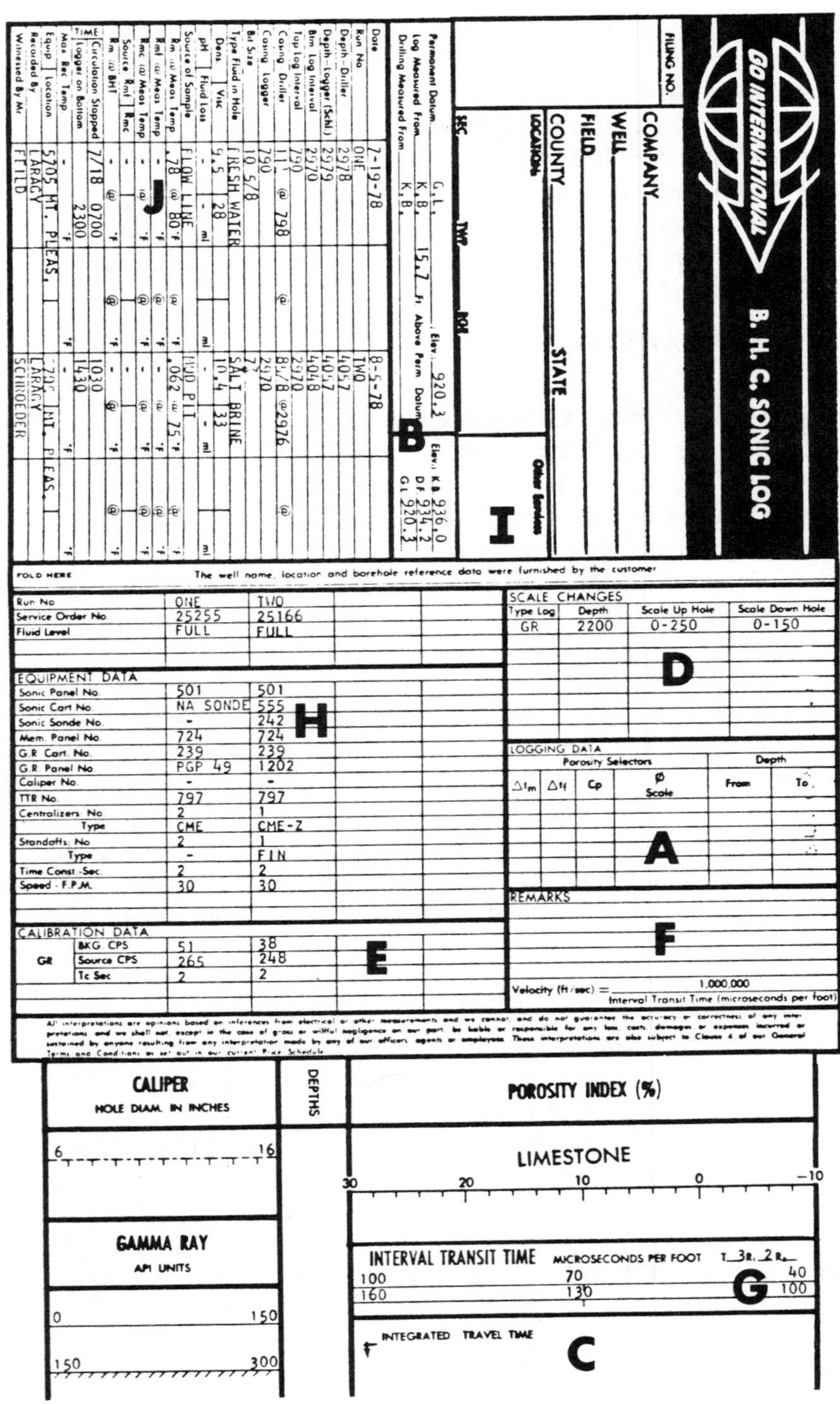

GO INTERNATIONAL — B. H. C. SONIC LOG

FILING NO. | COMPANY | WELL | FIELD | COUNTY | STATE | LOCATION | SEC. | TWP. | RGE. | Other Services

Permanent Datum G.L. Elev. 920.2
Log Measured From K.B. 15.7 Ft. Above Perm. Datum
Drilling Measured From K.B.
Elev.: K.B. 936.0, D.F. 934.2, G.L. 920.3

Date	7-19-78	8-5-78
Run No.	ONE	TWO
Depth—Driller	2978	4057
Depth—Logger (Schl.)	2979	4057
Btm. Log Interval	2970	4048
Top Log Interval	790	2970
Casing—Driller	11" @ 798	8 5/8 @ 2976
Casing—Logger	790	2970
Bit Size	10 5/8	7 7/8
Type Fluid in Hole	FRESH WATER	SALT BRINE
Dens. / Visc.	9.5 / 28	10.4 / 33
Source of Sample	FLOW LINE	MUD PIT
Rm @ Meas. Temp.	.78 @ 80°F	.062 @ 75°F
Time Circulation Stopped	7/18 0700	1030
Logger on Bottom	2300	1430
Equip. / Location	5705 / MT. PLEAS.	726 / MT. PLEAS.
Recorded By	LARACY	LARACY
Witnessed By	FIELD	SCHROEDER

The well name, location and borehole reference data were furnished by the customer

Run No.	ONE	TWO
Service Order No.	25255	25166
Fluid Level	FULL	FULL

EQUIPMENT DATA

Sonic Panel No.	501	501
Sonic Cart No.	NA SONDE	555
Sonic Sonde No.	-	242
Mem. Panel No.	724	724
G.R. Cart. No.	239	239
G.R. Panel No.	PGP 49	1202
Caliper No.	-	-
TTR No.	797	797
Centralizers No.	2	1
Type	CME	CME-Z
Standoffs No.	2	1
Type	-	FIN
Time Const.-Sec.	2	2
Speed-F.P.M.	30	30

CALIBRATION DATA

GR	BKG CPS	51	38
	Source CPS	265	248
	Tc Sec	2	2

SCALE CHANGES

Type Log	Depth	Scale Up Hole	Scale Down Hole
GR	2200	0-250	0-150

LOGGING DATA

Porosity Selectors				Depth	
Δt_m	Δt_f	Cp	ø Scale	From	To

REMARKS

Velocity (ft/sec) = 1,000,000 / Interval Transit Time (microseconds per foot)

All interpretations are opinions based on inferences from electrical or other measurements and we cannot, and do not guarantee the accuracy or correctness of any interpretations, and we shall not, except in the case of gross or willful negligence on our part, be liable or responsible for any loss, costs, damages or expenses incurred or sustained by anyone resulting from any interpretation made by any of our officers, agents or employees. These interpretations are also subject to Clause 4 of our General Terms and Conditions as set out in our current Price Schedule.

FIG. 3-10. Heading information: (A) The matrix or lithology, and fluid assumptions used to convert acoustic traveltime or bulk density to porosity, (B) log depth measurement reference elevation, (C) log scales, (D) scale changes which should also be marked where they occur on the log, (E) radiation tool calibration count rates, (F) any additional engineer comments about the log, (G) acoustic log spacing and receiver span, (H) sonde type or model, (I) other logs run (see Appendix A) and (J) mud properties.

the neutron reading is less than the density from 3 967 to 3 989 ft which normally would be an indication of gas for that interval. However, the first valid neutron reading is at 3 971 ft. Because the pickup-memory distance was not taken into account, completion was attempted in this interval and only water was produced.

Another false reading occurs when the tool sticks to the formation. The tool can stop, but the cable stretches as it continues to be reeled up at the surface. This false reading continues until the tool is pulled free and almost instantaneously returns to the proper depth relationship with the cable, or until the weak-point of the head is exceeded and the cable is pulled free from the stuck tool. This cable stretching condition is indicated by an increase of the cable tension (Figure 3-16, item A). The sharp decrease in tension indicates when the tool pulls free and begins coming uphole again.

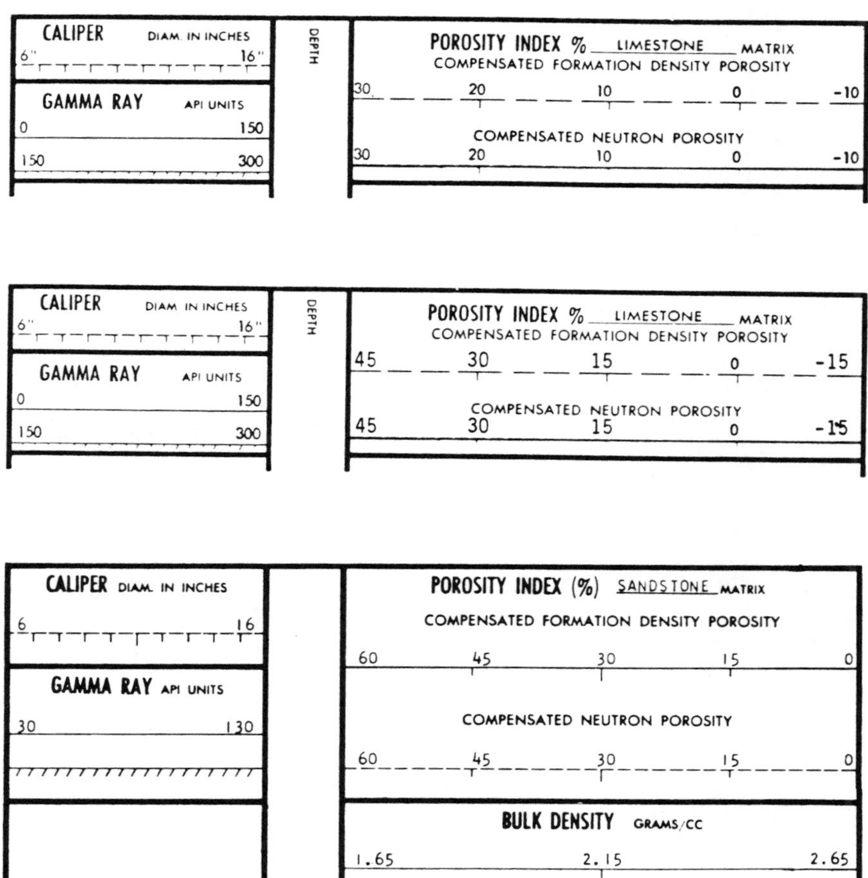

FIG. 3-11. Porosity scale formats. The top two formats predominately present carbonate environments. The bottom format is for sandstone environments. Note that the bottom format allows one curve to present both bulk density and density-derived sandstone porosity.

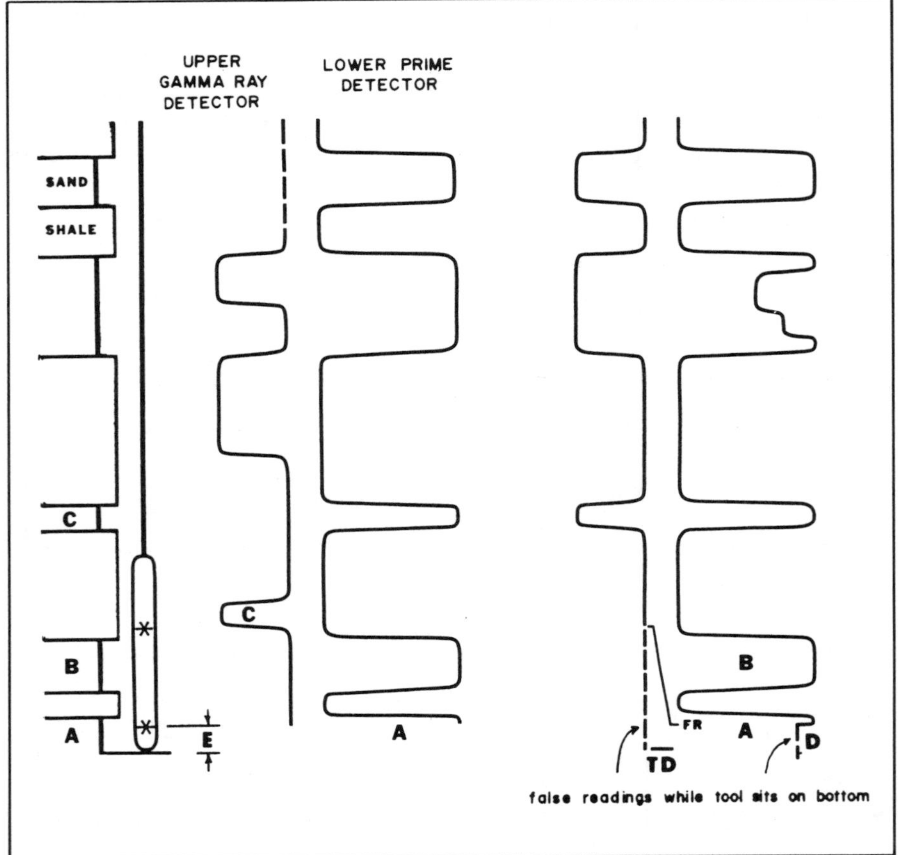

FIG. 3-12. Depth memorization where the upper detectors enter a formation before the lower detector. On the left, measurements are presented as they are detected; on the right, they are presented in proper depth relation by memorizing or delaying the gamma ray to the prime detector depths. The first reading (FR) of the lower detector as the tool moves up from sitting on the bottom, is indicated by initial movement of the curve (D) if the measurement is not statistical. Log total depth (TD) is the depth of the distance from the prime measurement point to the bottom of the tool (E) below the tool FR.

FIG. 3-13. Proper memorization. Note that there is not always a correspondence of movement between the correlation gamma ray and primary density curves, but enough points correspond to indicate proper memorization (A and B).

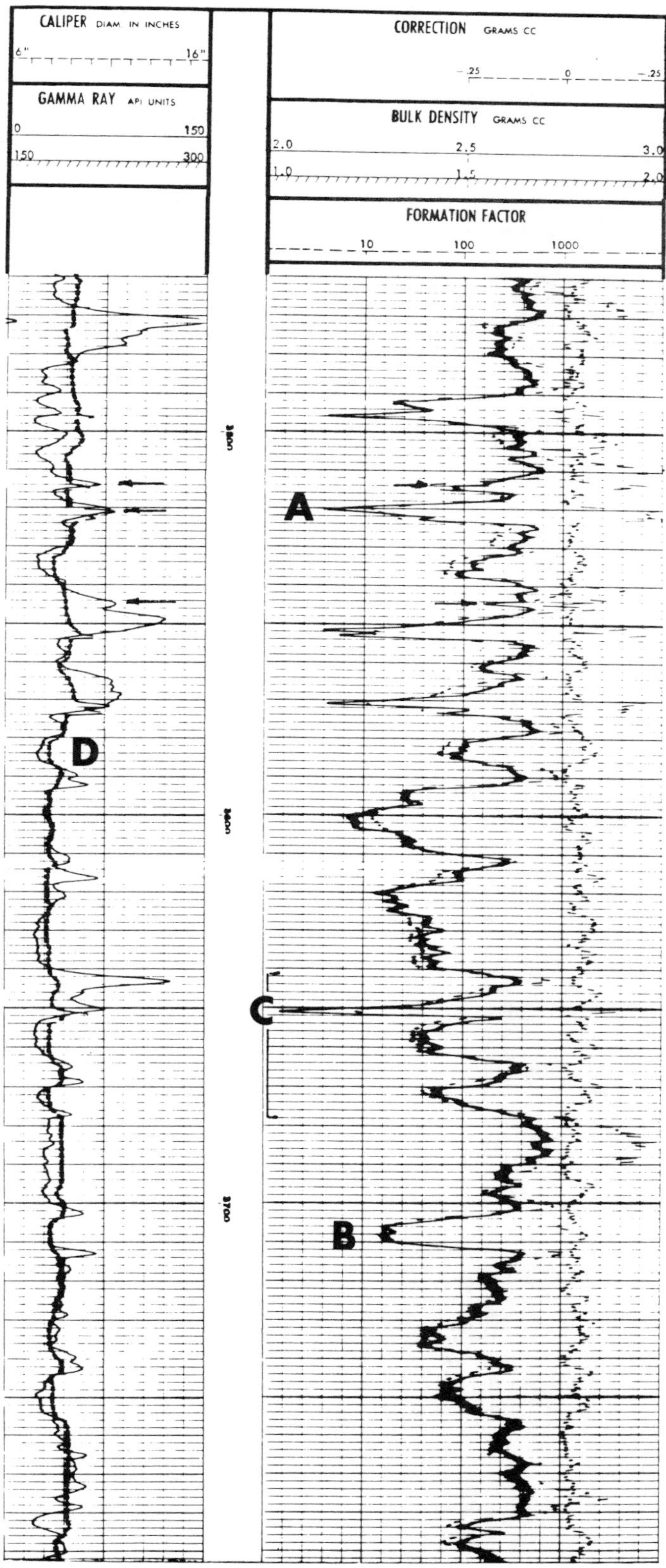
CALIPER DIAM IN INCHES
6"
16"
GAMMA RAY API UNITS
0
150
150
300
CORRECTION GRAMS CC
-.25
0
-.25
BULK DENSITY GRAMS CC
2.0
2.5
3.0
1.0
1.5
2.0
FORMATION FACTOR
10
100
1000
A
D
C
B

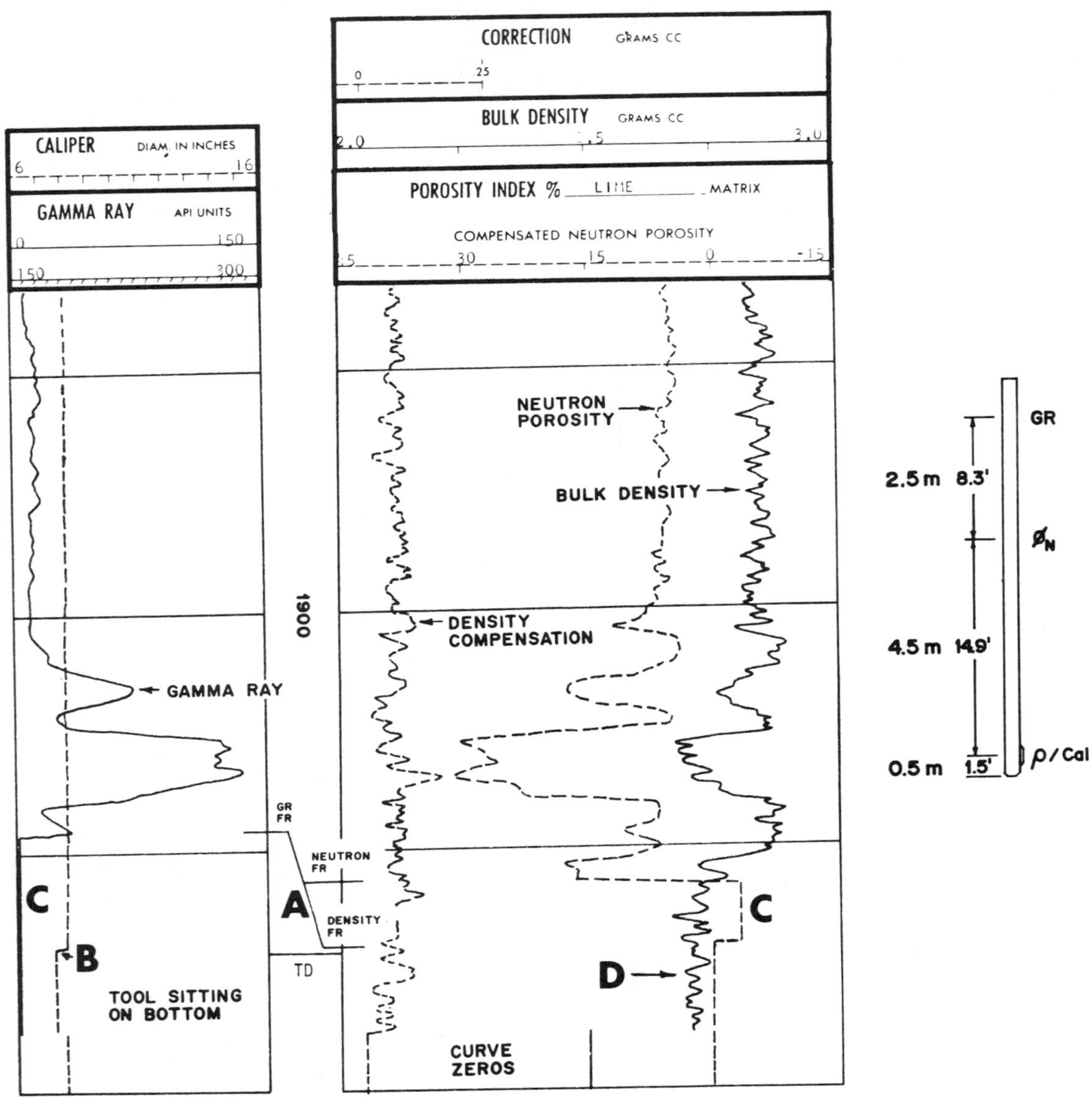

FIG. 3-14. False readings when the tool is sitting on the bottom of the borehole. The depths of the first valid readings of each measurement are usually marked in the depth track (A). Pickup is indicated by first movement of the caliper (B). The engineer switched in the memorizer at pickup, giving straight line readings (C) for memorized gamma ray and neutron. If the engineer had turned on the memorizer as winching up the cable began, these curves would have statistical readings like the density below its first reading (D).

For a stopped tool the detectors output a single value. The nonstatistical resistivity and acoustic measurements are recorded as straight lines (Figure 3-16, item B) and indicate the stopped tool situation. However this measurement for a stopped tool is not as obvious for statistical measurements (gamma ray, density, and neutron). Because of memorization, these false readings occur at different depths on the log and can be deceptive especially for the nearly always memorized gamma ray (Figure 3-16, item C).

If the gamma-ray portion of the tool stops in the space just below a sand and rebounds above the sand interval, the clean interval will not show on the gamma-ray presentation (Figure 3-17). Conversely, if the tool sticks while the gamma ray is opposite a sand interval, it incorrectly appears thicker (Cooley, 1974).

REFERENCES

Cooley, B. B., 1974, The delta tension curve for better log quality: Trans. 15th Ann. Soc. Prof. Well Log Analysts Logging Symposium, McAllen, Paper F.

Elliott, H. W. Jr, 1983, Some "pitfalls" in log analysis: Log Analyst, **24**, 1, 10-24.

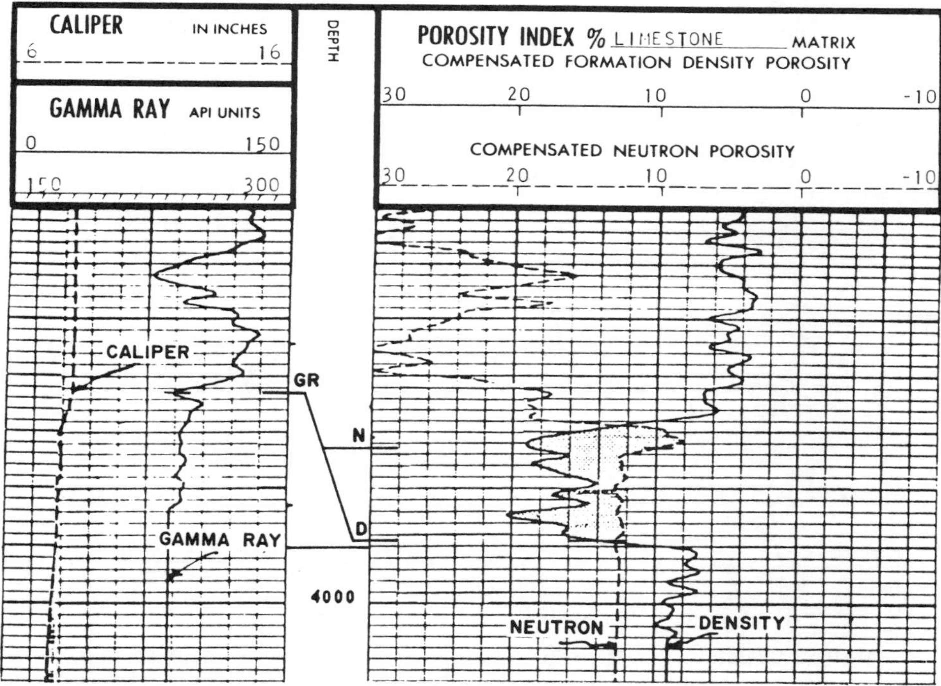

FIG. 3-15. False reading as the tool sits on the bottom of the borehole. The apparent gas cross-over of the neutron measurement from 3 969 — 3 939 ft is false, as the lowest valid neutron measurement is at 3 971 ft. The FR marks in the depth track were not on the original log (Elliott, 1983).

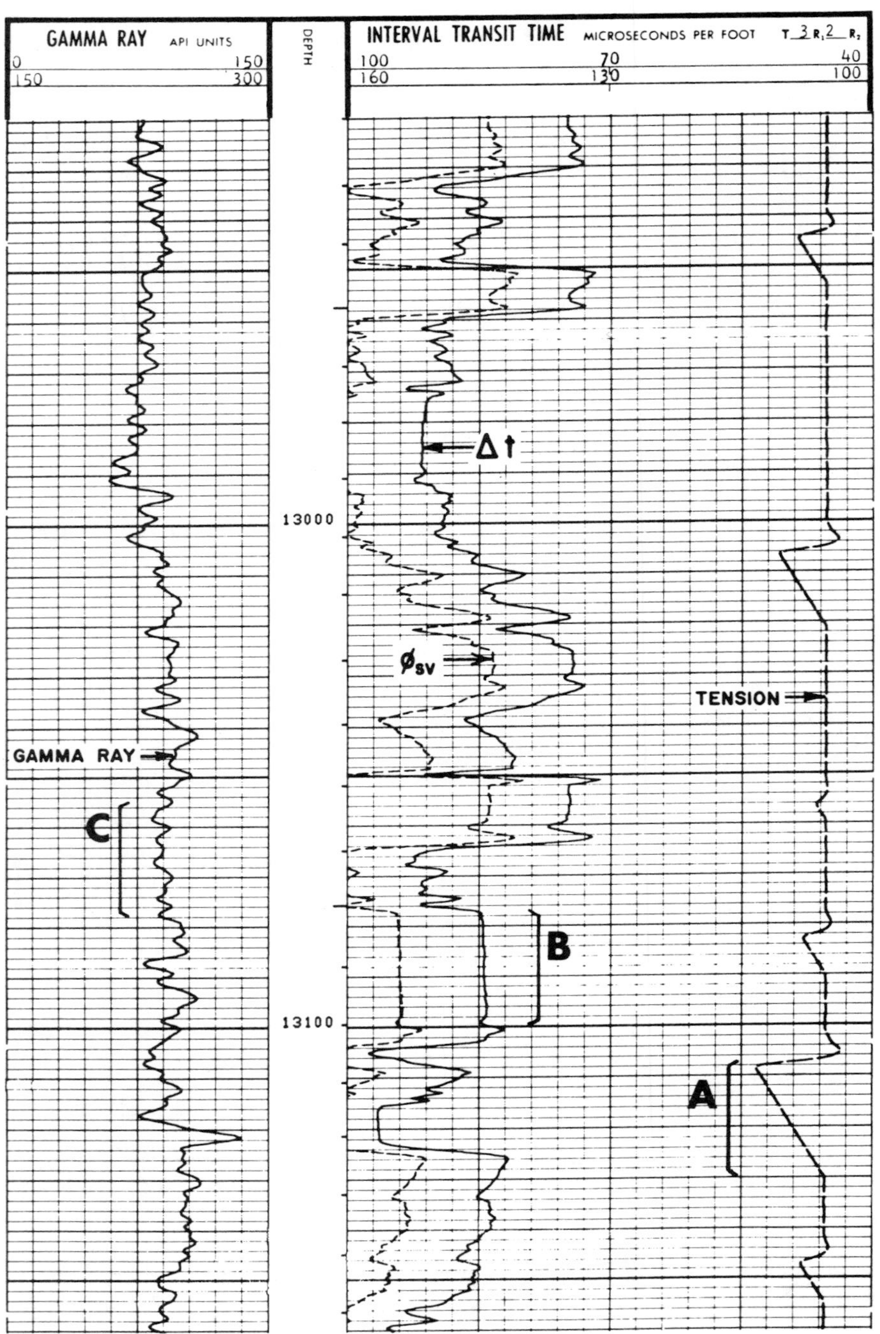

FIG. 3-16. False readings due to tool sticking, as shown by the increase in cable tension curve (A). The stopped tool is indicated by constant Δt readings (B). The gamma ray is also stopped, but due to the statistical nature of the measurement, it does not have a constant reading (C).

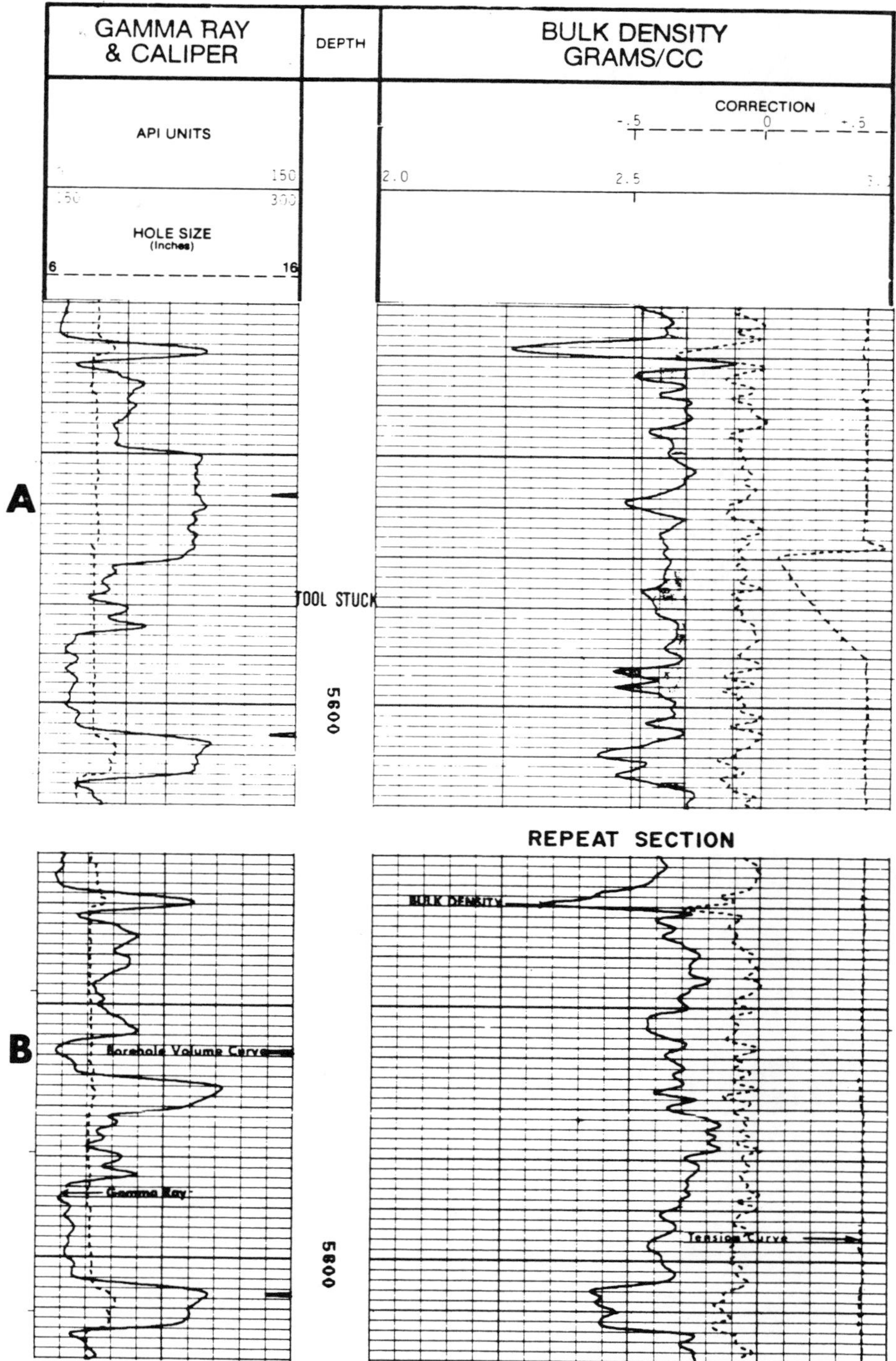

FIG. 3-17. False gamma-ray readings due to tool sticking. The false shale interval (A) on the main recording pass completely "covers" the clean interval (B) that is shown on the repeat section and on other logs.

Chapter 4

LOGGING OPERATIONAL ASPECTS

Operational aspects are events that occur during creation of the log that effect its usefulness. Individual measurement quality and depth compatibility of measurements to the formation as well as between sequential logs are dealt with. A basis for the editing of measurements between different logging runs over the borehole is also provided.

TOOL CALIBRATION PRINCIPLES

Tool calibration to known formation response is a multipart process. Analog systems use a three-part process: (1) adjusting the panel-to-recorder signal to a standard response, (2) checking the panel function-formers, and (3) calibrating the downhole detectors. Digital systems operate on the assumption that the recorder response and function former transforms are done by the computer program and therefore are correct. Thus, they present primarily the downhole detector calibration steps.

For analog systems the recorder adjustments (Figure 4-1, steps 1 and 2) set the mechanical zero and galvanometer deflections, without downhole signals present, for a standard response. Radiation tool panels use an internal test oscillator of several fixed count rates or ratios, duplicating the downhole measurement signals, to check function-former transform (Figure 4-1, steps 3 through 7).

The downhole tool is then calibrated by introducing a signal at the detector and adjusting the panel for proper galvonometer response. Resistivity tools use a precision calibration resistor switched into the measuring circuit when the tool is in the borehole. Radiation tools use secondary standards traceable to tool response in industry-agreed-upon primary standard formations (Figure 4-2). The secondary standards

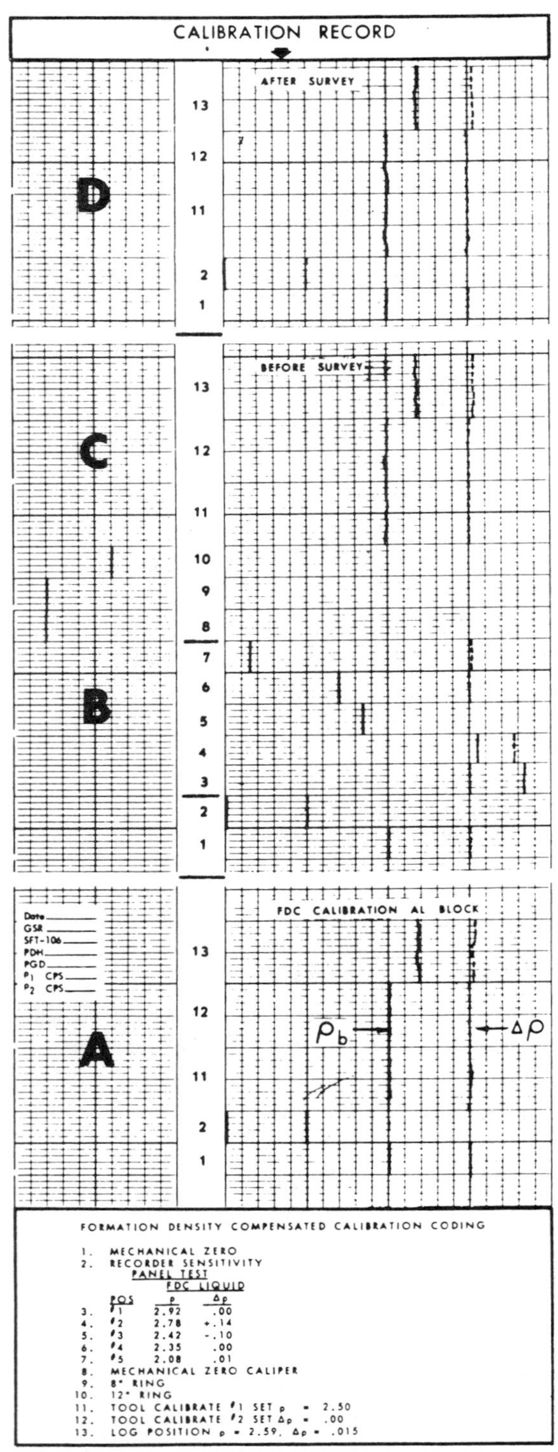

FIG. 4-1. Complete compensated density log calibration steps are shop calibration (A), panel test (B), before density survey (C), and after survey (D). The shop calibrations are done in the aluminum block secondary standard using the logging source. Additional steps 11, 12, and 13 may be shown to confirm the setting of the wellsite calibration jig. The panel-test checks the function formers using an internal test oscillator with no downhole input. The before-survey field calibration uses the wellsite calibration jig. Steps 11 and 12 may be done with a variable panel test osc. and then there would be no statistical variations. The after-survey is a check of key steps for calibration drift (Schlumberger, 1974).

FIG. 4-2. The Schlumberger compensated density tool secondary shop calibration standards are an aluminum and a sulfur block. When the density tool is calibrated to 2.59 g/cm³ in the aluminum block secondary standard, it would be able to measure correctly the industry agreed upon primary density calibration standards — Vermont Marble (2.675 g/cm³), Bedford Limestone (2.420 g/cm³), and Austin Limestone (2.211 g/cm³) — maintained by the A.P.I. in Austin. A low density check is provided by the sulfur secondary standard (1.90 g/cm³) for measurement "linearity." Varying thicknesses of artificial mudcakes, on top and inside left block, can be inserted between the tool and the secondary standards to check proper operation of the compensation system. After calibration in the secondary aluminum standard, the logging source is removed and a wellsite calibration jig is placed over the density detector and adjusted to duplicate the aluminum block response.

FIG. 4-3. The compensated density-compensated neutron tool (left) with density (A) and neutron (B) wellsite calibrators in place during field calibration. Also shown are the 3 — 5 ft borehole compensated acoustic tool (middle) and the dual induction tool (right), in its protective sleeve.

normally provide at least two reference values that bracket the expected formation range.

The compensated-density and compensated-neutron tools require massive secondary calibration standards that cannot be taken easily to the wellsite. Using their logging source they periodically are calibrated in one shop secondary standard and the response checked in a second standard. Then the logging source is removed and a portable wellsite calibration jig is fastened over the detectors. The shields for the low-level sources inside the jig are adjusted to make the count rates and curve responses exactly the same as if the detector were still in the first secondary standard (Figure 4-1, aluminum block steps). The field calibration jig, source, and detector then become a matched set. At the wellsite, the field standard is placed on the detectors (Figure 4-3) and the tool is calibrated to this response point (Figure 4-1, steps 11-3). Each step then is recorded for presentation with the log. The calibration jig is removed, with the logging source inserted just before the tool is lowered into the borehole.

After completion of the logging run, the field calibration jig is placed back on the tool and an after-calibration sequence is recorded on film without changing any calibration knobs to check for calibration drift. After-calibrations occasionally have been run before the logging run. Often in the analog system there is some drift of curve zeros during the logging run. A preliminary check should be made to compare the log mechanical zero at the top of the log with that on the after-calibration.

The increased accuracy of the digital surface systems eliminates the need for verification of tool signal-to-recorder transform response checks and thus eliminates the need for a digital equivalent of the panel test. Digital calibration is reduced to compensation of downhole tool variation. Calibration takes two forms; (1) digitization of the previous analog procedure or (2) direct use of the basic measurements.

Schlumberger uses the basic indirect method discussed earlier, a shop and wellsite calibrator sequence. At the wellsite the computer normalizes the wellsite calibration jig's response toward the count rates of the shop calibration (Figure 4-4). The results of this normalization are presented in print-out format and attached to the bottom of the log (Figure 4-5).

In contrast, Gearhart's digital system inputs the count rates of their two secondary shop calibration standards (aluminum and magnesium blocks) that are used to set up mathematical constants to convert the digital count rates directly to density values (Figure 4-6). These shop standard count rates are recorded in the log heading, with the calibration evidence reduced to a galvonometer response check (Figure 4-7).

OPEN-HOLE DEPTH CONTROL

Accurate depth control is maintained through the combination of an adjustable diameter measure wheel and magnetic marks every 100 ft along the cable. The moving logging cable turns the measure wheel that through a linkage system drives the recorder depth odometer (Figure 3-1). Because

(Text continued on page 56)

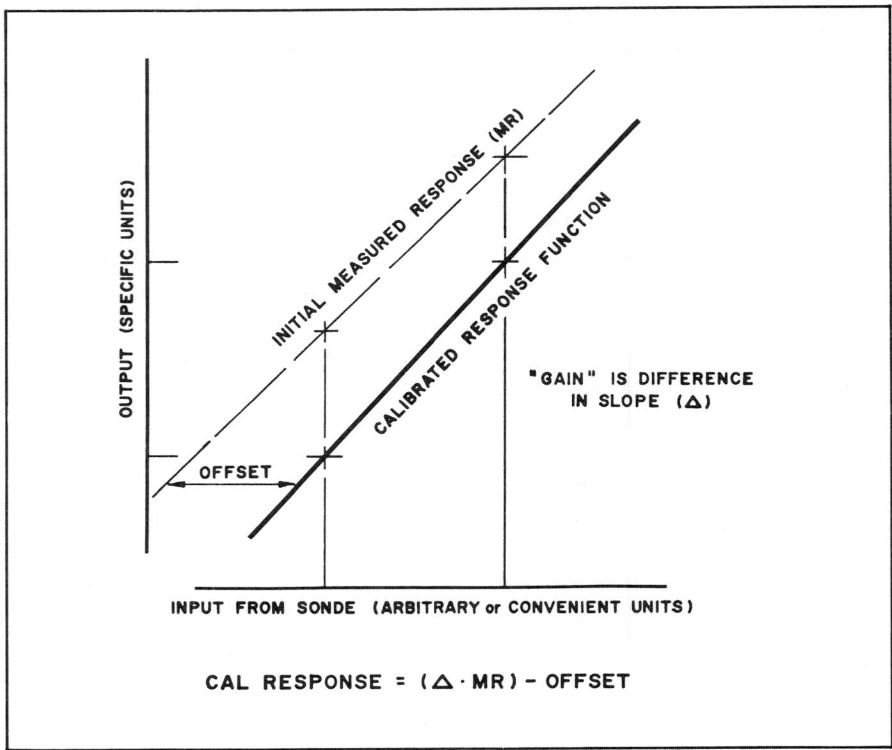

FIG. 4-4. Schlumberger digital calibration normalization principle. Known downhole inputs create an initial response line and with predetermined restraints the computer selects an offset and gain (Δ slope) to determine a calibrated response function that makes these inputs correspond to the required output values.

SCHLUMBERGER

AFTER SURVEY TOOL CHECK SUMMARY

PERFORMED: 78/09/26
PROGRAM FILE: NUC (VERSION 10.52 78/ 6/28)

PGTK TOOL CHECK

	JIG		
	BEFORE	AFTER	UNITS
FFDC	330	331	CPS
NFDC	538	536	CPS

BEFORE SURVEY CALIBRATION SUMMARY

PERFORMED: 78/09/26
PROGRAM FILE: NUC (VERSION 10.52 78/ 6/28)

PGTK DETECTOR CALIBRATION SUMMARY

	BLOCK	JIG		
	CALIBRATED	MEASURED	CALIBRATED	UNITS
FFDC	336	407	330	CPS
NFDC	527	647	538	CPS

SHOP SUMMARY

PERFORMED: 78/09/19
PROGRAM FILE: SHOP (VERSION 10.52 78/ 6/28)

PGTK DETECTOR CALIBRATION SUMMARY

	BLOCK		JIG		
	MEASURED	CALIBRATED	MEASURED	CALIBRATED	UNITS
FFDC	415	336	411	333	CPS
NFDC	634	527	632	526	CPS

(PGS:25 , PGC:78 , SFT:1324)

FIG. 4-5. Schlumberger digital processed compensated density log calibrations. The count rate normalization printout replaces the analog calibration film of Figure 4-1. However, the principles of calibration remain the same — a single point field calibration jig.

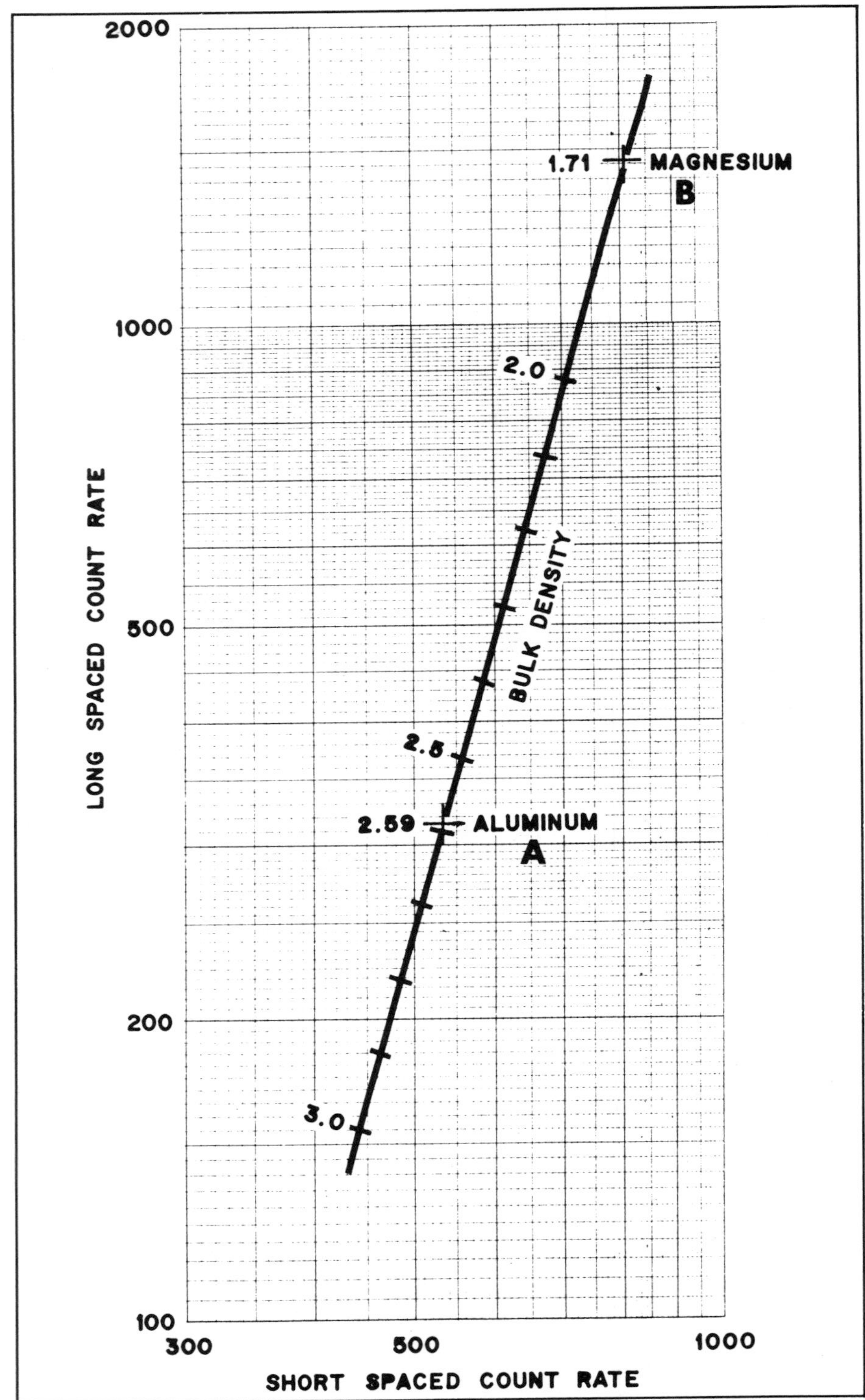

FIG. 4-6. Digital calibration of the Gearhart compensated density log. Using count rates established from two shop calibration standards (A and B). The tool then uses the equations established during this calibration to compute density directly from the tool near- and far-count rates during the logging run. See Head (1980).

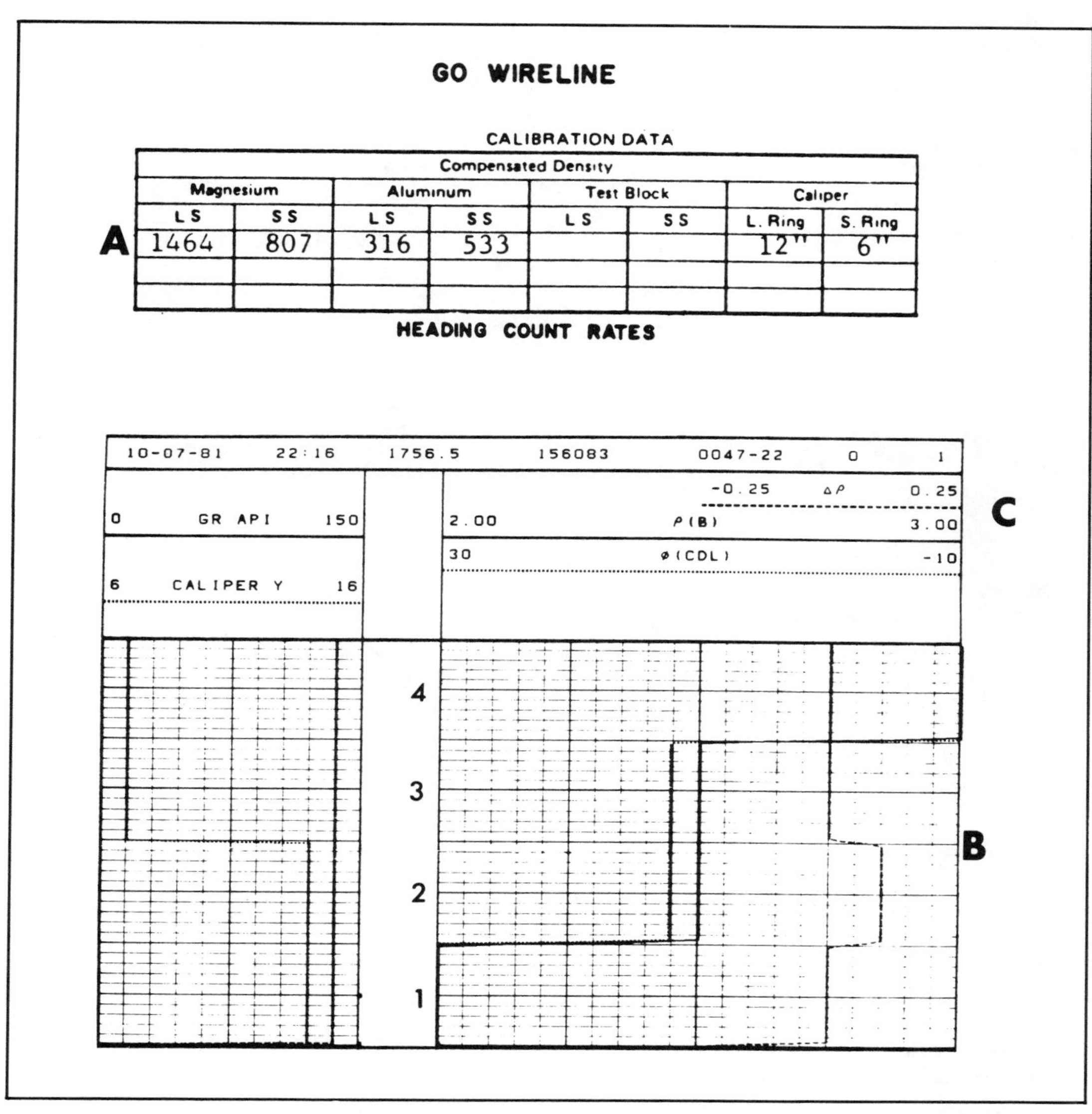

FIG. 4-7. Gearhart's direct digital calibration of their compensated density tool. The count rates for the two-field calibrator are in the heading (A). These count rates are input into the calibration process shown in Figure 4-6. The galvonometer check is included in the calibration tail (B). Note that the computer printing of the log scales positively denote the actual scales used (C).

FIG. 4-8. The driller's measured depth reference point is the top of the kelly bushing (KB). The kelly (A) is attached to the drill string and is turned by the kelly bushing (B) which is rotated by the rotary table (C). This system allows the drill string to turn the bit while allowing the drill string to lower as the bit cuts deeper. The elevation of the top of the KB is shown in the log heading (Figure 3-10, item B).

slight changes in the adjustable wheel circumference make a big difference in the measured footage, magnetic cable marks are needed as a correction. The linkage system crank is used to make the marks occur with the same last two digits showing on the recorder odometer — i.e., keeping the marks at 100 ft odometer intervals.

The tool is picked up into the derrick and slowly lowered until the primary tool measuring point is stopped at the drilling reference point, normally at the top of the kelly bushing (KB) (Figure 4-8, item B). Using the crank, the odometer is then set to zero and the tool is run into the hole past the first reliable mark detected at the measure wheel. The mark is kept at these last two digits as the tool is run to bottom.

A few hundred feet from the bottom an operator will be on the rig floor with a portable mark detector. The tool is brought up, taking slack

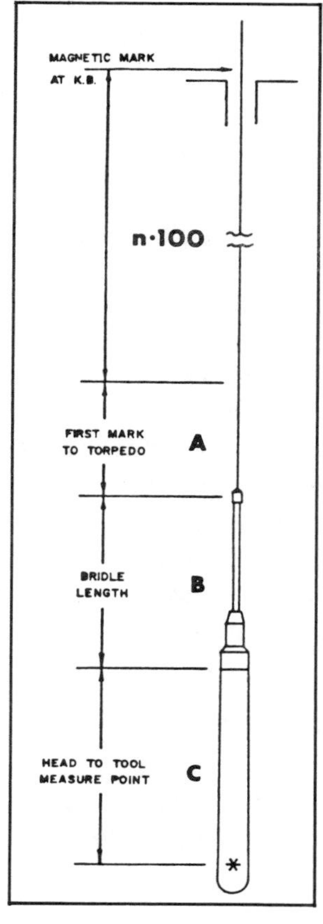

FIG. 4-9. Odometer corrections before logging for accurate depth control: with a magnetic mark at the drillers depth reference point (top of KB) the measure point of the tool is the distance for KB to the first (deepest) magnetic mark, plus the distance from the first mark to the tool measure point, or (n × 100) + A + B + C. If the measure wheel is approximately correct all that is needed is to correct the last two digits of the recorder odometer to the last two digits of the sum of A+B+C. For example, if A = 235 ft, B = 50 ft, C = 26 ft, then the total is 311 ft. Thus, with a magnetic mark at KB, the recorder odometer should be made to read xxx11. A correction for cable stretch (Figure 4-9) is then made. All the distances plus the date the cable was last marked should be posted inside the recorder cab.

out of the cable, and the mark detector is used to stop a mark at the drilling reference point. The odometer is then corrected to the last two digits of the sum of: the distance from the tool measuring point to the head (Figure 4-9, item C), the bridle length (Figure 4-9, item B), and the distance from the end of the cable to the first reliable mark (Figure 4-9, item A). These measurements are posted in the recording cab, along with the date the cable was last magnetically marked. Because the cable can undergo permanent stretch during use, periodically it needs to be remarked. A final correction due to downhole cable stretch is based on: the particular type and size of cable, cable tension, depth, and tool weight (Figure 4-10). After this correction the tool is brought up to detect the next mark at the truck and this mark becomes the starting mark for the logging run. During the logging pass, the engineer watches the mark detector and makes corrections when the odometer gets off this mark, while also evenly taking out the stretch correction over the borehole length. If it is necessary to correct more than one foot per 100 ft of hole, problems with the wheel or with the cable should be suspected.

Logs are run coming uphole to maintain accurate depth control. When the tool is run into the hole, there is slack in the cable and the odometer cannot indicate accurately the depth to the tool measuring point, as the tool drifts down through the mud above the indicated depth. By logging up, cable slack is removed, the resultant stretch accounted for, and the depth to the tool measuring point is determined accurately.

On subsequent logs, the tool is initially zeroed at the drilling reference and then run into the hole. A new bottom mark is calculated by starting with the original log mark and adding the difference between the tool lengths along with a correction for the difference in the tool weights and their borehole drag. Then a short section of log is run and optically checked for measurement deflections compared to the first-run log. Often the gamma-ray deflections are used for depth checks. This is all right as long as the gamma ray is correctly memorized during the check. If not, the result inevitably will be similar to the log in Figure 4-11. Apparently this log was run with the memorizer off, as the gamma ray is on depth with other prior logs but the prime density measurement is presented at least 20 ft too shallow. The engineer should make the same crank corrections, in amount and at the same depth points, as the first log to keep the logs on-depth with one another; if not, there will be periodic jumps in the log-to-log depth correlations.

REPEAT SECTION

The repeat section is run to check that the tool measurements can be duplicated. In older logs the memorizer is switched both out and in during

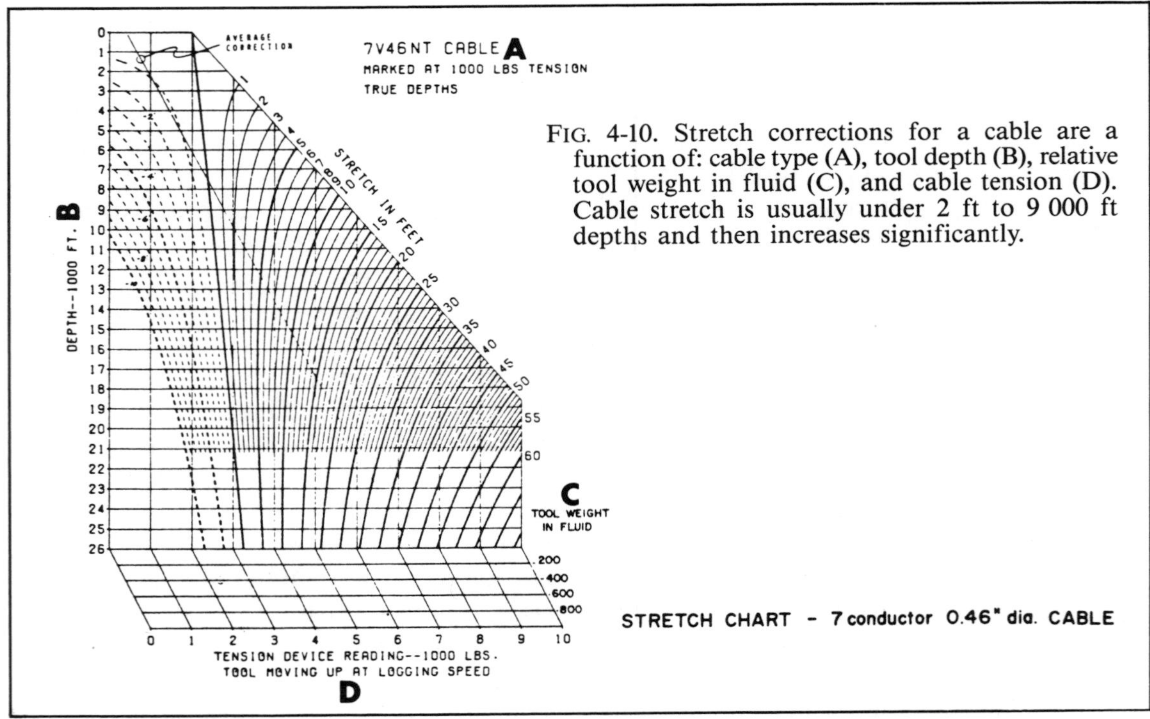

FIG. 4-10. Stretch corrections for a cable are a function of: cable type (A), tool depth (B), relative tool weight in fluid (C), and cable tension (D). Cable stretch is usually under 2 ft to 9 000 ft depths and then increases significantly.

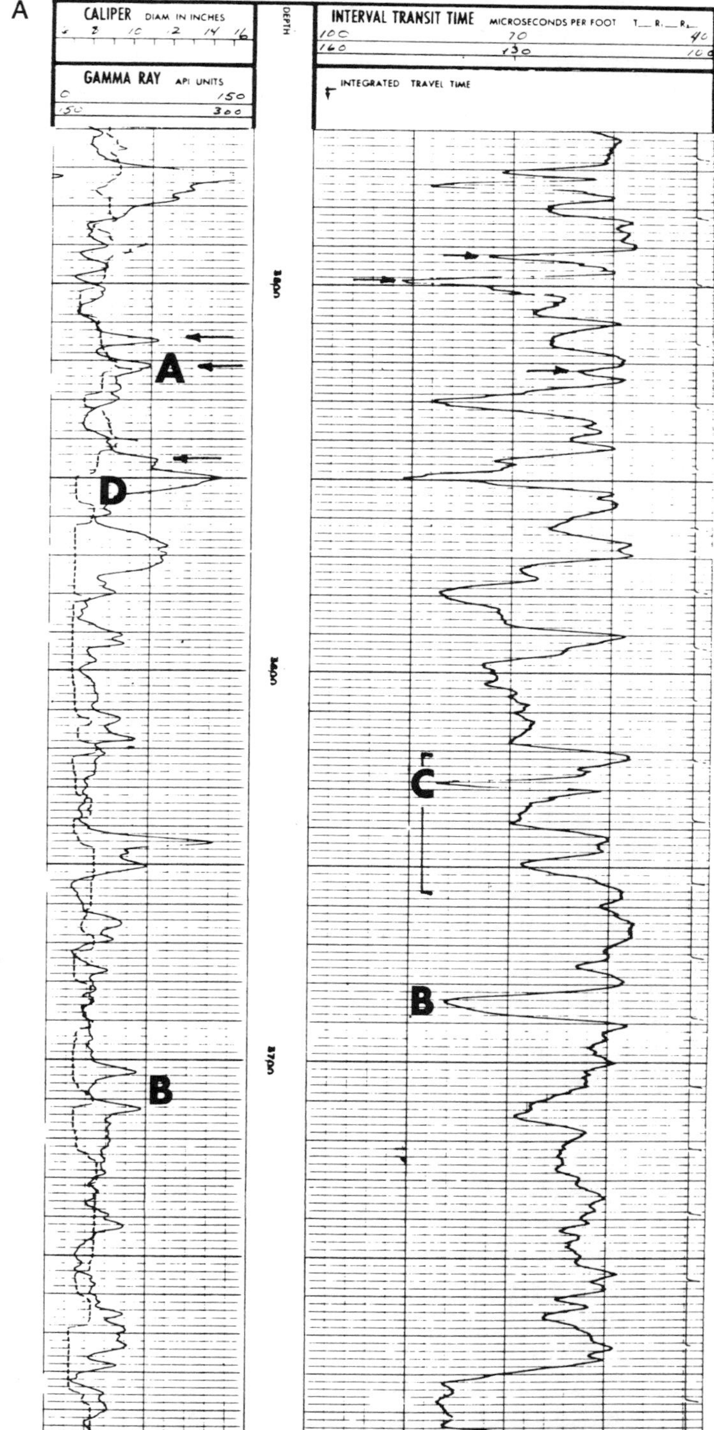

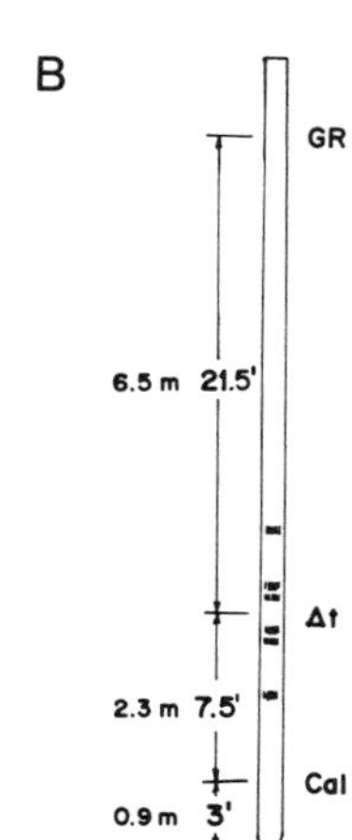

FIG. 4-11. Poor log-to-log depth control. The acoustic Δt measurement is at least 20 ft shallow as compared to the bulk density measurement (Figure 3-13, points B and C). However, the acoustic caliper differs by almost 30 ft (D) from the density caliper. The acoustic log probably was run with the memorizer off, normally gamma ray and Δt memorized to caliper. Instead of using the primary measurements for depth control, the two logs were depth tied using the gamma ray curves alone.

the repeat section to verify that it only changes depth of the readings and does not shift the readings. The repeat is normally limited to ±200 ft of hole and will normally be run on the bottom of the borehole *unless* the company representative specifies differently.

Some engineers use the repeat section to fine tune interlog depths which can cause a difference in depths between the repeat and the main logging pass. This should be noted on the repeat section and heading, but is often omitted.

RUN OVERLAP

Most deep wells are drilled in segments, with the casing run and cemented into the borehole before drilling deeper. Two problems in borehole measurements result: (1) a discontinuity in depth control, and (2) missing measurements at the bottom of each intermediate logging run.

With each logging run independently depth checked, depth discrepancies between logging runs are possible. To check for these discrepancies, the deeper logging run must be overlapped with the shallower run and the measurements to do this must be capable of detecting formation changes behind or through the cemented casing. An overlap section should be attempted on all measurements that can detect the formation through casing.

The easiest and most common overlap measurement is the natural gamma ray (GR). The presence of the casing and cement decreases the absolute count rates but the relative deflections are adequate to check depth control between logging runs.

The next best group of measurements is from the various neutron logs. The compensated neutron system allows the measurement of porosity through casing with reasonably good accuracy, compared with open-hole measurements (Figure 4-12). If a single-detector, uncompensated, epithermal neutron log is run, the neutron-porosity curve response cannot be corrected for the presence of the casing. However, the neutron-detector count rate yields an excellent correlation curve of the formation behind casing. This curve, similar in response to the older thermal neutron (gamma ray of capture tool), can be calibrated for qualitative analysis.

Under certain conditions the compensated density and acoustic logs can be used through casing. These conditions are discussed in Chapter 8 and Chapter 10, respectively.

An overlap section also allows filling the gap of the missing measurement below the shallower-run, first-tool readings. With combined tools this distance is significant especially for the acoustic log on top of a simultaneous resistivity-acoustic log (Figure 4-13).

In practice, the through-casing measurements should be continued for at least 200 ft above the casing shoe, or until some definitive markers have been logged. This ensures adequate information on both run-to-run depth control and on missing sections.

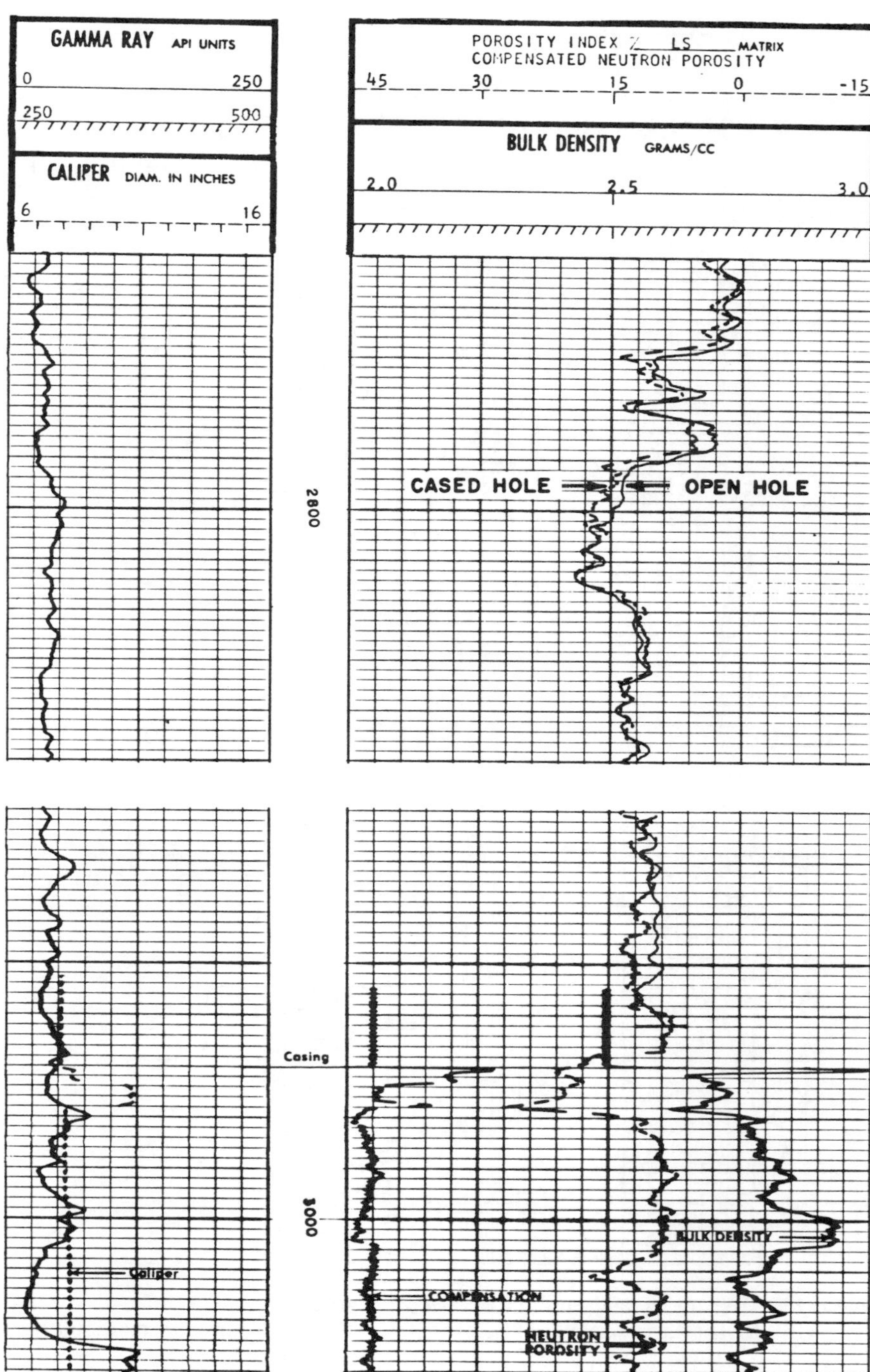

FIG. 4-12. The compensated neutron log run in open-hole (solid curve) and in cased-hole (dashed curve), which shows very good measurement agreement between the runs.

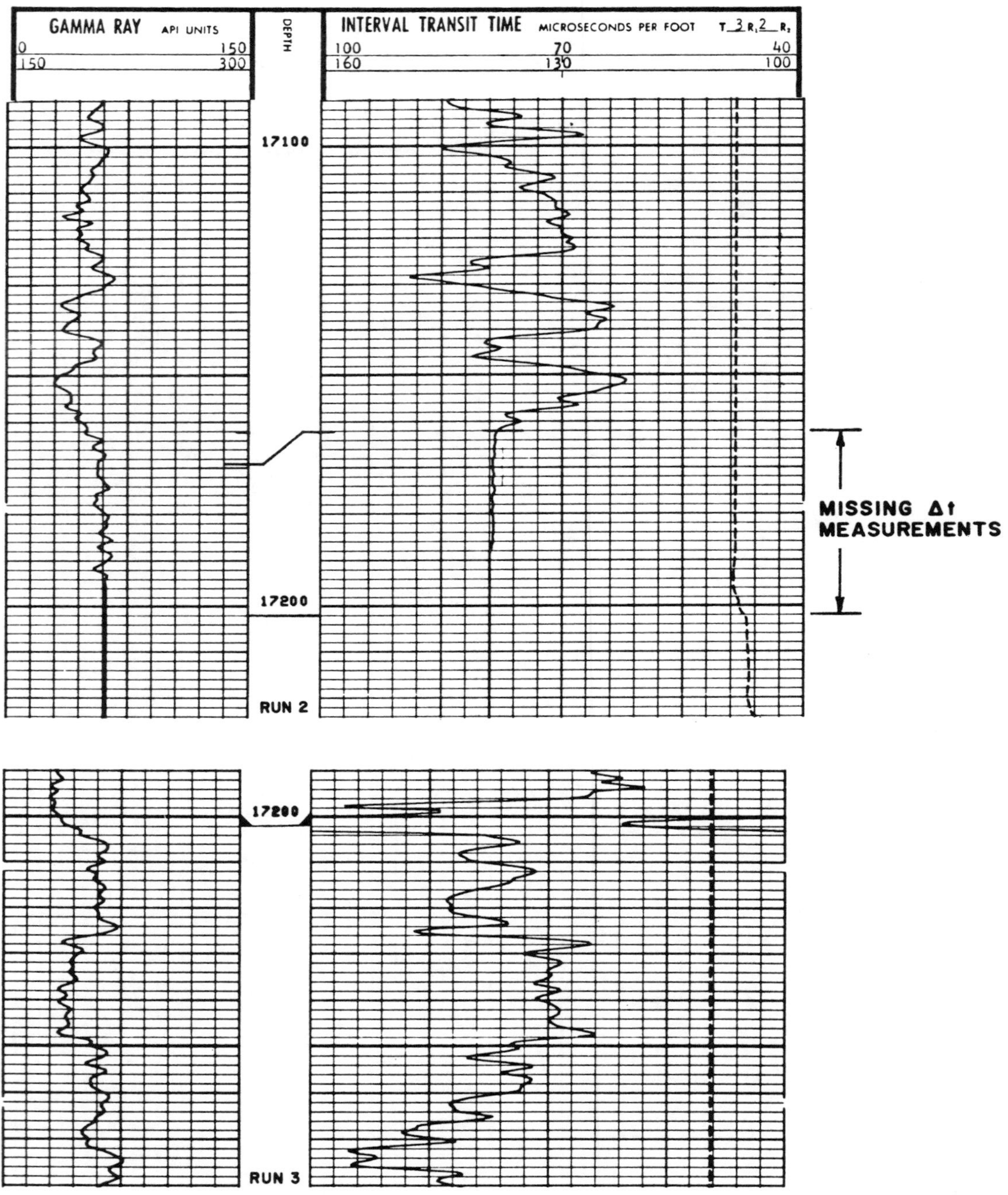

FIG. 4-13. Missing measurements between logging runs. 40 ft of the borehole is not recorded due to the large Δt FR-TD distance on the simultaneous resistivity-acoustic tool.

REFERENCES

Head, M. P., and Barnett, M. E., 1980, Digital log calibrations: The compensated density log: 55th Ann. Soc. Petr. Eng. Conference, Dallas, SPE-9343.

Schlumberger Well Services, 1974, Calibration and Quality standards.

REFERENCES FOR GENERAL READING

Bosworth, A. F., 1972, Log calibrations surface and downhole: Can. Well Logging Soc. J. **5**, 1, 39-68.

Evans, M. T., and Frost, E. Jr., 1983, Digital calibration of well log instruments: 57th Ann. Soc. Petr. Eng. Conference, San Francisco, SPE-12047.

Neinast, G. S., and Knox, C. C., 1973, Normalization of well log data: Trans. 14th Ann. Soc. Prof. Well Log Analysts Logging Symposium, Lafayette, Paper I.

Waller, W. C., Cram, M. E., and Hall, J. E., 1975, Mechanics of log calibration: Trans. 16th Ann. Soc. Prof. Well Log Analysts Logging Symposium, New Orleans, Paper GG.

SERVICE COMPANY REFERENCES

Schlumberger Well Services, 1975, Cased Hole Applications.

Schlumberger Well Services, CSU Calibration Guide.

Chapter 5

FORMATION EVALUATION PRINCIPLES

Borehole geophysical measurements are quantitatively used in obtaining indirectly a hydrocarbon indicator, which along with porosity (percent of pore volume) and pay thickness gives the hydrocarbon amount in place. The most common hydrocarbon indicator is water saturation (S_w), the percentage of the porosity that contains water. Use of hydrocarbon saturation is logical, but water saturation is the convention. With an S_w of 100 percent, the porosity contains all water and no hydrocarbons. Adding hydrocarbons into the formation porosity drives out some of the water and S_w decreases. Water saturation never goes to zero, because some fraction of water is always present as a very thin, tightly bonded coating on the pore surface.

ARCHIE'S EQUATION METHOD

Archie (1942) published an empirical relationship between S_w and core measurements that is the primary basis for quantitative formation evaluation:

$$S_w^n = F_R \, R_W \, R_t^{-1} \tag{5-1}$$

where F_R is formation resistivity factor, R_w is the connate (native) water resistivity, and R_t is true resistivity of the core and its fluids. This water saturation equation is often presented for graphical solution (Figure 5-1). Formation resistivity factor F_R is related to porosity through a second empirical relationship,

$$F_R = K_R \, \phi^{-m} \tag{5-2}$$

where K_R (traditionally "a") is an empirical constant that is one or less,

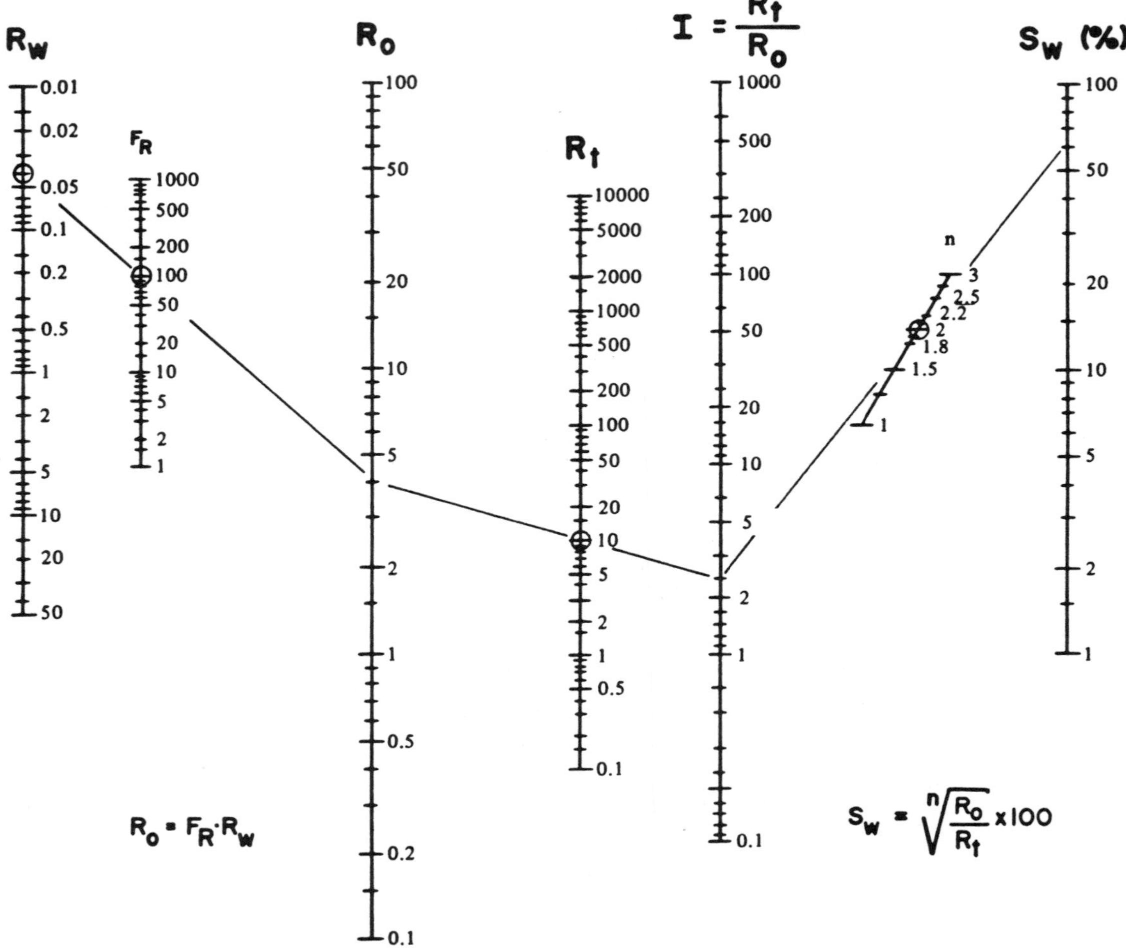

FIG. 5-1. Graphic solution of Archie's water saturation equation. If R_w is .04 Ω•m and F_R is 100 (approximately 10 percent porosity) and if the formation were water wet it would measure 4 Ω•m (R_o). If it actually measured 10 Ω•m (R_t) and n was 2.0, then the water saturation of the interval would be 63 percent.

m is an empirical constant related to cementation with a common range from one to three, and ϕ is porosity. Several combinations of K_R and m (Figure 5-2) are used depending on the degree of formation compaction and lithology.

In its simplest form, with $K_R = 1$, $m = 2$, and $n = 2$, Archie's water saturation equation reduces to

$$S_w^2 = R_w \, \phi^{-2} \, R_t^{-1}. \tag{5-3}$$

This procedure works better in the laboratory, where R_w, ϕ, and R_t can be measured independently and accurately. However, in the borehole a single measurement to exactly define each parameter in equation (5-1) is not probable. As a result, a number of log measurements aimed at determining the same parameters are run to eliminate additional unknowns introduced by the borehole environment.

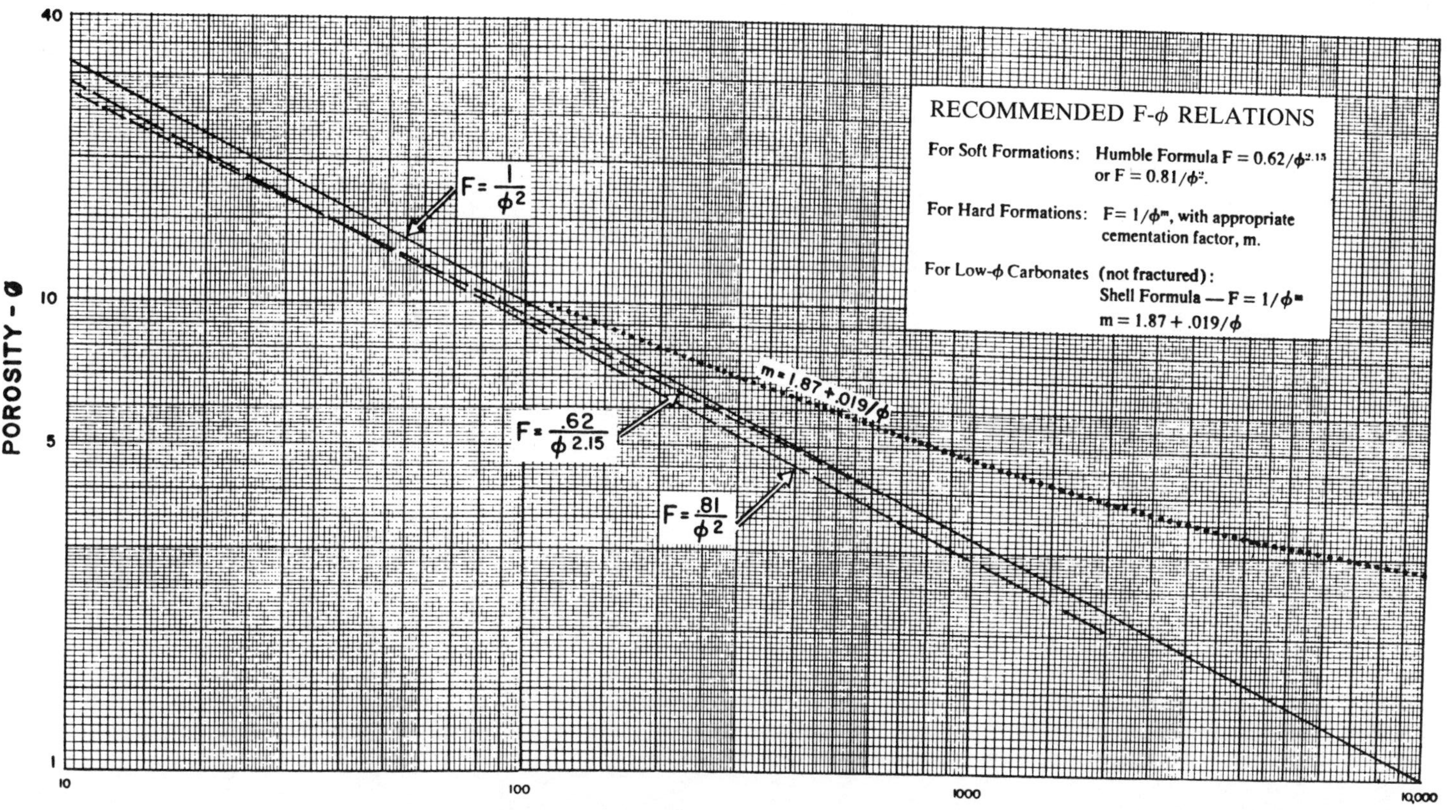

FIG. 5-2. Relationship between formation resistivity factor (F_R) and porosity (ϕ): the exact relationship to use depends primarily on the degree of formation compaction.

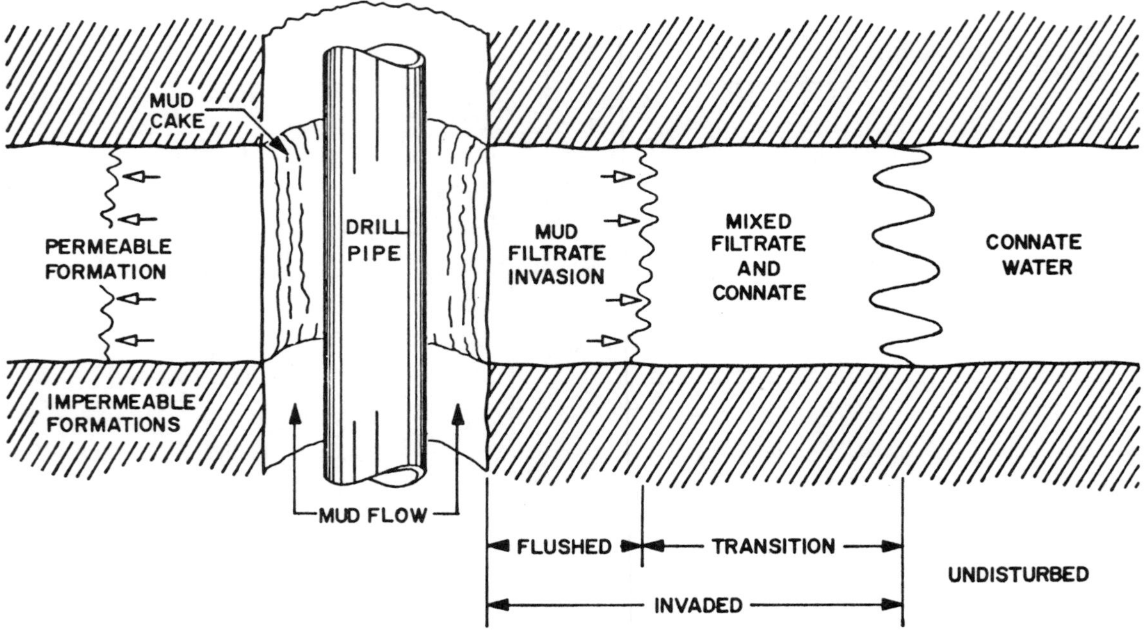

FIG. 5-3. Invasion of a water-filled formation. During drilling mud filtrate invades permeable formations displacing native connate water and forming a mudcake on the borehole wall. Water saturation is 100 percent in both the invaded and undisturbed zones. Adapted from Scientific Software Logging Manual.

CONNATE WATER

Connate water resistivity, R_w is determined either by measuring the resistivity of a drill stem or production test water sample from the wet part of the formation being evaluated, or by using a water catalog tabulation of measurements made in the same formation in the same geographic area. R_w can also be calculated from a nearby wet formation through use of the spontaneous potential (SP) measurement to calculate (R_{we}) or by back-calculating Archie's equation (R_{wa}). R_w normally can be predicted for each formation from surrounding wells and this value is commonly assumed in routine log calculations. On rank wildcat wells, however, R_w can be an elusive quantity.

FORMATION RESISTIVITY

The resistivity tools react to a combination of pore fluid content, porosity, interconnecting pore geometry, and lateral variation of pore fluid about the borehole. When solving equation (5-3) for R_t,

$$R_t = R_w \, \phi^{-2} \, S_w^{-2}, \qquad (5\text{-}4)$$

note that as S_w decreases (hydrocarbon content increasing) the formation resistivity increases. But resistivity can also increase for a porosity

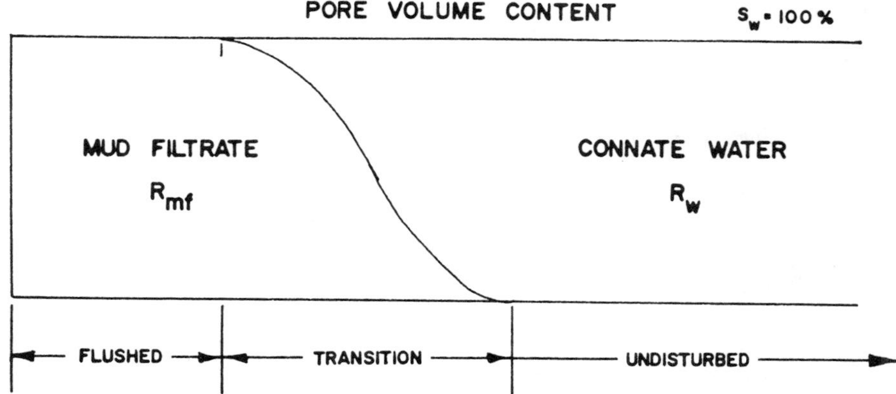

FIG. 5-4. Changing pore volume content with distance from the borehole, of a water-filled interval, as a result of mud filtrate flushing.

decrease (a tighter interval) or a R_w increase (a fresher connate water interval). Therefore, high resistivity alone cannot be used as a hydrocarbon indicator.

Lateral variations in pore fluid content greatly complicate the measuring of formation resistivity. To prevent blowouts wells are drilled with a positive hydrostatic pressure, i.e., with more pressure in the borehole than in the formation. The result is a flushing action along the borehole sides during drilling (Figure 5-3). The mud filtrate, the liquid part of the mud, invades the *permeable* formations surrounding the borehole and leaves the filtrate solids in a mudcake in the borehole walls with all or part of the original connate water driven away from the borehole. This flushing creates an invaded or flushed zone around the borehole with a transition zone between the flushed and undisturbed zone (Figure 5-4). The depth of invasion is determined by porosity, differential drilling pressure, permeability, water loss of the drilling fluid, and time.

For the general case equation (5-4) becomes

$$R_a = F_R \, R_f \, S^{-n} \tag{5-5}$$

where R_a is apparent or measured resistivity of a zone away from the borehole, R_f is resistivity of the fluid in that zone, and S is water saturation of that zone. If the saturation is 100 percent and if F_R is constant, the apparent resistivity of a zone is a function of the resistivity of the pore fluid(s). Because the mud filtrate (R_{mf}) rarely has the same resistivity as the displaced connate water (R_w) there is a resultant lateral variation of the apparent resistivity from the borehole into the formation (Figure 5-5). Because of this lateral resistivity variation, it is difficult to obtain R_t with a single resistivity measurement. Even though a resistivity device may be "tuned" to read as deep as possible beyond the invaded zone (Figure 5-6), it cannot compensate for the unknown depth of invasion and the nature of the transition zone. A solution is to use several resistivity measuring devices with each responding predominately to a different depth

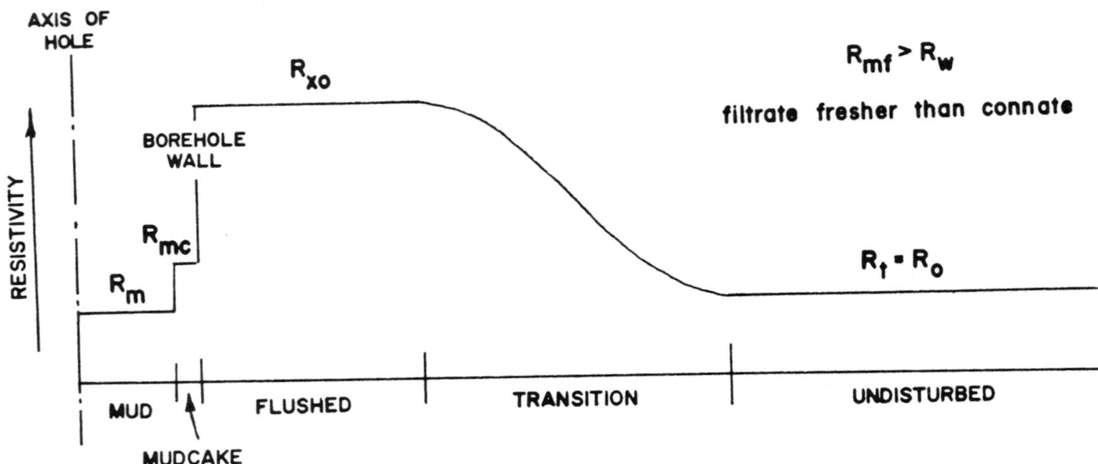

FIG. 5-5. Lateral variations of formation resistivity as a result of filtrate invasion of a water-bearing interval. If S_W is 100 percent and F_R is constant, the apparent resistivity varies as the resistivity of the pore fluids. Because $R_{mf} > R_w$, the common fresh drilling mud condition, then $R_{xo} > R_t$.

of investigation: deep, medium, shallow, and in some cases very shallow or micro (Figure 5-7). After assuming an invasion profile, the shallower reading devices are used to correct the deep device toward an undisturbed R_t value (Figure 5-8).

There are three types of resistivity measuring devices: (1) electrical (normals and laterals), (2) induction, and (3) focused resistivity (laterolog). Because each device has advantages and limitations, they are often used in combinations for the multiple resistivity tools.

Electrical devices. — Electrical resistivity measuring devices are the oldest of the three types. Electrodes are placed on the sonde and bridle, with current passing between one pair of electrodes and the resultant voltage drop measured across the second pair (Figure 5-9). The electrode spacing and configuration determines the depth of investigation. Greater spacing gives greater depth of investigation. Conductive fresh mud must be in the borehole for the system to work, while low-resistivity salt muds short circuit the current flow. High-resistivity formations force the current close to the borehole and therefore greatly reduce the recorded value and the depth of investigation.

The most common electrical survey (ES) includes a 16 inch short normal, a 64 inch long normal, and a 18 ft 8 inch lateral (Figures 5-10 and 5-11). The lateral is the deepest reading device. However, unlike most logs it has a nonsymmetrical measurement with unusual bed-boundary effects (Figure 5-12). Electrical logs have been superseded for most logging operations, except in high temperature hostile borehole environments where downhole electronic equipment is prone to fail. Because the electrical log was for a long time the only resistivity tool, well log files contain many of these logs: Complete explanations are given in Hilchie (1979) and Ross (1979).

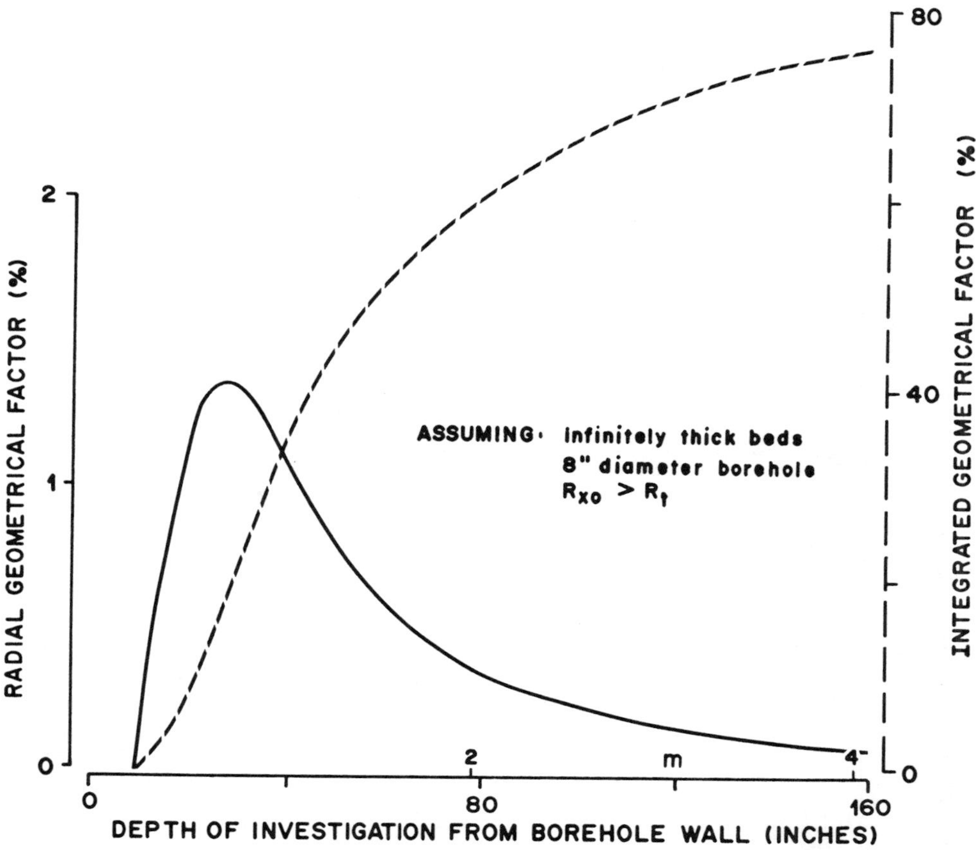

FIG. 5-6. Depth of investigation of Schlumberger's 6FF40 Deep Induction device. The radial geometrical factor (solid curve) indicates the signal contribution of each zone away from the borehole. The first 10 inches from the borehole do not influence the resultant resistivity measurement of this deep induction device. The same information is more commonly presented as an integrated geometrical factor (dashed curve), the summed contribution of all zones from the borehole to a given depth.

Induction devices. — Induction or focused conductivity devices induce a current into the formation about the borehole and measure a back-induced current that is proportional to formation conductivity, the reciprocal of resistivity (Figure 5-13). This device has a deeper depth of investigation and better vertical bed resolution than electrical devices. The primary disadvantage is that it is short circuited by low-resistivity salt mud. However, because the induction device does not use electrodes, it can be used in nonconductive mud, air, or gas-filled boreholes. Because it measures conductivity, it does not accurately determine resistivity above 100 Ω•m.

The induction-electrical survey (IES) contains a deep-reading induction device and a shallow-reading 16 inch short normal to indicate invasion (Figure 3-5). It uses a linear format with x10 scaled back-up resistivity curves. The dual induction tool contains an improved deep-reading

(Text continued on page 78)

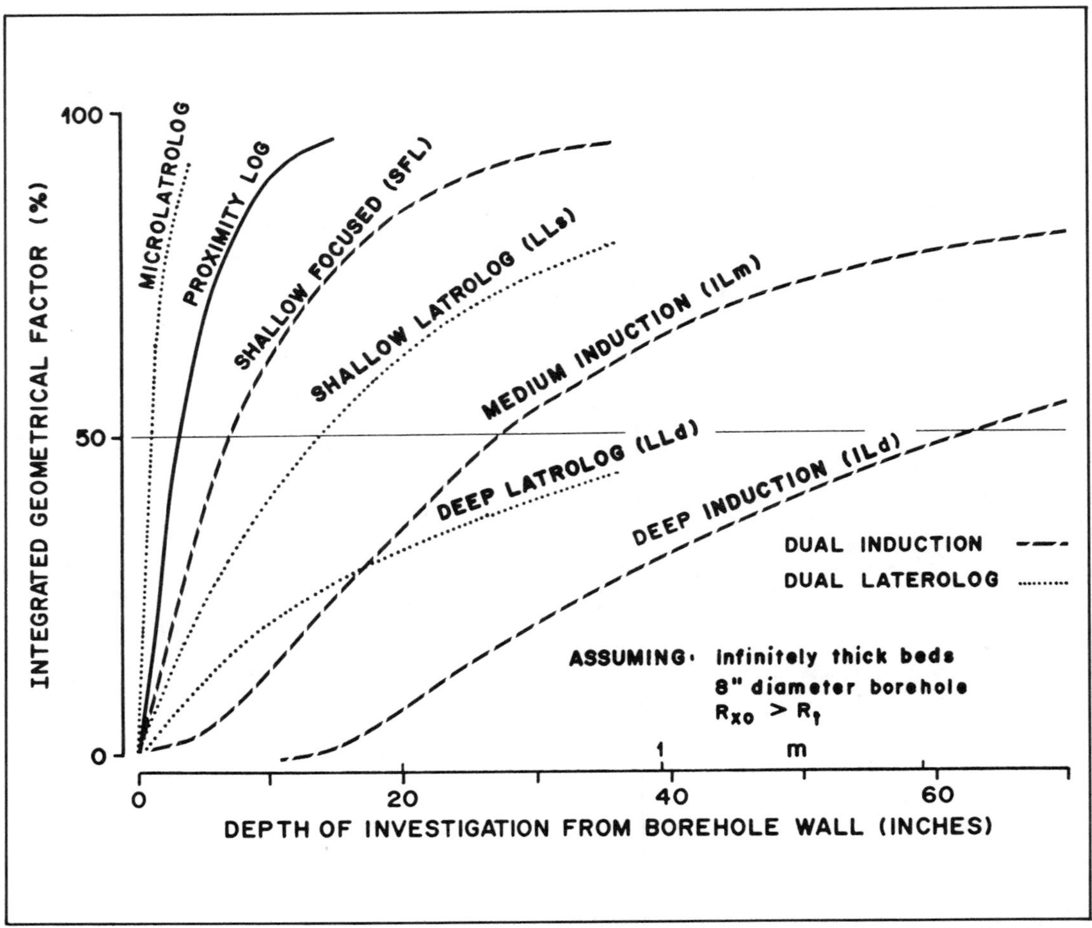

FIG. 5-7. The relative depths of investigation of modern resistivity devices as indicated by their integrated geometrical factor. The 50 percent points indicate that 50 percent of the resultant measurement is coming from the interval from the borehole to that depth. Adapted from Schlumberger (1972).

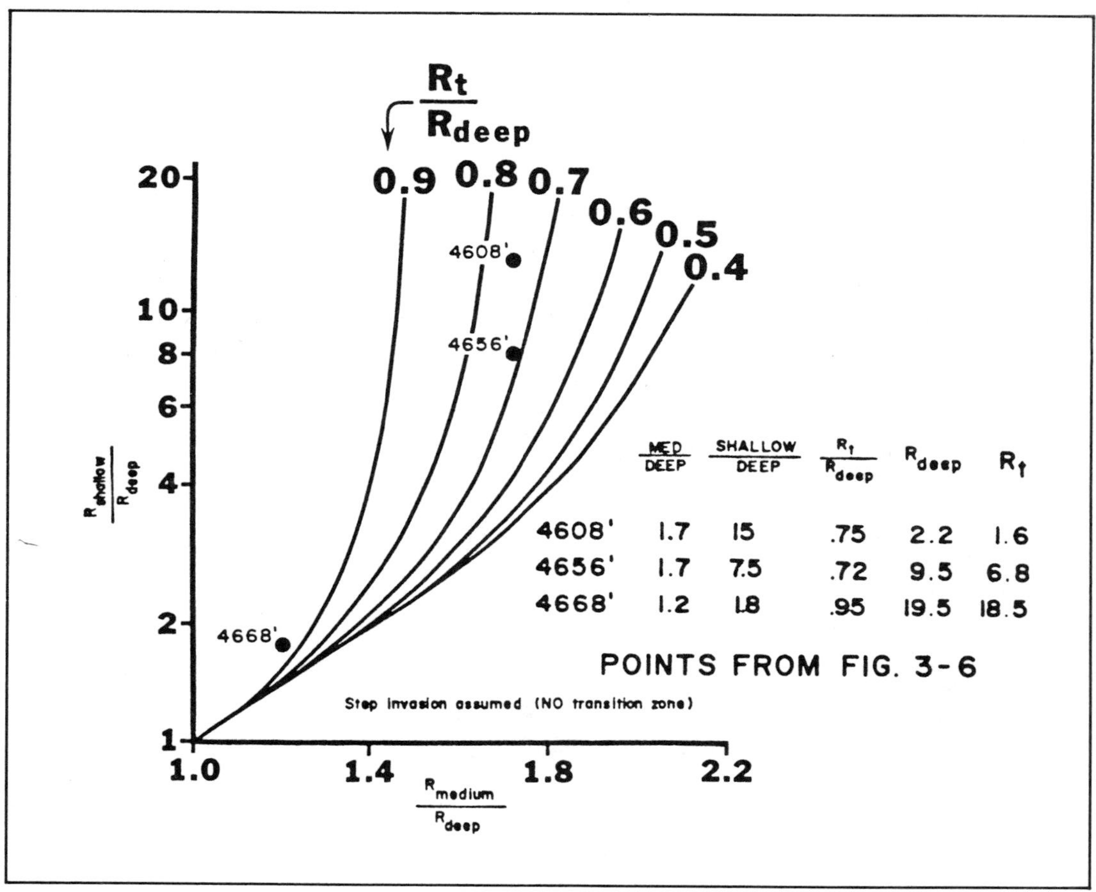

FIG. 5-8. Overcoming invasion influence to obtain an uncontaminated zone resistivity (R_t). The ratios of the three resistivity measurements are used to project R_{deep} to R_t. Note that the more the medium-deep separation, a larger ratio, the larger the correction applied to the deep resistivity. For medium-deep ratios of less than 1.2, R_t is approximately R_{deep}. There is a different correction chart for each service company tool. Also, a R_{xo}/R_t ratio and diameter of invasion estimate is obtainable using this crossplot technique.

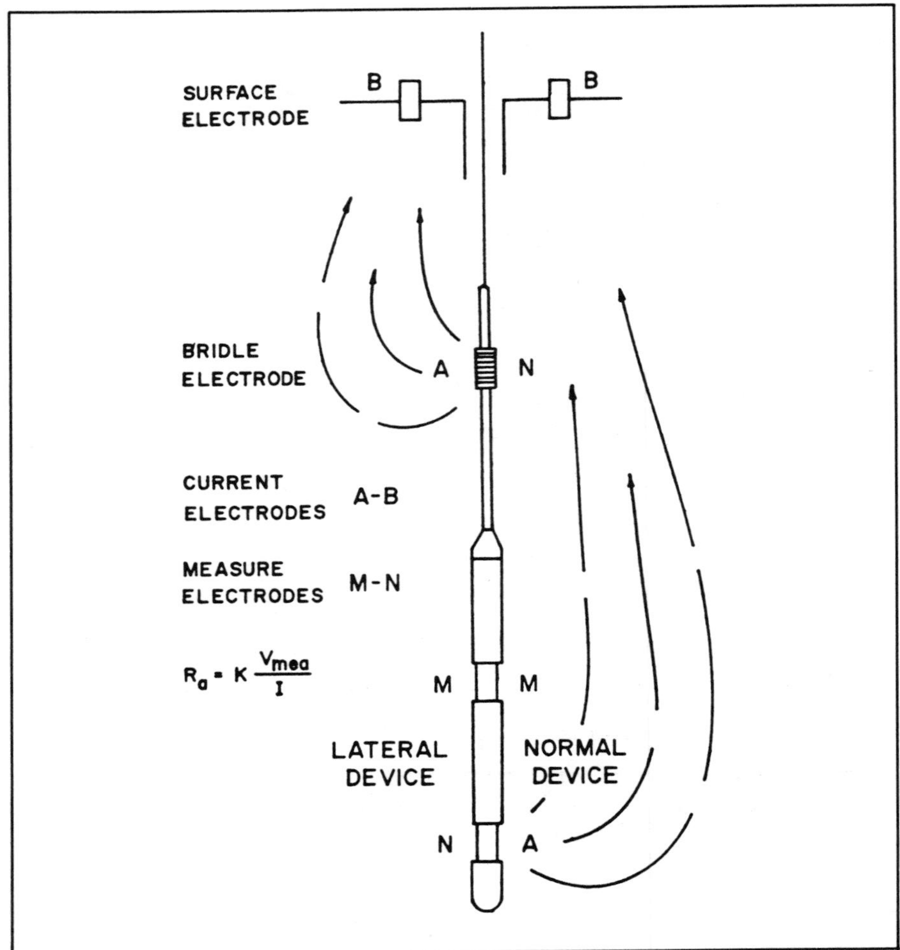

FIG. 5-9. Electrical resistivity measurement principles: normal electrode configuration to right and lateral configuration to left. Current flows between A and B electrodes with the resultant potential drop measured between M and N. K is a constant dependent on the electrode configuration. Spacing is the A to M distance for the normal and A to O (midway between M-N) for the lateral. Depth of investigation is twice AM for the normal and approximately AO for the lateral.

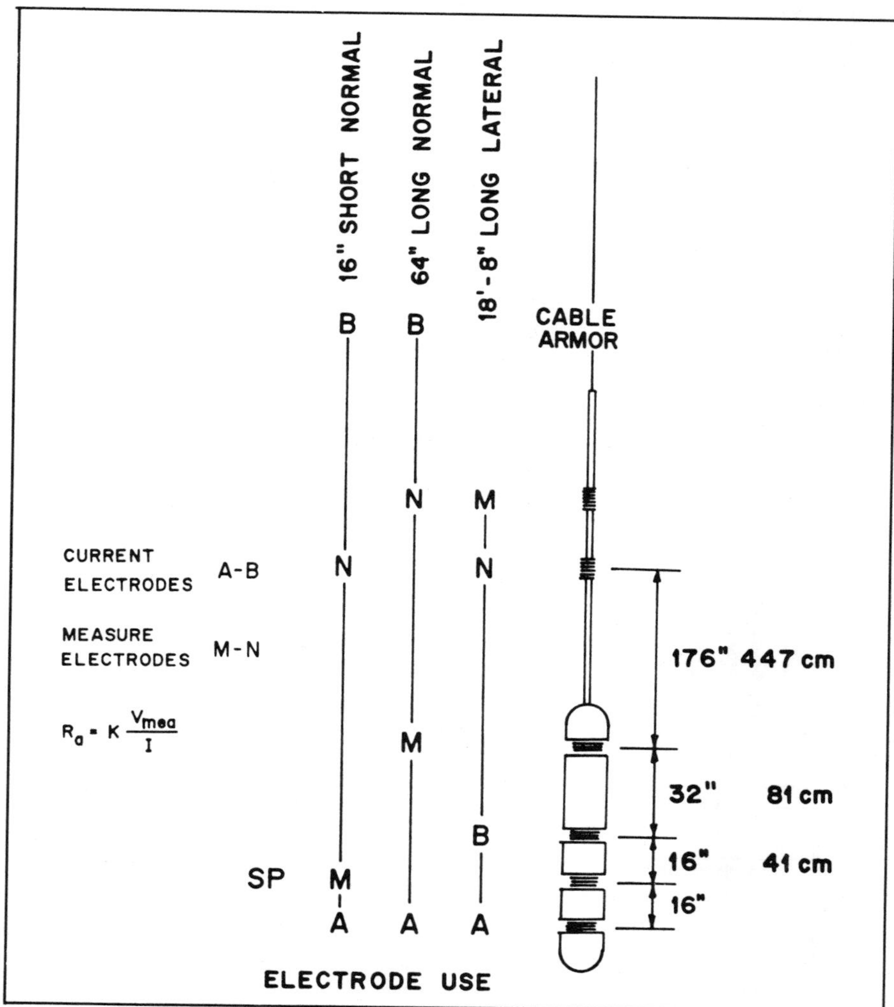

FIG. 5-10. The time sequential electrical log with seven multiple-use electrodes are connected in time sequence to record SP and three different resistivity measurements. The lateral electrode configuration is inverted from Figure 5-9 to be compatible with the normals; i.e., so each electrode is used only for current flow or voltage measurement.

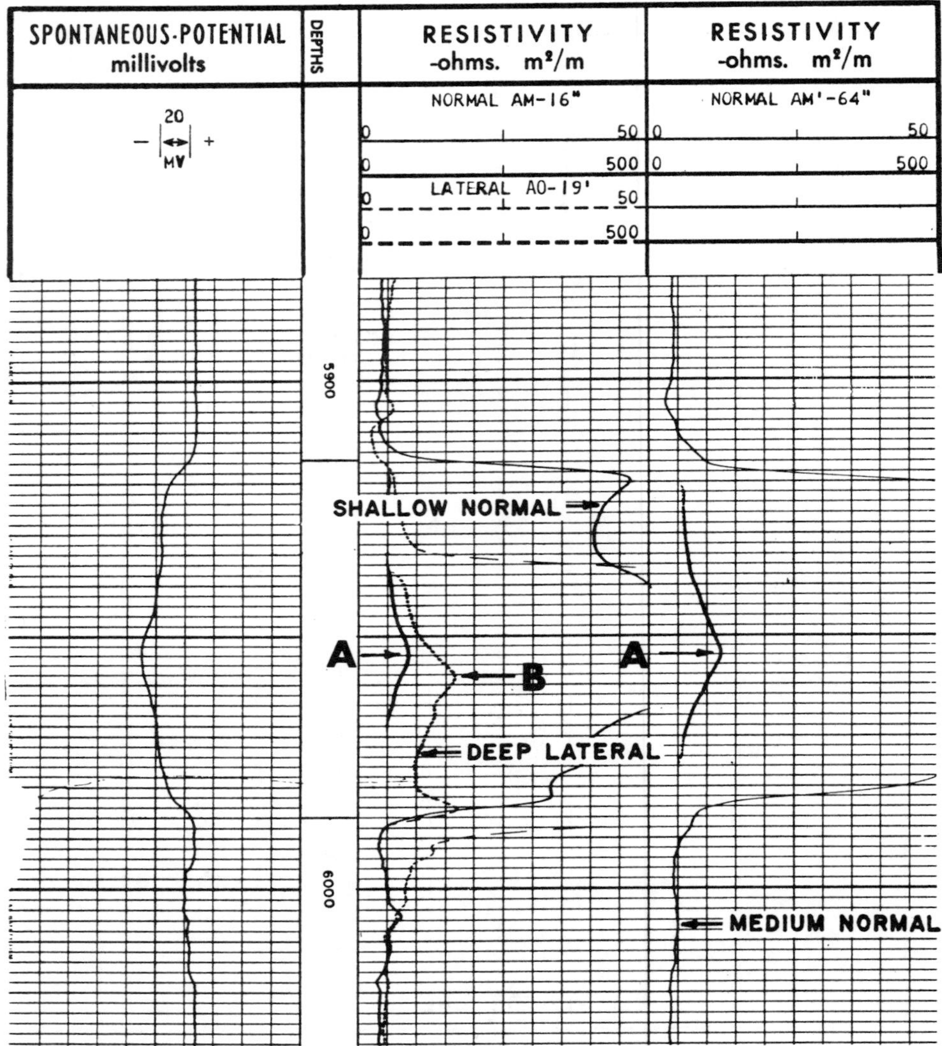

FIG. 5-11. Electrical survey presentation where the shallow-reading 16-inch normal and deep-reading lateral are presented in Track II. Some area presentations have both normals in Track II and the lateral in Track III. All curves are presented in resistivity units. Note the vertical displacement between (A) normal and (B) lateral resistivity peaks.

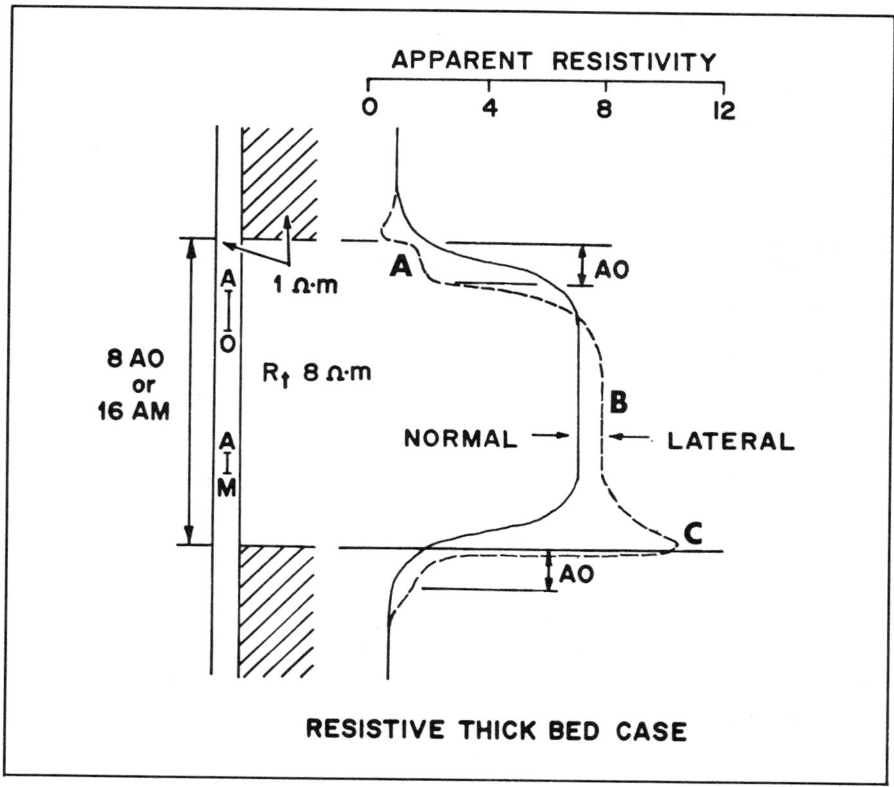

FIG. 5-12. Bed boundary effects of the asymmetrical lateral-resistivity device (dashed curve) as compared with the symmetrical normal device (solid curve). The normal never quite measures the true formation resistivity but the lateral does for this resistive thick-bed case. Note the significant displacement (A) of the upper bed boundary for the lateral. For a lateral with the typical 18 ft 8 inch AO spacing, this bed would have to be at least 150 ft thick. For beds of less than 6 AO thickness the lateral measurement never flat tops (B) at R_t, and the lower curve feature (C) predominates indicating a too high resistivity.

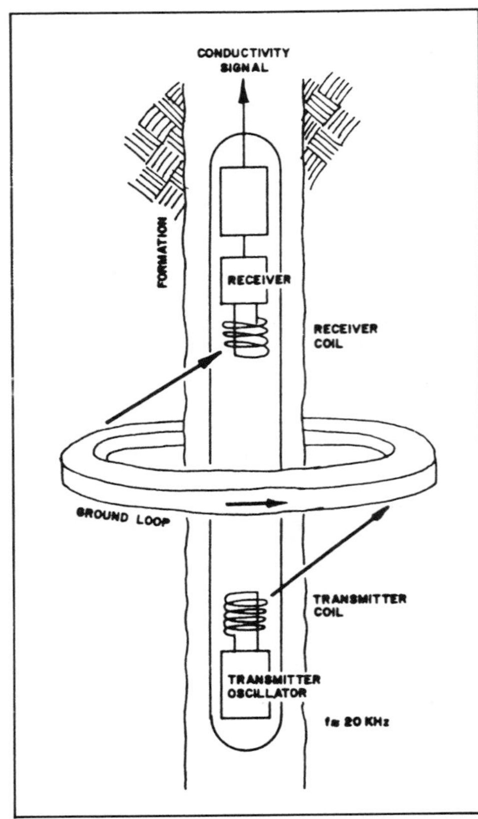

FIG. 5-13. Induction (focused conductivity) principles: the transmitter induces a ground loop about the borehole which in turn induces a current in the receiver coils that is proportional to the formation conductivity.

induction, a medium induction, and a shallow-measuring focused log (Figure 3-6). Invasion effects are determined by using three different depths of investigation. An approximation of R_{xo}/R_t can be included as a hydrocarbon indicator. The induction logs have moderate vertical bed resolution so a shallow focused log was added to better define bed boundaries. Either a logarithmic or a linear resistivity format for the 2 inch/100 ft correlation log is available.

Focused-resistivity devices. — Focused resistivity or laterologs are electrode devices that force a measuring current into the formation (Figure 5-14). They work poorly in fresh mud conditions when the mud resistivity becomes an appreciable part of the measured signal. However, they measure high resistivities better than the other two devices and they also have better vertical bed resolution.

The dual laterolog contains deep and shallow focused logs along with an optional pad, microfocused resistivity device (Figure 5-15). The shallow device is focused to be strongly influenced by the flushed zone. In conditions where the mud filtrate is equal to connate water resistivity, as it often is in salt-mud conditions, the apparent resistivity of a zone is a function of the water saturation of that zone for constant porosity [equation (5-5)]. For wet zones the saturations are 100 percent, and the measured resistivities at shallow and deep depths of investigation are the

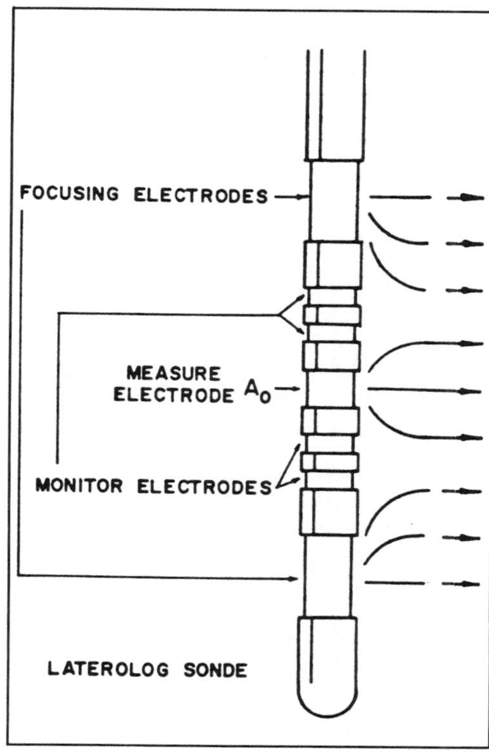

FIG. 5-14. Focused-resistivity principles: the current from the focusing electrodes is controlled by the monitor electrodes so that the measure current from electrode A_o flows out in a thin sheet into the formation even for resistive beds.

same. Because in an invaded hydrocarbon-bearing interval, the saturations are less than 100 percent and continue to decrease farther from the borehole (Figure 5-16), the shallow measurement can be used as a hydrocarbon indicator where it separates from, and is less resistive than, the deep laterolog (Figure 5-15).

The choice between induction and focused laterolog is based on mud resistivity, connate water resistivity, and porosity (Figure 5-17). Normally, induction logs are used in fresh mud and focused logs in salt mud. However, if the formation resistivity is high, over 200 Ω•m, the focused log is preferred.

The electrical and focused-resistivity principles are also used for very shallow or microdepths of investigation aimed at measuring the flushed zone resistivity (R_{xo}) and detection of permeable intervals. To achieve this very shallow depth of investigation the spacings are small and they must be mounted on a liquid filled pad which has been hydraulically forced against the borehole wall. The most common R_{xo} tool is the combination microlog-microlaterolog (Figures 5-18 and 5-19). The nonfocused microlog detects permeable intervals using a microinverse (microlateral) and a slightly deeper reading micronormal device. In a permeable interval a mudcake forms that is detected when the microinverse reads less than the micronormal, termed "positive separation." The flushed zone resistivity is measured by a shallow measuring microlaterolog, proximity log, or microspherical log that is only focused for a few inches into the

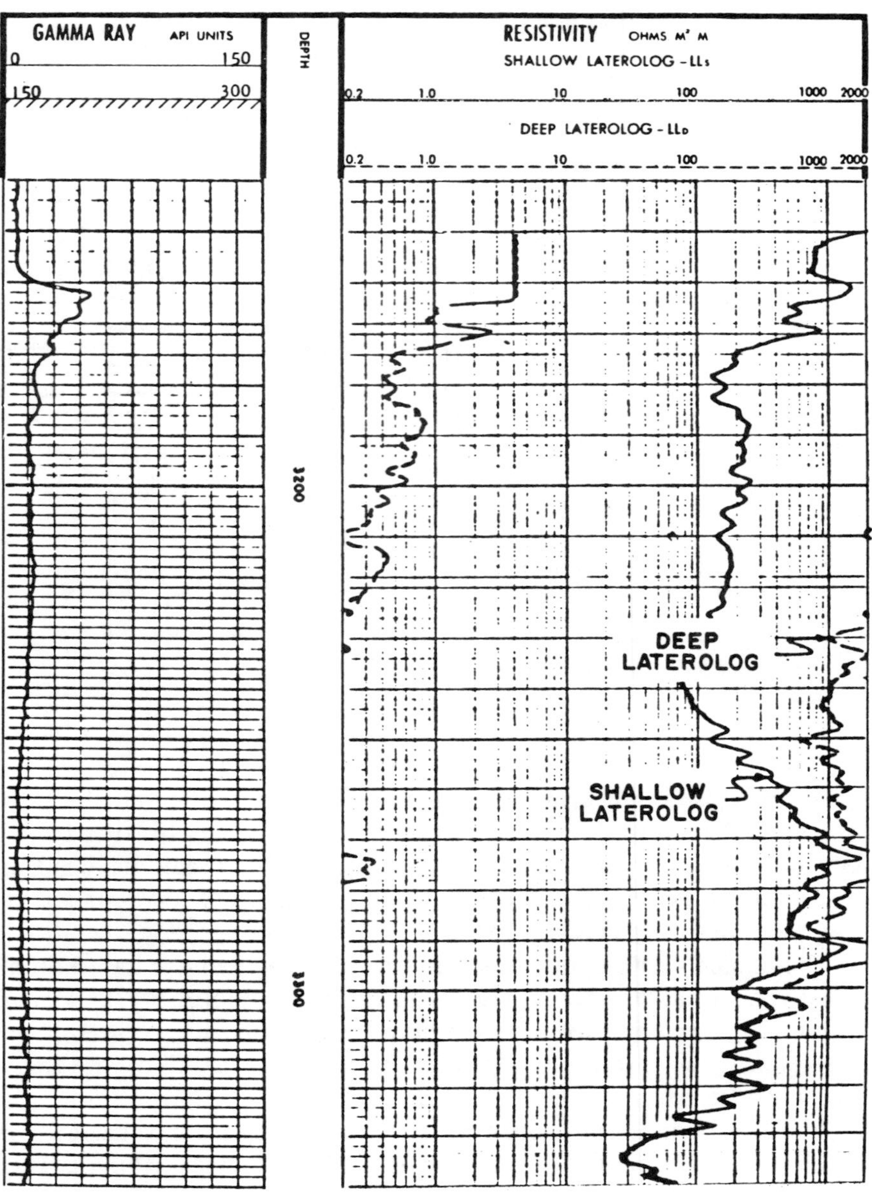

FIG. 5-15. Dual laterolog presentation: Schlumberger's deep reading device measures up to 40 000 Ω•m, the backup curve enters from the left on a 2 000 to 20 000 Ω•m scale. In salt mud conditions R_{mf} and R_w are often equal, so mud filtrate flushing does not change the measured resistivity of the deep and shallow laterologs, i.e., they track in wet intervals. In hydrocarbon-bearing intervals the mud filtrate flushing changes the water saturation near the borehole causing the curves to separate (equation (5-5) and Figure 5-16) Because the filtrate is less resistive than the displaced hydrocarbons, the shallow will read less in the hydrocarbon-bearing intervals — from 3 150 to 3 306 ft in this example. A microresistivity device can be added to the tool for a better R_t determination using a crossplot technique as shown in Figure 5-8.

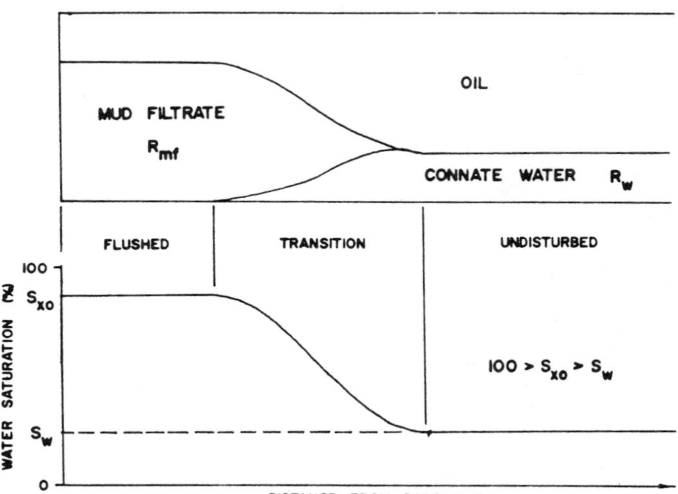

FIG. 5-16. Invasion of a hydrocarbon-bearing zone: The water saturation changes laterally due to mud filtrate flushing of hydrocarbons away from the borehole. This variation of water saturation can indicate the presence of hydrocarbons (Figure 5-15 and Figure 6-8).

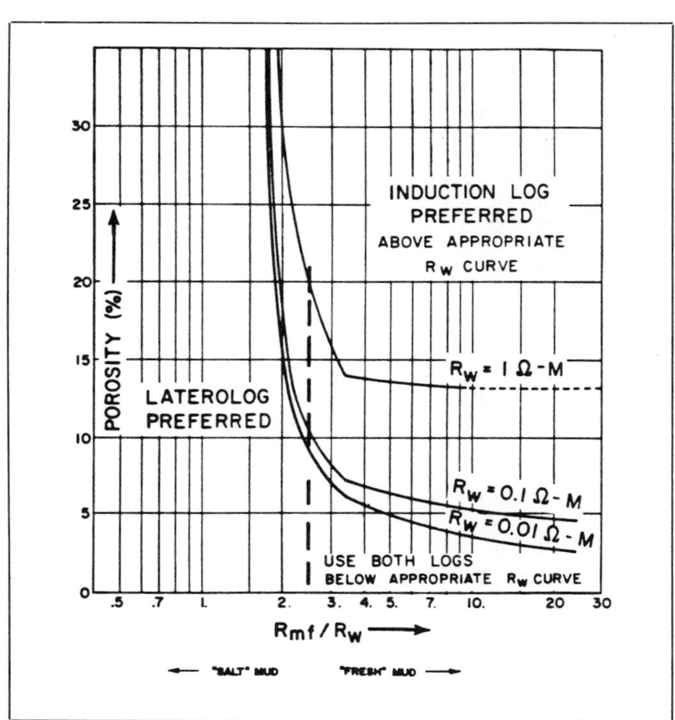

FIG. 5-17. The choice of induction or laterolog for resistivity measurement is mainly based on contrast between the mud filtrate (R_{mf}) and connate water (R_w) resistivities, with some influence of porosity. (Schlumberger, 1972.)

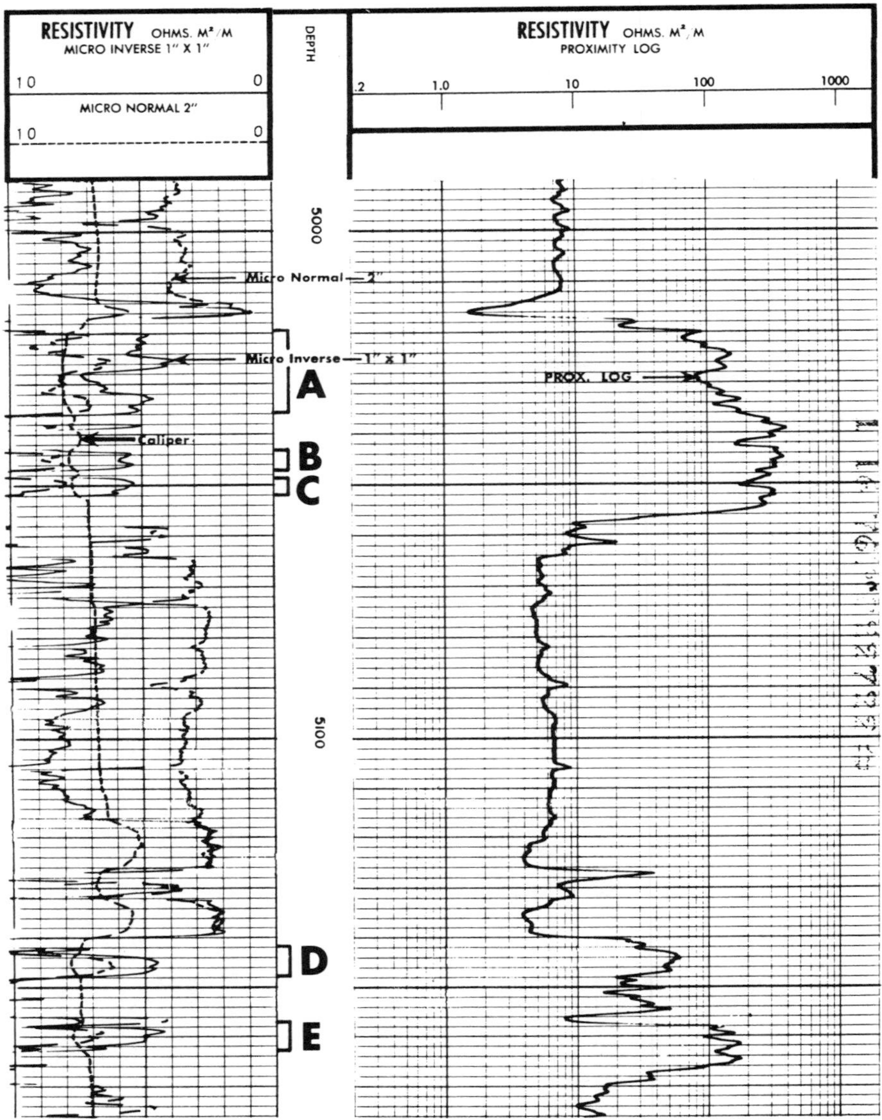

FIG. 5-18. The shallow reading microlog-proximity tool uses two pad-mounted sets of electrodes on opposite sides of the tool (Figure 5-19). The intervals A through E display "positive separation" of the microlog, i.e., the microinverse reading less than the micronormal, indicating mud-cake-forming permeable intervals. A shallow-reading microlaterolog is used in salt mud conditions in place of the proximity log.

formation. The R_{xo} device is mounted on a second pad on the opposite side of the tool from the microlog pad (Figure 5-19).

The presence of metallic casing makes the measurement of formation resistivity impossible. Thus, a method other than Archie's water saturation equation must be used to determine S_w for formations behind casing.

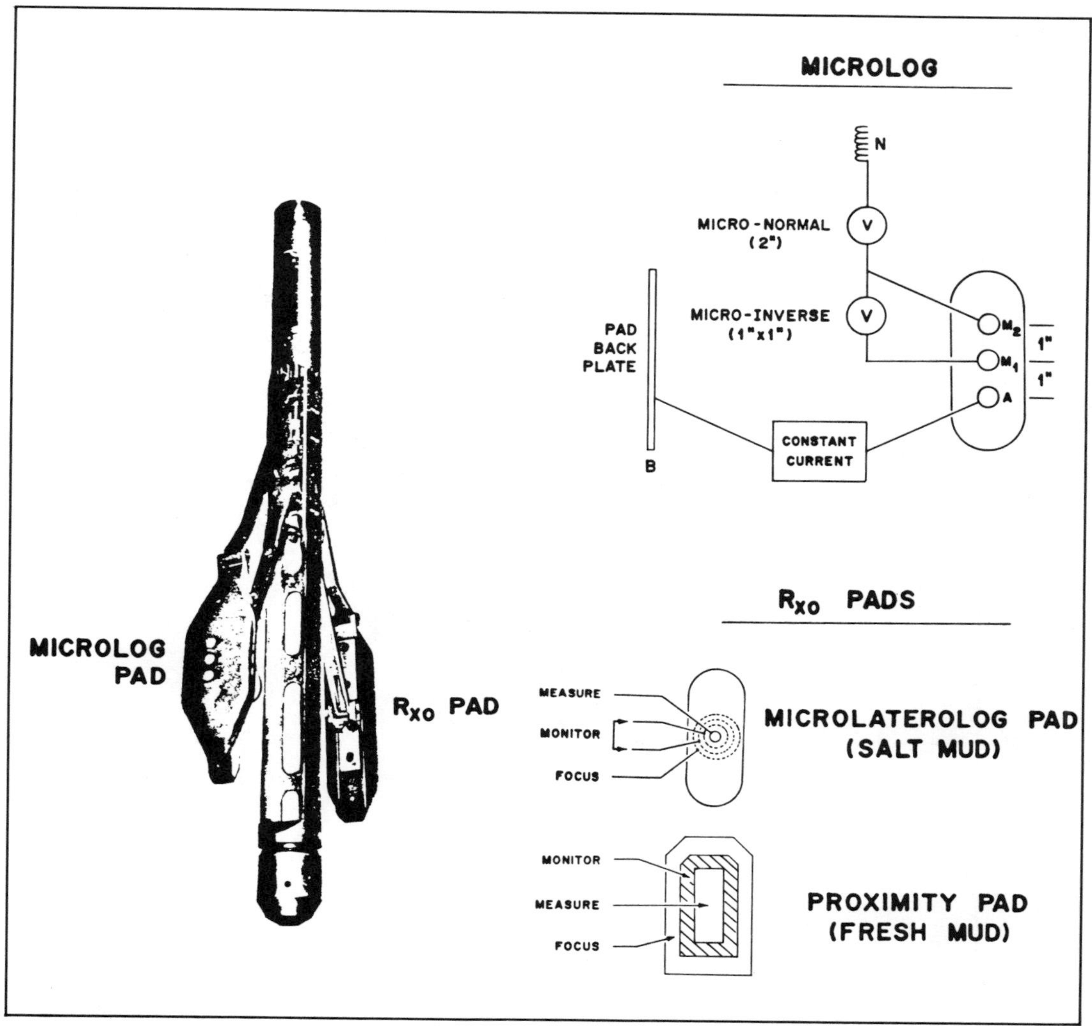

FIG. 5-19. The microlog-R_{xo} tool. Because of the tool's very shallow depths of investigations, the electrodes must be placed on pads and forced against the borehole wall to measure the formation. The two liquid-filled pads contain the microlog electrodes, on the near pad, and the microlaterolog or proximity R_{xo} electrodes are on the back pad. The proximity pad provides better R_{xo} measurements in fresh muds where the mudcakes are usually thicker.

POROSITY

The so-called "porosity" measuring methods — acoustic, density, and neutron — react to a combination of type of porosity, pore fluid, formation matrix (rock type), the presence of shale and for gas. Each tool reacts differently to these variables, with porosity and matrix being predominant. If a formation is composed of a uniform matrix and is "clean" (nonshaley), a single porosity tool may be sufficient to determine porosity. The accuracy of the porosity value depends, however, on the correctness

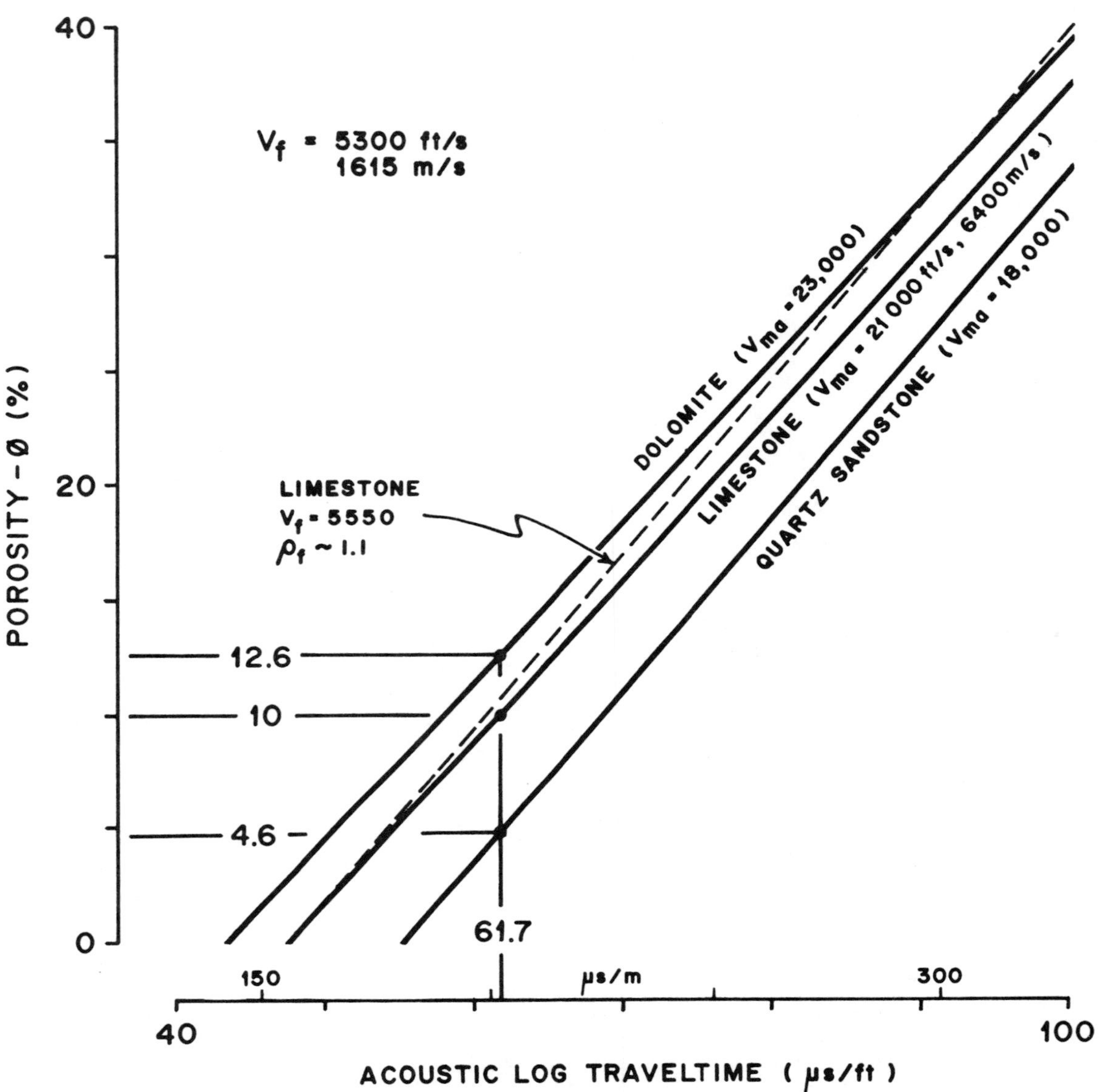

FIG. 5-20. Obtaining a porosity value from a single "porosity" measurement by assuming the formation matrix or rock type. For a 61.7 μs/ft value the porosity is 10 percent if the formation is limestone. However, if the formation is really a dolomite, its porosity is 12.6 percent. With a single porosity tool it is difficult to determine directly which is the proper formation matrix.

of the matrix assumption (Figure 5-20). Density and acoustic tools are discussed in detail in Chapters 8 and 10, respectively.

A formation evaluation resistivity-porosity crossplot (Hingle, 1959; or Fertl, 1979a) can also be used to determine the formation matrix, when the interval is thick enough and of sufficiently varying porosity (Figure 5-21).

Multiple porosity measurements that react differently to the unknowns

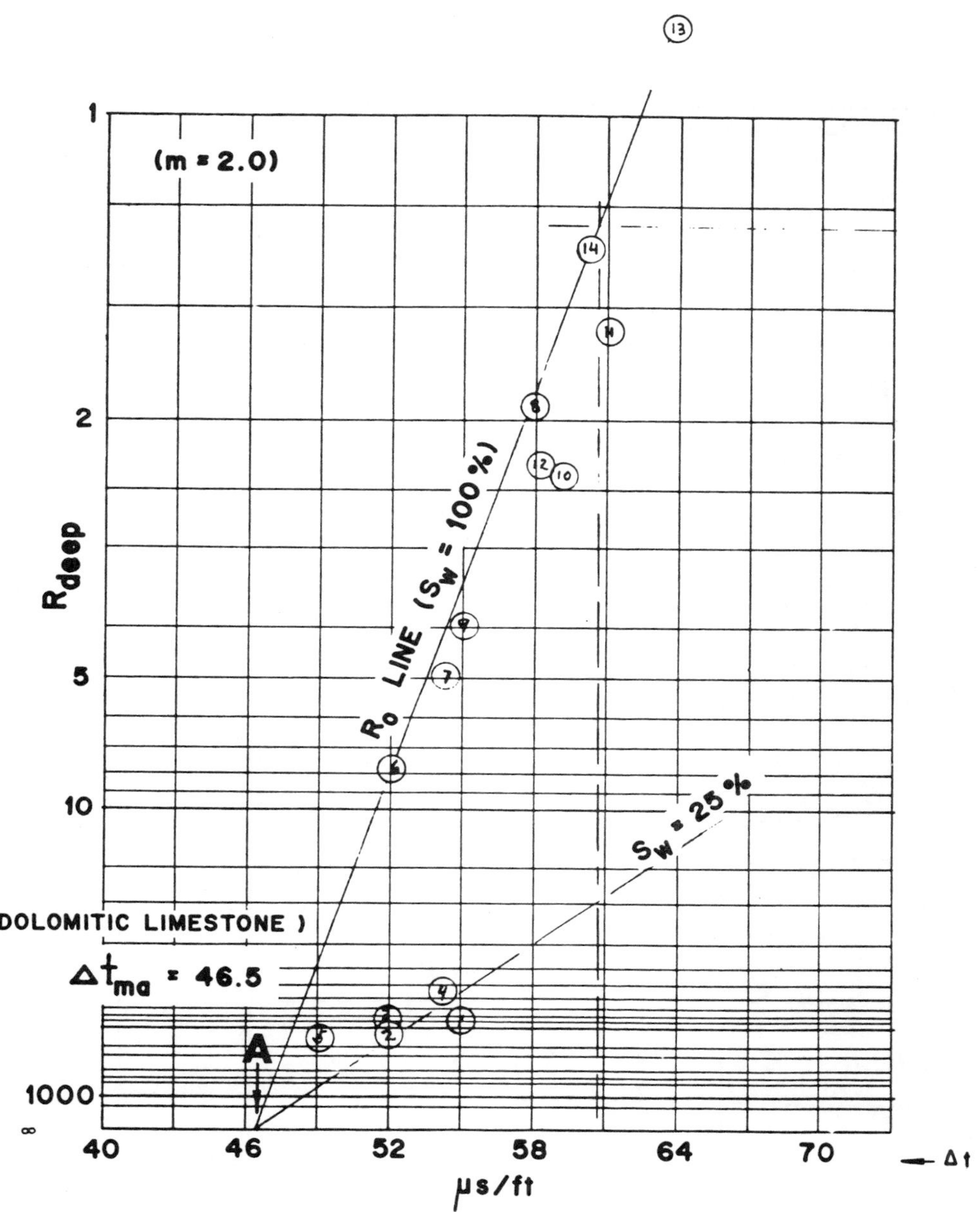

FIG. 5-21. The Hingle Plot crossplots porosity against deep resistivity. *Provided* the interval is thick enough and has an adequate variation in porosity, the formation matrix is determined at the R_o line infinite resistivity intercept (A). Points 1-5 are hydrocarbon bearing, while deeper points 6-15 are wet.

can be used to determine both porosity and formation matrix. Combining two different porosity devices on one tool string allows easy visual determination of the changing formation lithology (Figure 5-22). By crossplotting neutron porosity (ϕ_n) against bulk density (ρ_b) both lithology and porosity are accurately determined (Figure 5-23).

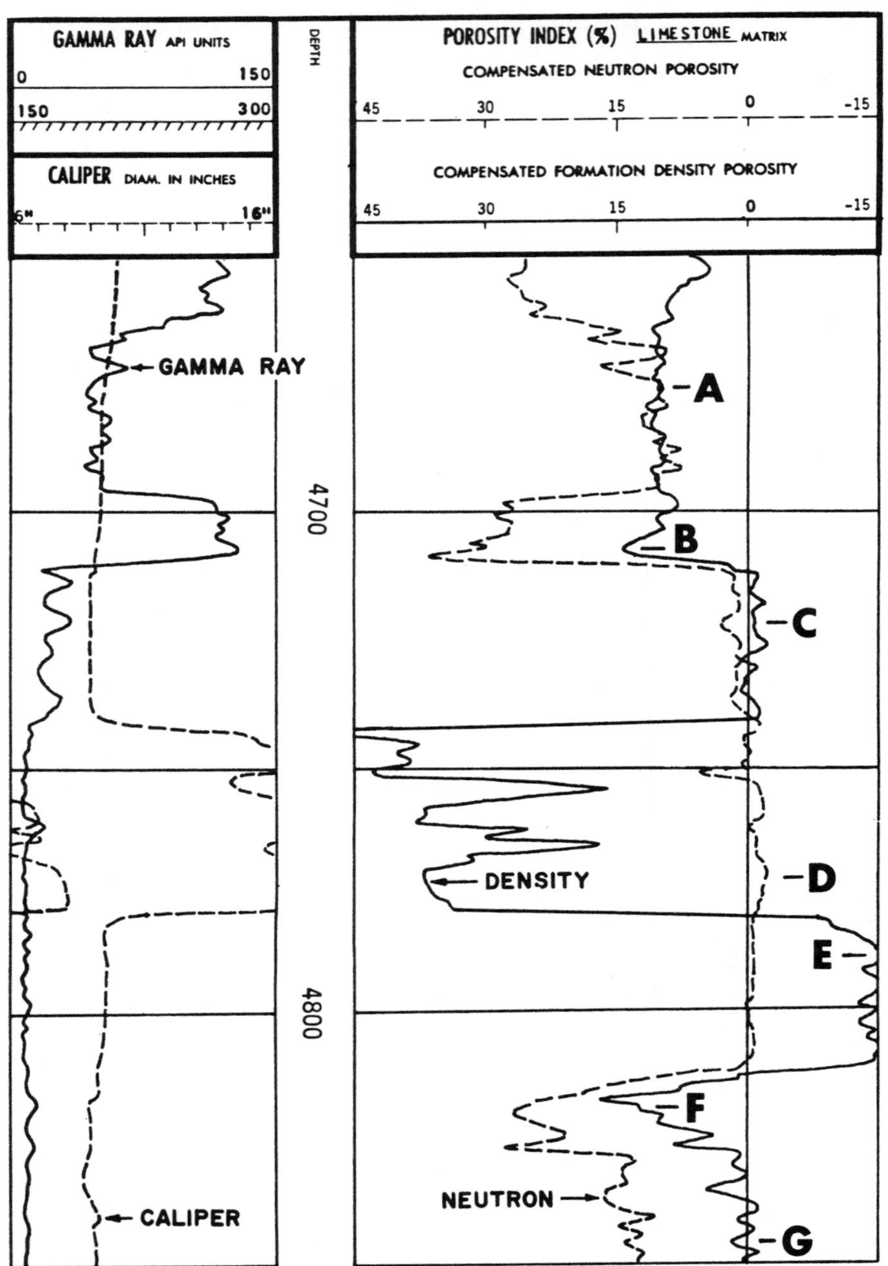

FIG. 5-22. By combining two "porosity" measurements on a tool it is possible to determine both formation matrix and porosity (Figure 5-23): A — limestone, B — shale, C — low porosity dolomitic limestone, D — salt, E — anhydrite, F — porous dolomite, and G — limy dolomite. The porosity value read directly from the log is correct only when the two curves track (A).

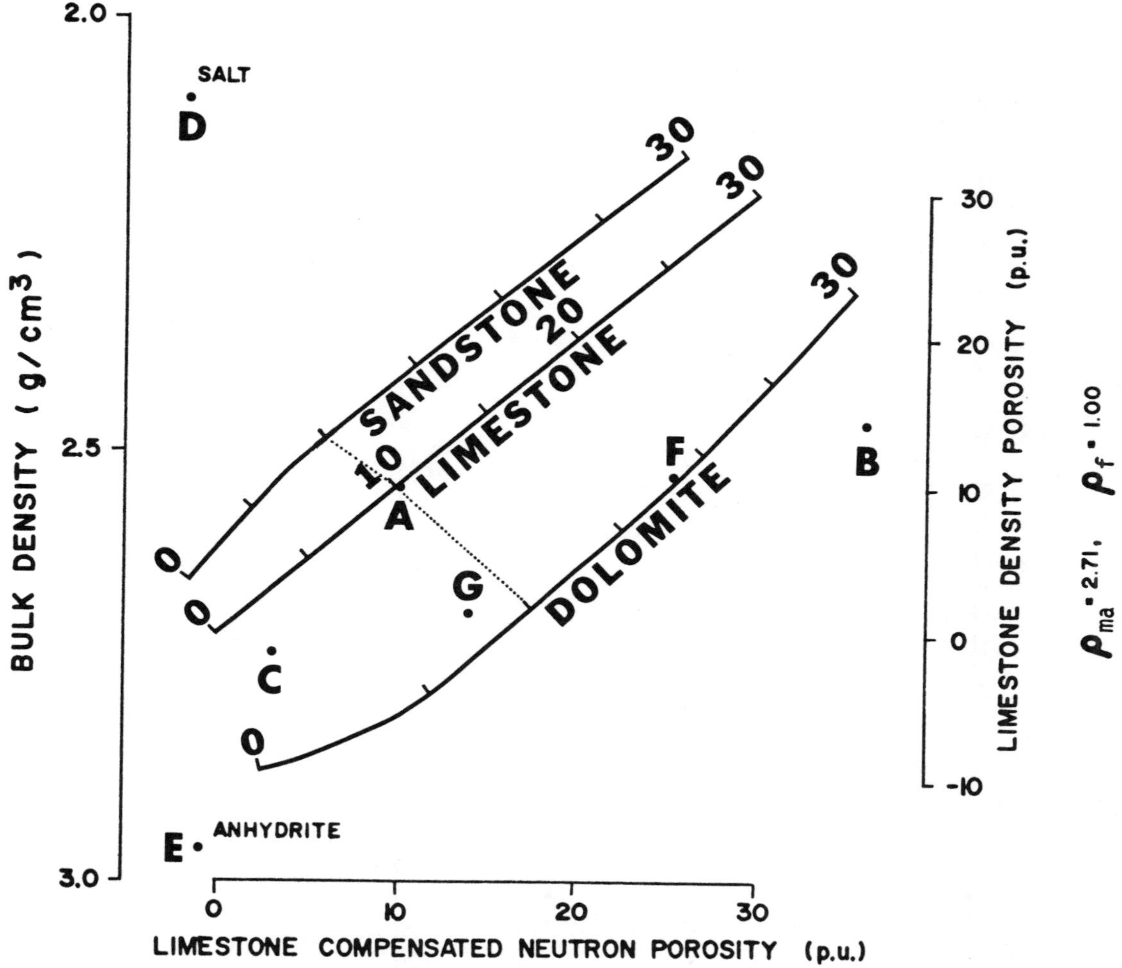

FIG. 5-23. Obtaining porosity and lithology by crossplotting two "porosity" measurements. The density measurement can be either in bulk density (ρ_b), using the left scale, or in limestone density-porosity units (ϕ_D) units, using the right scale. The points are from the log in Figure 5-22. Notice the near straight lines of equal porosity. Thus, if a sand and dolomite mixture is the real lithology at A, the matrix may be determined incorrectly but the porosity will be nearly correct. Compare this with the lithology-porosity crossplots of Figures 8-33, 10-46, and 10-48.

RESISTIVITY RATIO METHOD

It is possible to calculate S_w without determining porosity by using only resistivity measurements. For the undisturbed zone

$$S_w^n = F_R\, R_w\, R_t^{-1} \tag{5-1}$$

where S_w is water saturation of the undisturbed zone well away from the borehole, R_w is resistivity of the undisturbed formation connate water, and R_t is resistivity of the undisturbed zone away from the borehole. For

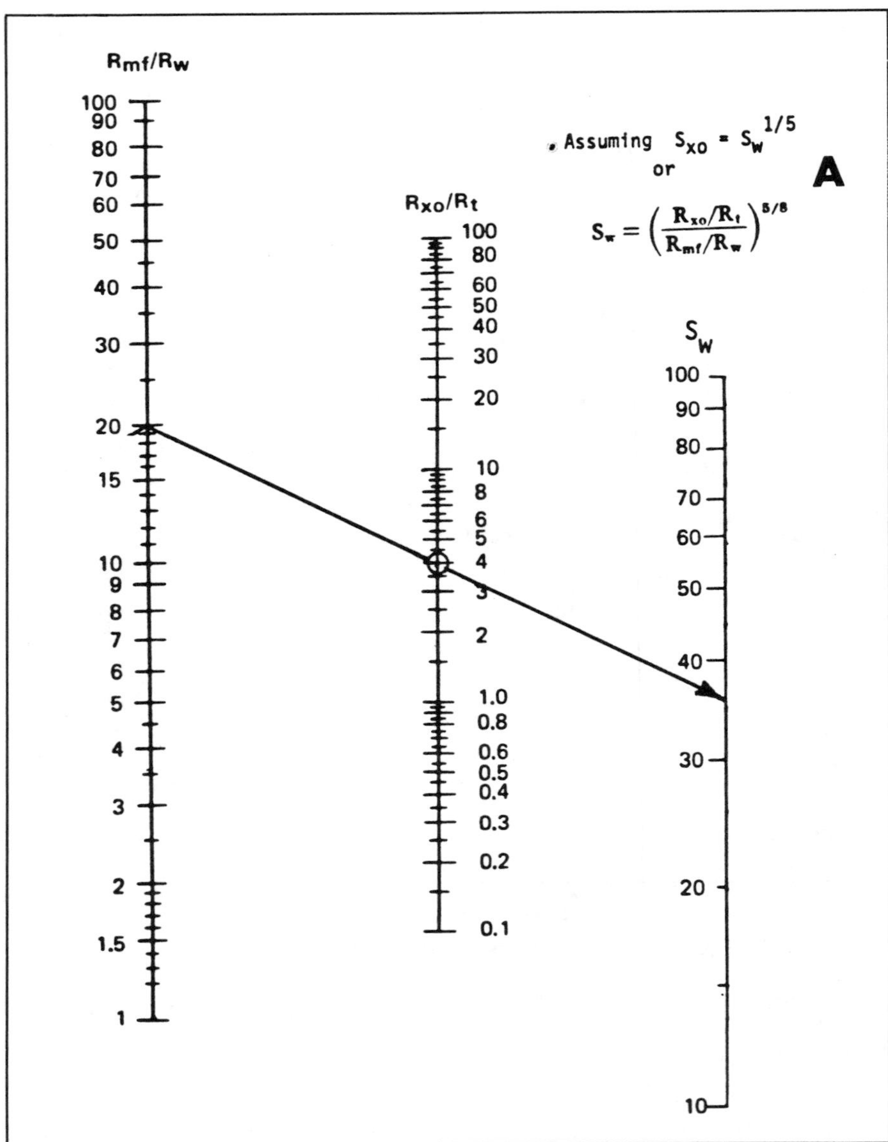

FIG. 5-24. The resistivity ratio method determines water saturation through resistivity measurements alone by assuming a uniform porosity from borehole to deep in the formation, and a relationship between water saturations near the borehole and in the undisturbed zone (A). In practice R_{mf}/R_w stays nearly constant in an interval and R_{xo}/R_t changes.

the completely flushed invaded zone immediately around the borehole

$$S^n_{xo} = F_R \, R_{mf} \, R^{-1}_{xo} \qquad (5\text{-}6)$$

where S_{xo} is water saturation of the completely flushed zone very near the borehole, R_{mf} is resistivity of the invading mud filtrate, and R_{xo} is measured resistivity of the flushed zone near the borehole. If in both the flushed and undisturbed zones the formation resistivity factor F_R is the

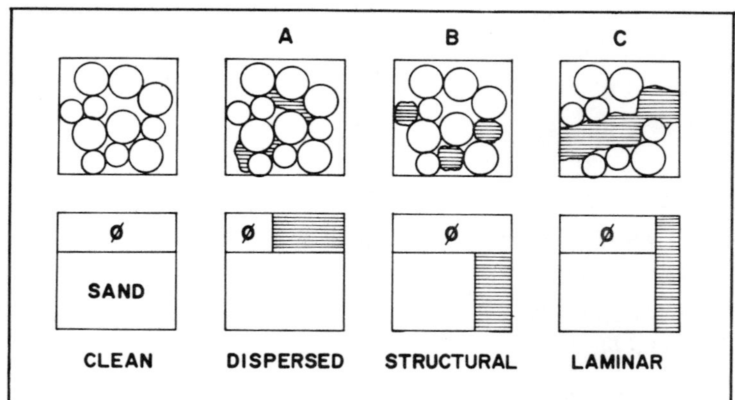

FIG. 5-25. The type of shale determines whether the measurement should be part of the porosity (A), part of the matrix (B), or an additional matrix (C).

same, then equations (5-1) and (5-6) can be solved for F_R and equated as

$$R_{xo}\, S^{n}_{xo}\, R^{-1}_{mf} = R_t\, S_w\, R^{-1}_w \tag{5-7}$$

$$(S_w/S_{xo})^n = (R_{xo}/R_t)\,(R_{mf}/R_w)^{-1}. \tag{5-8}$$

For homogeneous formations such as sandstones the above relationships are exact. However, they are only approximate for less homogeneous carbonates.

If it is possible to determine an "average" relationship between S_w and S_{xo}, then it is possible to solve for S_w using resistivity ratios. Empirical studies show that

$$S_{xo} = S_w^{1/5} \tag{5-9}$$

is a reasonable relationship. Thus, if n is 2, then equation (5-8) becomes (Figure 5-24)

$$S_w = \left[\frac{R_{xo}/R_t}{R_{mf}/R_w}\right]^{5/8} \tag{5-10}$$

SHALE

Shale is a complicating factor in formation evaluation. Because shale has a low resistivity it causes a low reading of the resistivity tools that can mask the increasing resistivity effect of hydrocarbons. It also complicates the porosity measurements because it can act as a part of the porosity, part of the matrix, or as an additional matrix (Figure 5-25), depending on the type of shale. Porosity transform equations (3-1) and (3-2) become

$$\phi_D = [(1 - V_{sh})\, \rho_{ma} + (V_{sh})\, \rho_{sh} - \rho_a]\, [\rho_{ma} - \rho_f]^{-1} \tag{5-11}$$

and

$$\phi_{SV} = [\Delta t_a - (1 - V_{sh})\, \Delta t_{ma} - (V_{sh})\, \Delta t_{sh}]\, [\Delta t_f - \Delta t_{ma}]^{-1} \tag{5-12}$$

where V_{sh} is bulk volume fraction of shale or clay, and Δt_{sh} and ρ_{sh} are log shale properties. Thus, porosity log measurements are effected by the presence of shale and gas (Figure 5-26) and log interpretation is made considerably more difficult with these additional unknowns in clastic sequences.

REFERENCES

Archie, G. E., 1942, The electrical resistivity log as an aid in determining some reservoir characteristics: J. Petr. Tech., **5**, 1, TP-1422.

Fertl, W. H., 1979a, Hingle crossplot speeds long-interval evaluation: Oil and Gas J., **77**, 3, 114-118.

Hilchie, D. W., 1979, Old electric log interpretation: Tulsa, IED Exploration.

Hingle, A. T., 1959, The use of logs in exploration problems: 29th Ann. Soc. Explor. Geophys. Meeting, Los Angeles, paper 29.

Ross, D., Ed., 1979, The art of ancient log analysis: Houston: Soc. Prof. Well Log Analysts.

Scientific Software, Well logging manual.

Schlumberger Well Services, 1972, Log interpretation/principles.

TRUE POROSITY (30 20 10 0)	PORE CONTENT	PORE TYPE	LITHOLOGY
SONIC HIGH - DUE TO SHALE DENSITY LOW - DUE TO SHALE NEUTRON HIGH - DUE TO SHALE	WATER		SHALE
SONIC HIGH - GAS & UNCOMPACTION DENSITY HIGH - GAS NEUTRON LOW - GAS	GAS	GRANULAR	UNCOMPACTED CLEAN SAND
SONIC HIGH - UNCOMPACTION DENSITY - O.K. NEUTRON - O.K.	OIL OR WATER		
SONIC HIGH - SHALE (possible gas) DENSITY HIGH - GAS NEUTRON LOW - GAS	GAS	GRANULAR	COMPACTED SHALY SAND
SONIC HIGH - SHALE DENSITY - O.K. - (slightly low? - shale) NEUTRON HIGH - SHALE	OIL OR WATER		
SONIC - O.K. DENSITY HIGH - GAS NEUTRON LOW - GAS	GAS	GRANULAR	COMPACTED CLEAN SAND
SONIC - O.K. DENSITY - O.K. NEUTRON - O.K.	OIL OR WATER		
SONIC - O.K. DENSITY - O.K. (slightly high? - gas) NEUTRON - O.K. (slightly low? - gas)	GAS	GRANULAR	CARBONATE DOLOMITE OR LIMESTONE
SONIC - O.K. DENSITY - O.K. NEUTRON - O.K.	OIL OR WATER		
SONIC LOW - (secondary porosity) DENSITY - O.K. NEUTRON - O.K.	GAS	VUGS OR FRACTURES	
SONIC LOW - (secondary porosity) DENSITY - O.K. (slightly low? - fractures) NEUTRON - O.K.	OIL OR WATER		

FIG. 5-26. The effects of shale and gas on porosity tool measurements. (Source unknown.)

REFERENCES FOR GENERAL READING

Burke, J., Campbell, R., and Schmidt, A., 1969, The litho-porosity crossplot: Log Analyst, **10**, November-December, 25-43.

Denoo, S., 1978, Neutron density log is a valuable open-hole porosity tool: Oil and Gas J., **76**, 39, 96-102.

Doll, H. G., 1950, The microlog — A new electrical logging method for detailed determination of permeable beds: Pet. Trans. Am. Inst. Min., Metall., Petr. Eng., **189**, 155-164, TP 2880.

———, 1951, The laterolog: A new resistivity logging method with electrodes using an automatic focusing system: Pet. Trans. Am. Inst. Min., Metall., Petr. Eng., **192**, 305-316, TP 3198.

———, 1953, The microlaterolog: Pet. Trans. Am. Inst. Min., Metall., Petr. Eng., **198**, 17-32, TP 3492.

Fertl, W. H., 1979b, Lithology and other effects on porosity logs: Oil and Gas J., **77**, 11, 68-70.

Guyod, H., 1944, Guyod's electrical well logging: Houston, Welex — a Halliburton Company.

Hilchie, D. W., 1978, Applied open hole log interpretation: Tulsa, IED Exploration.

Picket, G. R., 1973, Pattern recognition as a means of formation evaluation: Trans. 14th Ann. Soc. Prof. Well Log Analysts Logging symposium, Lafayette, paper A.

Poupon, A., Clavier, C., Dumanoir, J., Gaymard, R., and Misk, A., 1970, Log analysis of sand-shale sequences — A systematic approach: J. Petr. Tech., **22**, 867-881.

Poupon A., and Gaymard, A., 1970, The evaluation of clay content from logs: Trans. 11th Ann. Soc. Prof. Well Log Analysts Logging symposium, Los Angeles, paper G.

Ransom, R. C., 1975, Glossary of terms and expressions used in well logging: Houston, Soc. Prof. Well Log Analysts.

Raymer, L. L., and Biggs, W. P., 1963, Matrix characteristics defined by porosity computations: Trans. 4th Ann. Soc. Prof. Well Log Analysts Logging Symposium, Oklahoma City, paper X.

Sheriff, R. E., 1970, Glossary of terms used in well logging: Geophysics, **35**, 1116-1130.

Suau, J., Grimaldi, P., Poupon, A., and Souhaite, P., 1972, The dual laterolog — R_{xo} tool: 47th Ann. Fall Soc. Petr. Eng. Meeting, San Antonio, SPE 4018.

Tixier, M. P., Alger, R. P., Biggs, W. P., and Carpenter, B. N., 1963, Dual induction-laterolog: A new tool for resistivity analysis: 38th Ann. Fall Soc. Petr. Eng. Meeting, New Orleans, SPE 713.

Tixier, M. P., Morris, R. L., and Connell, J. G., 1968, Log evaluation of low-resistivity pay sands in the Gulf Coast: The Log Analyst, **9**, November-December.

Chapter 6

WELLSITE HYDROCARBON INDICATORS

Before introduction of computer processing, wellsite interpretation was not intended to compute precise porosity, water saturation, or hydrocarbon recovery for large intervals of the borehole. Rather, it was to point out which intervals needed careful formation evaluation and which intervals needed more testing with additional logs, drill stem tests, or wireline formation test.

Primarily only intervals with hydrocarbons need wellsite evaluation. With several thousand feet of hole to evaluate quickly, and as expensive idle rig time mounted, methods were needed to cull the nonhydrocarbon-bearing "wet" intervals and quickly locate possible hydrocarbon-bearing intervals.

Wellsite "quick look" indicator techniques tend to be optimistic which is fine, because it is better to examine and test a marginal or even anomalous nonhydrocarbon interval than to abandon a well with untested hydrocarbons. Continuous analog indicators are usually recorded over the entire logged section and the nonpermeable intervals are visually rejected. Then preliminary interpretation calculations, using improved crossplot porosities and corrected resistivities, are made to confirm the indicator (Figure 6-1).

Each quick-look technique is reliable for only certain porosity and invasion conditions and these limitations should be understood in much more detail than is given here. The five prime techniques are: (1) apparent water resistivity (R_{wa}), (2) $F_R - R_o$ overlays, (3) movable-oil plot (MOP), (4) the R_{xo}/R_t indicator, and (5) the neutron-density gas overlay. Most of these older techniques are used in current wellsite computer interpretations.

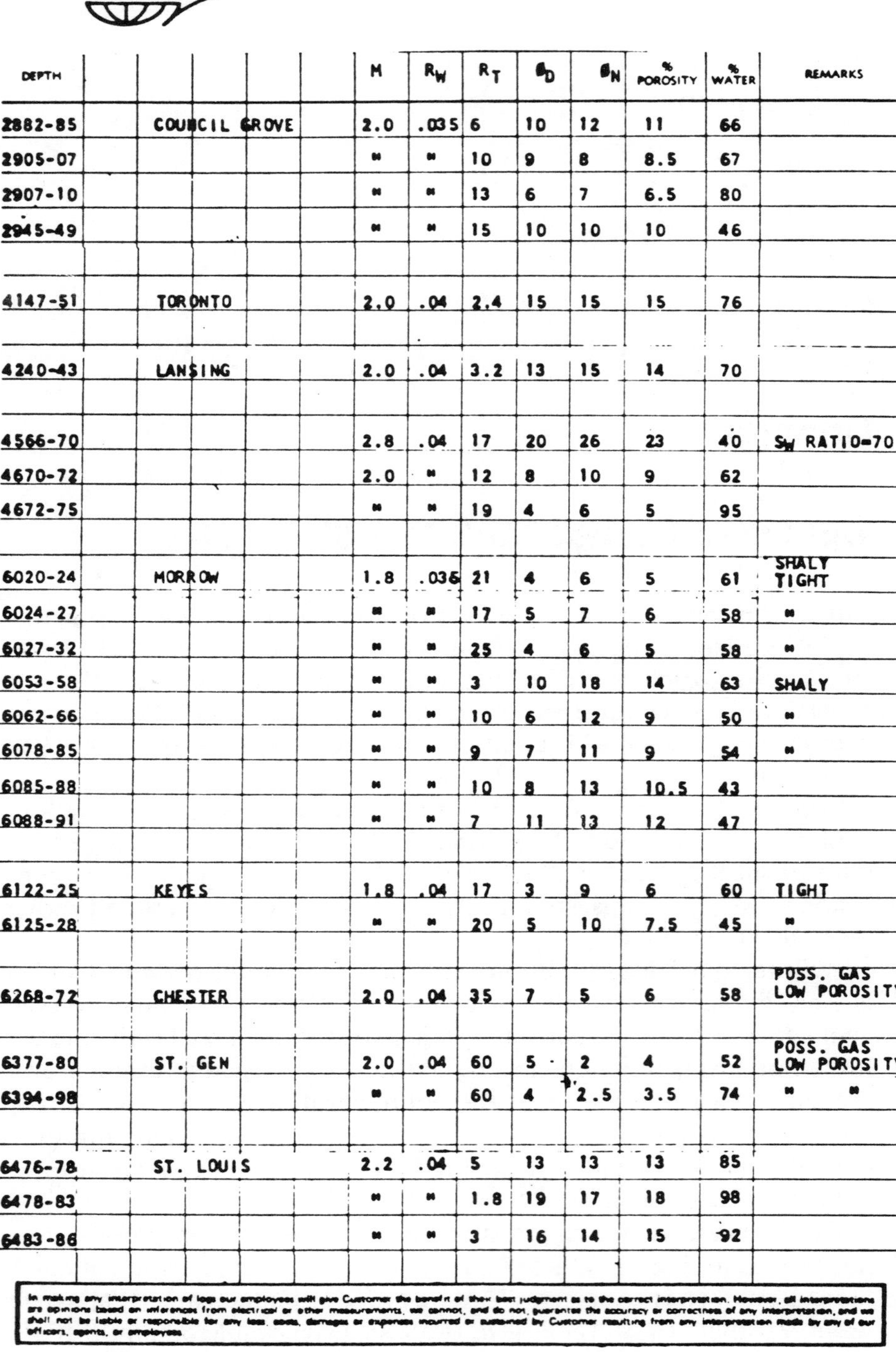

FORM 110

DEPTH		M	R_W	R_T	ϕ_D	ϕ_N	% POROSITY	% WATER	REMARKS
2882-85	COUNCIL GROVE	2.0	.035	6	10	12	11	66	
2905-07		"	"	10	9	8	8.5	67	
2907-10		"	"	13	6	7	6.5	80	
2945-49		"	"	15	10	10	10	46	
4147-51	TORONTO	2.0	.04	2.4	15	15	15	76	
4240-43	LANSING	2.0	.04	3.2	13	15	14	70	
4566-70		2.8	.04	17	20	26	23	40	S_W RATIO=70
4670-72		2.0	"	12	8	10	9	62	
4672-75		"	"	19	4	6	5	95	
6020-24	MORROW	1.8	.036	21	4	6	5	61	SHALY TIGHT
6024-27		"	"	17	5	7	6	58	"
6027-32		"	"	25	4	6	5	58	"
6053-58		"	"	3	10	18	14	63	SHALY
6062-66		"	"	10	6	12	9	50	"
6078-85		"	"	9	7	11	9	54	"
6085-88		"	"	10	8	13	10.5	43	
6088-91		"	"	7	11	13	12	47	
6122-25	KEYES	1.8	.04	17	3	9	6	60	TIGHT
6125-28		"	"	20	5	10	7.5	45	"
6268-72	CHESTER	2.0	.04	35	7	5	6	58	POSS. GAS LOW POROSITY
6377-80	ST. GEN	2.0	.04	60	5	2	4	52	POSS. GAS LOW POROSITY
6394-98		"	"	60	4	2.5	3.5	74	" "
6476-78	ST. LOUIS	2.2	.04	5	13	13	13	85	
6478-83		"	"	1.8	19	17	18	98	
6483-86		"	"	3	16	14	15	92	

In making any interpretation of logs our employees will give Customer the benefit of their best judgment as to the correct interpretation. However, all interpretations are opinions based on inferences from electrical or other measurements, we cannot, and do not, guarantee the accuracy or correctness of any interpretation, and we shall not be liable or responsible for any loss, costs, damages or expenses incurred or sustained by Customer resulting from any interpretation made by any of our officers, agents, or employees.

FIG. 6-1. A wellsite interpretation using corrected and crossplotted values, especially for intervals pointed out by the quick-look techniques.

APPARENT WATER RESISTIVITY

Connate water resistivity can be back-calculated using Archie's water saturation equation (5-1):

$$R_w = S_w^n R_t F_R^{-1} \tag{6-1}$$

If all intervals are assumed wet, S_W = 100 percent, an apparent water resistivity (R_{wa}) can be defined as:

$$R_{wa} \equiv R_{\text{deep}} F_R^{-1}. \tag{6-2}$$

For intervals that are actually wet, R_{wa} is nearly equal to R_w, and R_{wa} should be near a minimum. For hydrocarbon intervals, the deep resistivity will be higher and thus R_{wa} greater than R_w. In practice, R_{wa} must be at least three times R_w to indicate a prospective hydrocarbon interval. This figure of merit was chosen because water-bearing shaley sands can achieve a R_{wa} of two times R_w. The simultaneous induction-sonic tool provides both the deep resistivity and a sonic derived porosity, allowing presentation of a continuous R_{wa} curve, using the simple $F_R = \phi^{-2}$ relationship (Figure 6-2). The R_{wa} technique used is in preliminary wellsite computer interpretations to determine R_w.

F_R – R_o OVERLAY METHOD

The presence of oil, a S_w decrease, causes an increase in formation resistivity:

$$R_t = F_R R_w S_w^{-n}. \tag{6-3}$$

A simple technique is to compare what formation resistivity should be if it were wet versus what it actually measures. The formation resistivity, if 100 percent water saturated (R_o), is defined as:

$$R_o \equiv F_R R_w. \tag{6-4}$$

The deep resistivity and R_o should be nearly equal in wet zones but separated, with R_o less than R_{deep}, in hydrocarbon-bearing intervals. A continuous recording of F_R is made at the same time the porosity log is recorded (Figure 6-3). With F recorded in a logarithmic format it is easy to convert it to R_o for any R_w when tracing F_R onto a logarithmic resistivity presentation. To create an R_o for an R_w of .04 Ω•m, the 100 grid index line of the F Log is placed under the 4 Ω•m grid line of the resistivity log — then the F log curve is traced on the resistivity log in the nonshale intervals (Figure 6-4). A separation between R_o and R_{deep} indicates possible hydrocarbons. A lower S_w is indicated by a greater separation. Because R_o is a direct function of porosity, a scale is included to read it directly in porosity units. The same technique is done in the computer to create the "wet" resistivity curve in a wellsite computer interpretation (Figure 6-11).

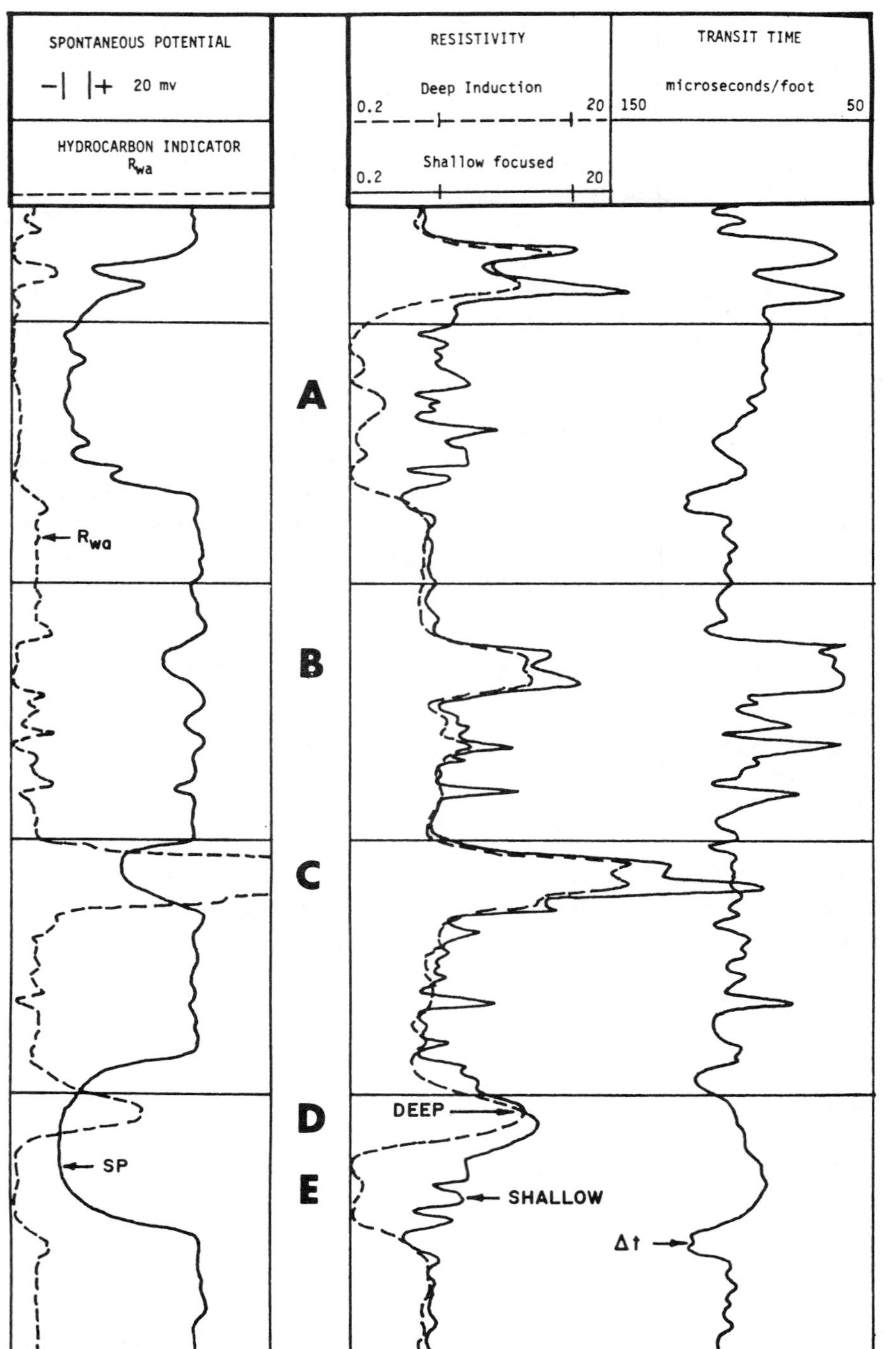

FIG. 6-2. The apparent water resistivity (R_{wa}) indicator. Intervals A, B, and E are wet. Intervals C and D need further examination due to their high R_{wa}.

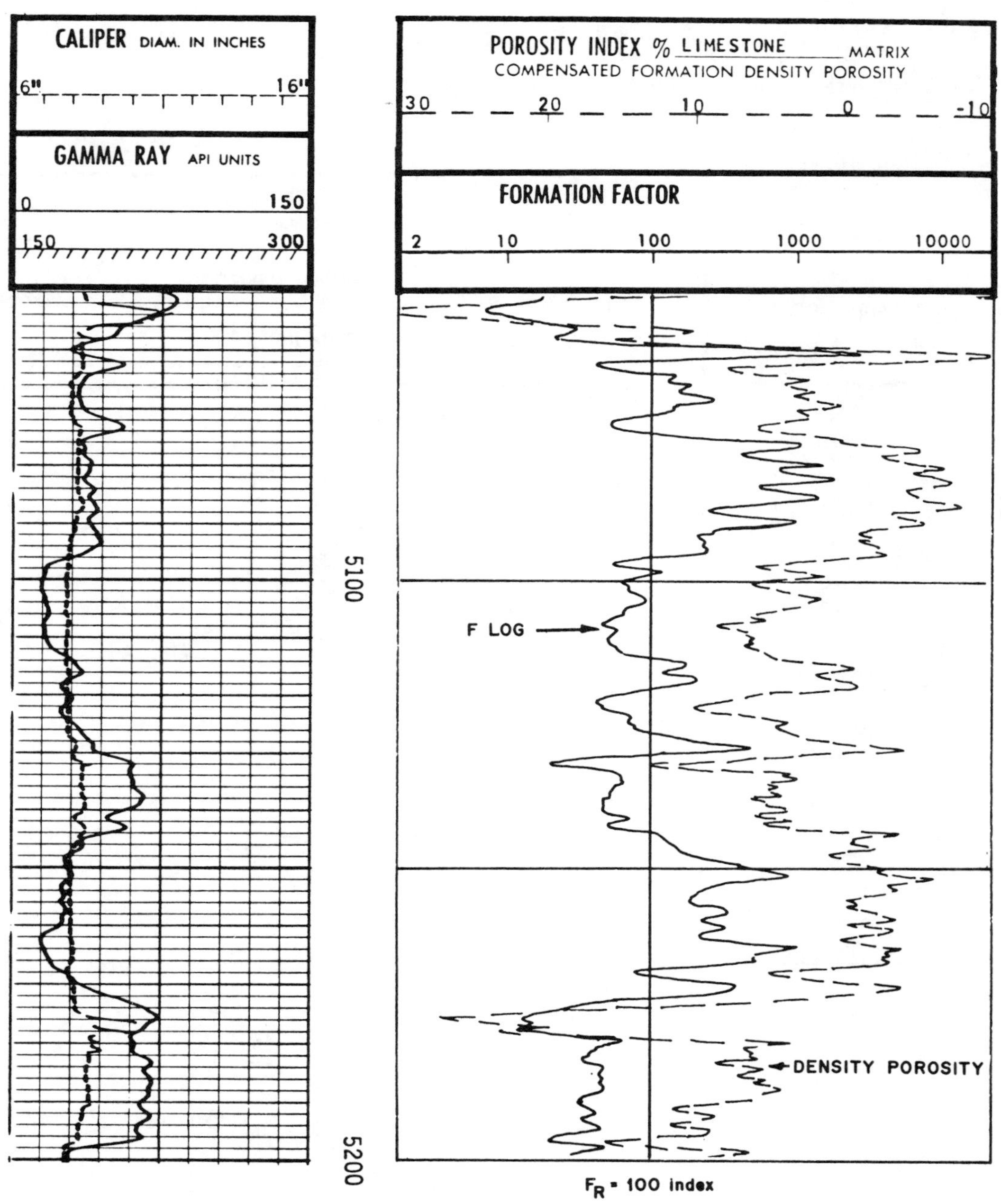

FIG. 6-3. Construction of an R_o curve: the F log curve is derived from density porosity (for a limestone lithology in this example) using $F_R = \phi^{-2}$. On Figure 6-4 the F curve is traced onto the resistivity log to read R_o for an R_w of .04 Ω•m.

This entire process can be inverted with everything done in porosity units. A porosity derived from a resistivity measurement, by again assuming everything wet, would be:

$$\phi_{\text{res}} = (K_R\, R_w\, R_{\text{deep}}^{-1})^n \qquad (6\text{-}5)$$

This result is then compared with the measured porosity. Again, the two

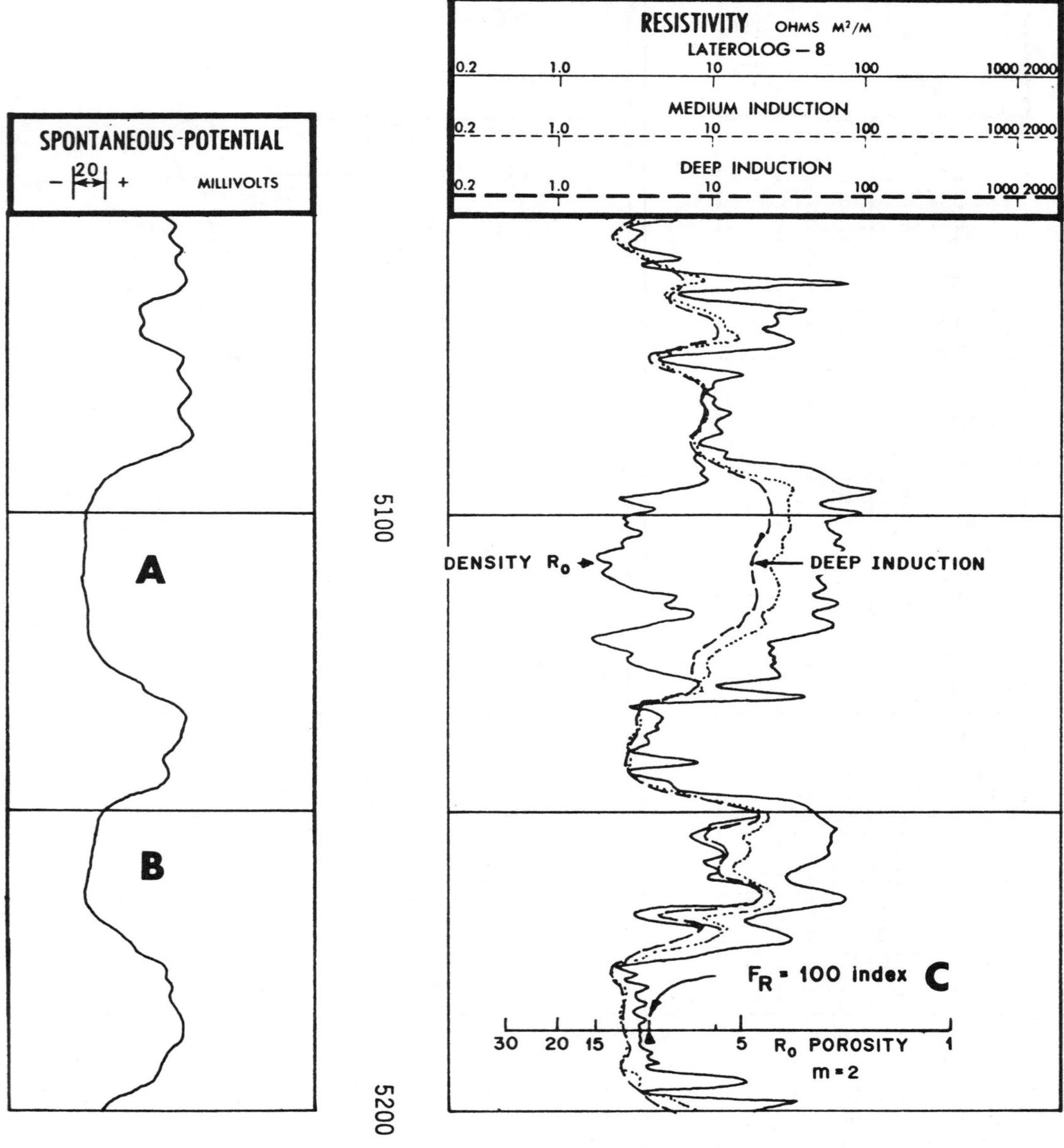

FIG. 6-4. R_o technique: the upper sand (A) needs further examination due to separation between R_o and R_{deep}. The relative S_W can be determined from the amount of separation, the greater the separation the lower the S_W. Porosity is read on the R_o scale. The lower sand (B) is probably wet because R_o and R_{deep} overlay. The R_w value is determined by taking the resistivity at the scale index (C) and dividing by 100, i.e., .04 Ω•m in this example.

porosities should be the same in wet intervals and will separate in potential hydrocarbon-bearing intervals.

MOVABLE OIL PLOT

The $F_R - R_o$ overlay technique indicates how much hydrocarbon is contained in the pore space — but the indication is misleading if that hydrocarbon cannot be produced. An example would be a very low-viscosity hydrocarbon, such as tar, in the pore spaces.

Most completion techniques depend substantially on water as the flushing media for moving hydrocarbons from the formation, through the perforated casing, and into the borehole. A similar, but reversed, action occurs during drilling — a portion of the hydrocarbon in an interval is flushed by the drilling mud filtrate away from the borehole. If the two conditions, flushing and production, are analagous, then a measure of the drilling-flushing action should indicate the movability, and thus producibility, of the hydrocarbons in an interval.

A comparison of the difference between R_t and R_{xo} is a comparison between undisturbed zone saturation and flushed, invaded zone saturation; S_w and S_{xo}, respectively. For the undisturbed zone

$$(S_w)^n = F_R \, R_w \, R_t^{-1}. \qquad (5\text{-}1)$$

Solving for R_t/R_w yields

$$R_t/R_w = F_R \, (S_w^{-n}), \qquad (6\text{-}6)$$

while for the completely flushed, invaded zone immediately around the borehole

$$(S_{xo})^n = F_R \, R_{mf} \, R_{xo}^{-1}. \qquad (5\text{-}6)$$

Solving for R_{xo}/R_{mf} yields

$$R_{xo}/R_{mf} = F_R \, (S_{xo}^{-1}). \qquad (6\text{-}7)$$

R_{xo} is obtained by a very shallow reading microresistivity device, such as the microlaterolog.

The ratios expressed in equations (6-6) and (6-7) are thus a measurement that differs from F_R by a reciprocal function of the saturation of the measured zone. If the ratios were plotted on a common scale along with F_R, their separations would indicate a difference in saturations as a result of the drilling-flushing movement of hydrocarbons away from the borehole — i.e., hydrocarbon movability under drilling conditions.

The ratios are obtained easily by rescaling the logarithmically displayed R_{deep} and R_{xo} resistivity measurements, where on a logarithmic grid, division is accomplished by a lateral shift of the grid scale (Figure 6-5). Knowing where the $F_R = 100$ index is for each of the resistivity measurements, they are then plotted on a common grid with the porosity F_R measurement (Figure 6-6). The separation between the porosity derived

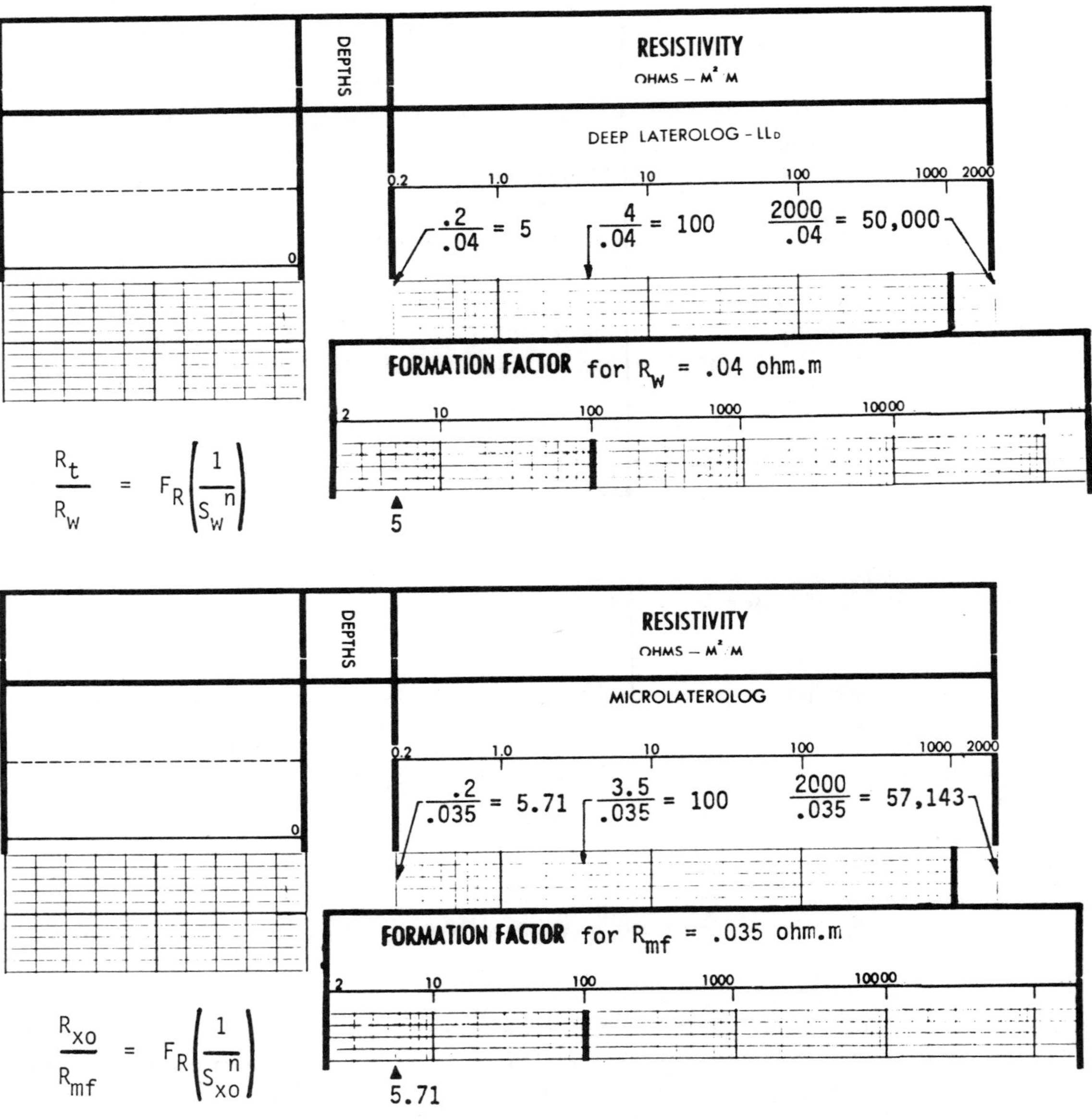

FIG. 6-5. Rescaling a deep resistivity curve (top) and microresistivity (bottom) into $F(S^{-n})$ units for use in a movable-oil plot, Figure 6-6. Division of both resistivity measurements which are in a logarithmic grid presentation involves a lateral shift of the grid scale.

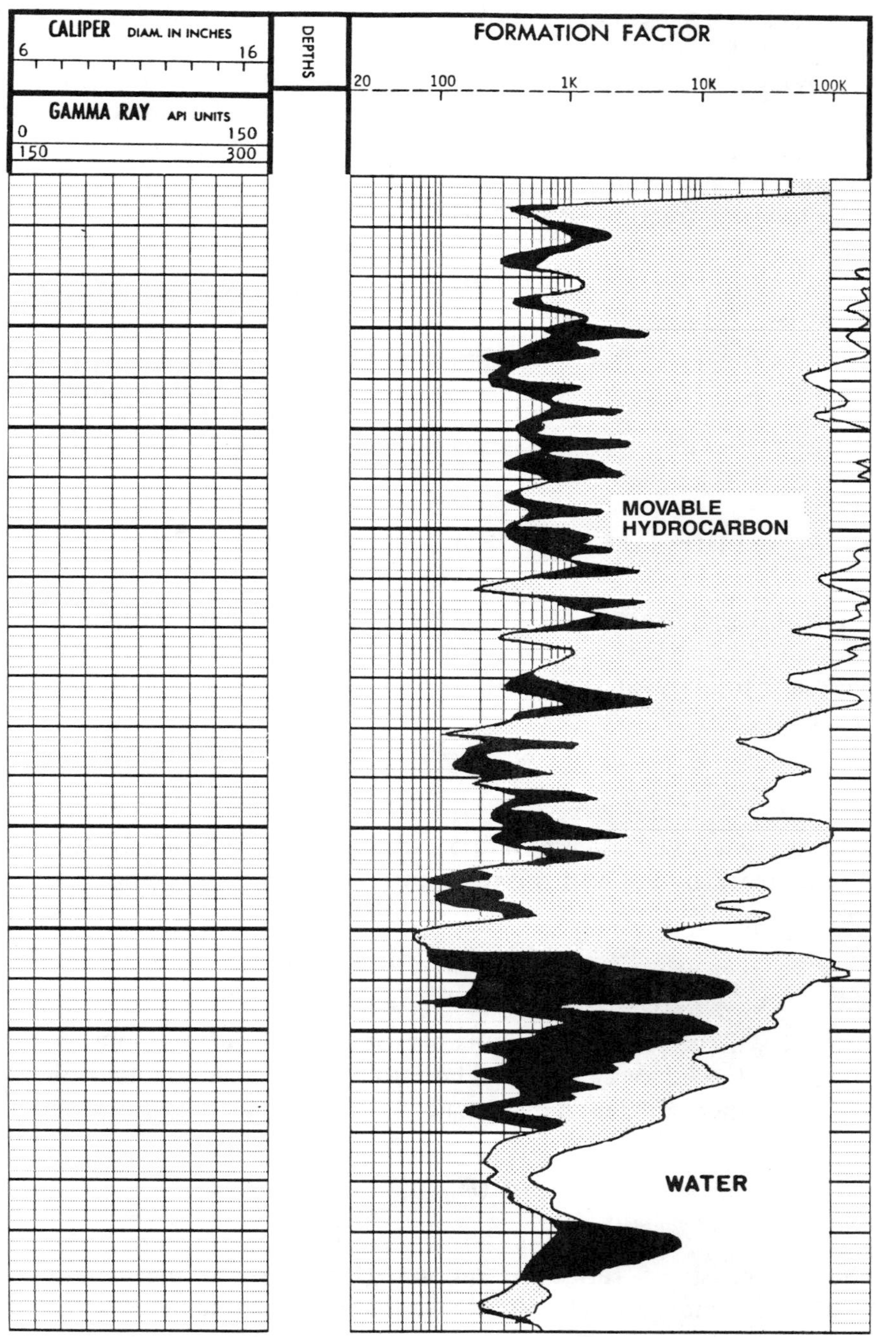

FIG. 6-6. Movable-oil plot (MOP) scaled in F units. The dark area is residual hydrocarbon, stipple area is movable hydrocarbon, and white area is water. The F scale is open-ended in terms of porosity, so a F of 100 000 or 0.5 percent porosity cutoff is used in this example.

F_R curve and the R_{xo} derived curve (Figure 6-6, dark region) is an indication of nonmovable residual hydrocarbons, and between the R_{xo} derived and R_{deep} derived curves (Figure 6-6, stipple region) indicates movable flushed hydrocarbons.

A related presentation is done for wellsite computer interpretation (Figures 6-7 and 6-11, track III, porosity analysis). The three fractions are shown as a partitioning of the porosity. This is determined using the calculated saturation from R_{xo} and R_{deep} and converting to a bulk volume water fraction (BVW), where

$$\text{BVW} = \phi \, S_w \tag{6-8}$$

indicates the water fraction (white fraction) and a similar relationship using S_{xo} is used to divide the hydrocarbon fraction into residual and movable hydrocarbon.

R_{xo}/R_t APPROXIMATION INDICATOR

The resistivity ratio water saturation method discussed in Chapter 5 can be used as a wellsite hydrocarbon indicator. In this case it is not necessary to know an "average" relationship between S_w and S_{xo}. From equation (5-8)

$$(S_w/S_{xo})^n = (R_{xo}/R_t)\,(R_{mf}/R_w)$$

if an interval is wet, then S_w and S_{xo} are both 100 percent because there is no hydrocarbon to be displaced by mud filtrate flushing and therefore

$$\frac{S_w}{S_{xo}} = 1, \text{ or } (R_{xo}/R_t) = (R_{mf}/R_w). \tag{6-9}$$

However, if the interval contains movable hydrocarbons some hydrocarbon is flushed away from the borehole by the invading mud filtrate, and S_{xo} is then greater than S_w and

$$\frac{S_w}{S_{xo}} < 1, \text{ or } (R_{xo}/R_t) < (R_{mf}/R_w). \tag{6-10}$$

The two resistivity ratios in equation (5-8) thus become a potential hydrocarbon indicator. The SP measurement is related to the R_{mf}/R_w ratio. Empirical studies provide a transform of the deep and shallow resistivity measurements of the dual induction log into an approximate R_{xo}/R_t ratio (Dumanoir, et al, 1972). With proper scaling of the R_{xo}/R_t approximation, it will overlay the SP in wet intervals and separate, with approximation having less deflection than SP in permeable intervals, from the SP in hydrocarbon-bearing intervals (Figure 6-8).

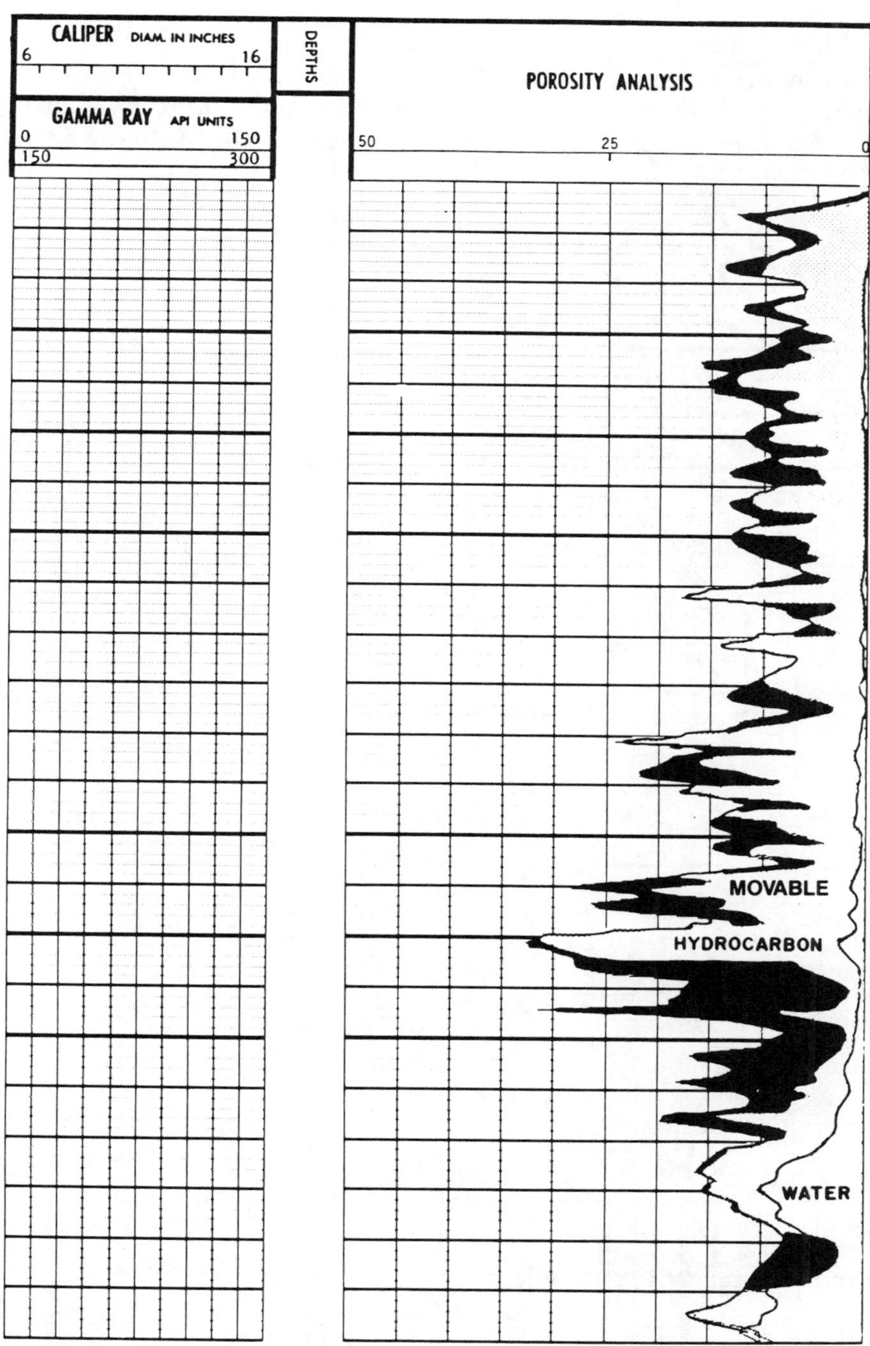

FIG. 6-7. Bulk volume water (BVW) presentation showing a linear porosity unit movable oil plot format. This is the same information given in Figures 6-6 and 7-19. S_{xo}, S_w, and equation 6-8 are used in partitioning of the porosity.

DENSITY-NEUTRON GAS OVERLAY

The compensated density-compensated neutron combination tool is used at the wellsite to determine gas bearing intervals. Neutron logs measure the presence of hydrogen and determine a porosity value by assuming the measured hydrogen is in water and oil filled pore spaces.

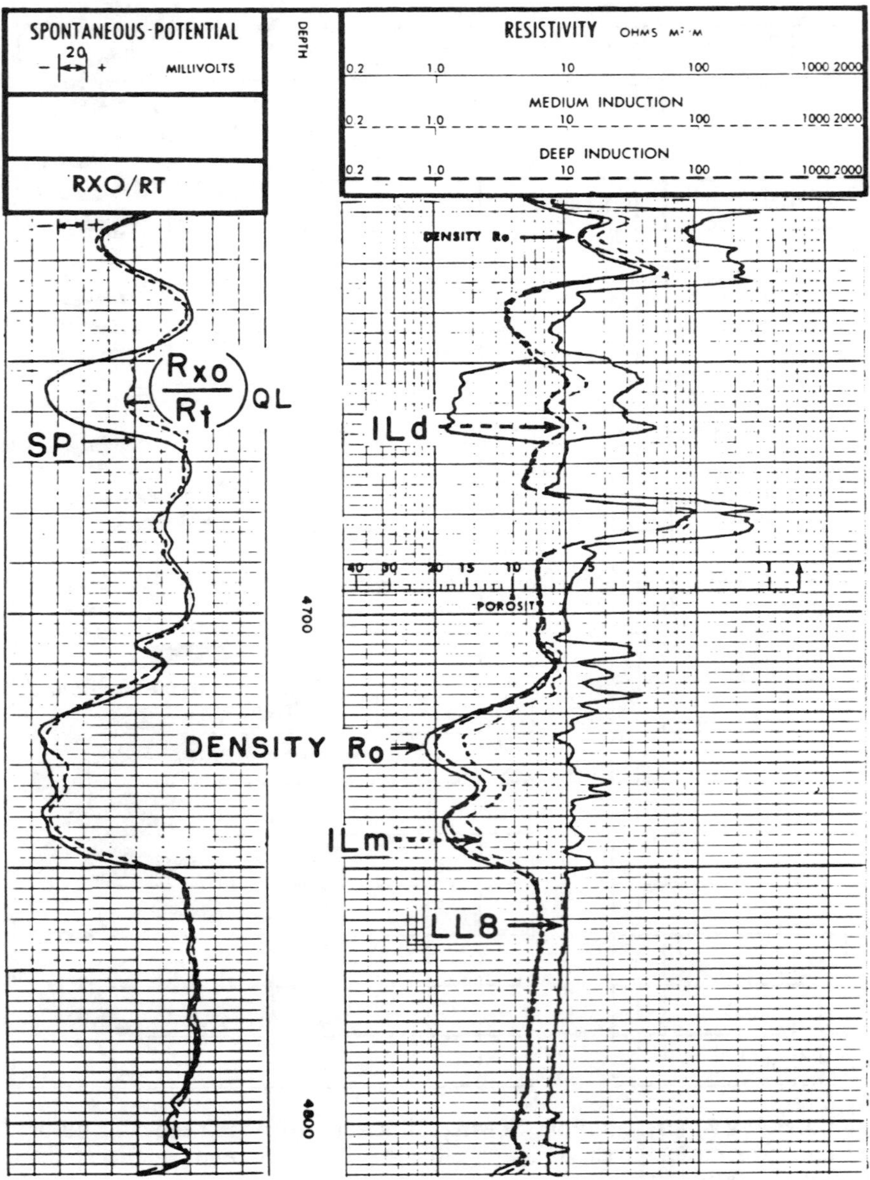

FIG. 6-8. R_{xo}/R_t indicator where the upper sand (A) needs further examination due to separation of the R_{xo}/R_t indicator and the SP measurement, plus $R_o - R_{deep}$ separation. The lower sand (B) is probably wet. (Dumanoir et al, 1972, courtesy of Schlumberger).

But for gas filled pores, because gas has a considerably lower hydrogen index or density than oil or water, the neutron tool presents this as a porosity decrease. This neutron "gas effect" is used as a gas detector when compared with the density log, which only has a slight gas effect. A change in the curve separation toward crossover (Figure 6-9) when lithology is constant indicates gas. This may also be presented in a separate overlay when the two porosity curves are overlain, or normalized for

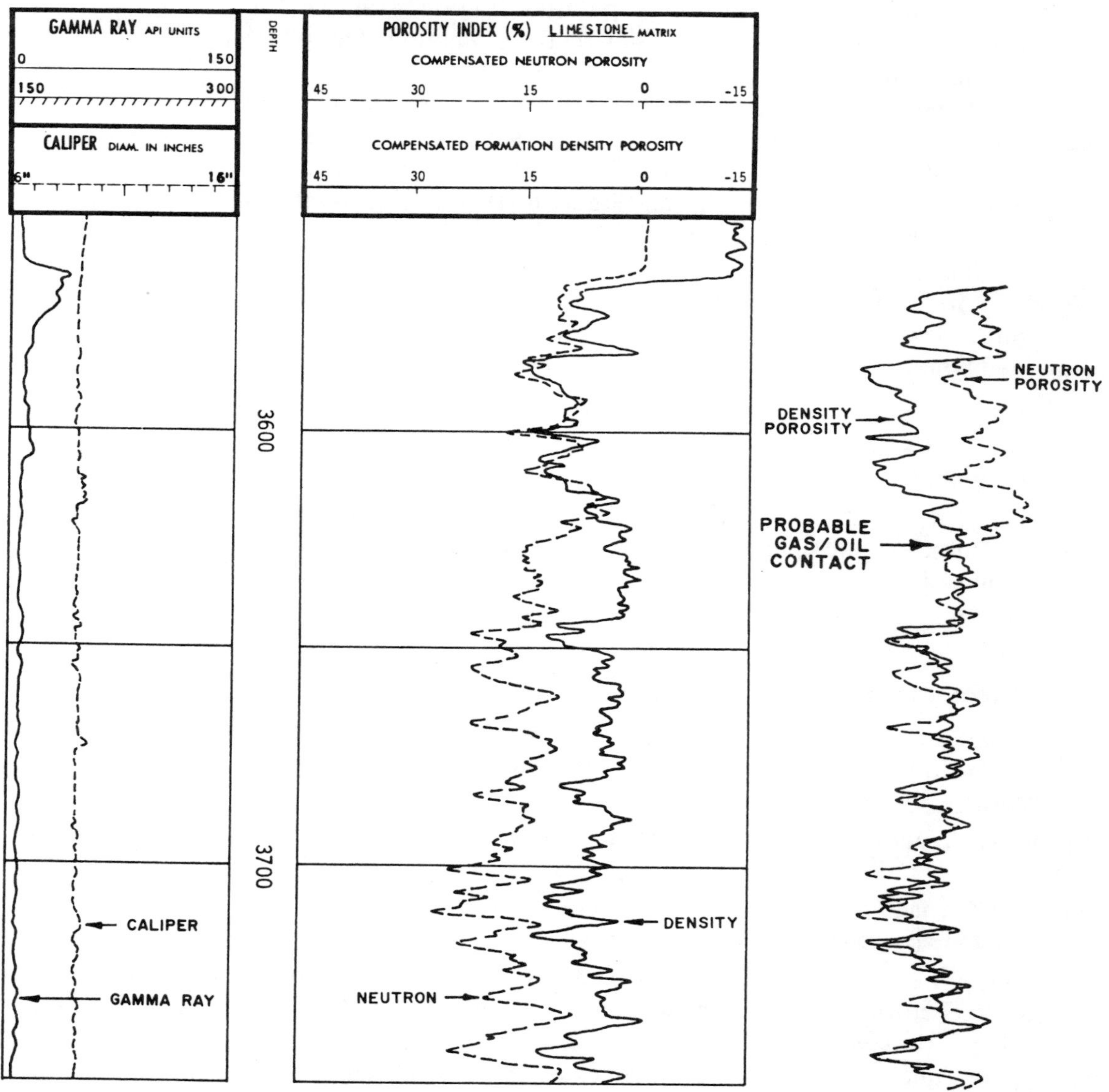

FIG. 6-9. A gas detection overlay (right) made from the density-neutron log (left) by assuming a constant lithology for the interval 3 558 to 3 750 ft. The two curves are normalized for lithology or made to overlay in the lower part of the interval. Gas is indicated by the separation in the 3 558 to 3 626 ft interval, with the neutron measurement going to the right or less porosity.

lithology, in the water or oil interval (Figures 6-9 and 7-18). Gas is indicated in the overlay where the neutron reads less than the density log.

WELLSITE COMPUTER INTERPRETATION

Use of digital computers at the wellsite makes possible more exact interpretation, over much longer intervals than can be done conveniently by hand. With more complete wellsite interpretation available, previously used indicators are becoming less important. Wellsite computer processed logs often have an initial parameter computation (Figure 6-10) to aid in selection of R_w and grain density for gas effect determination. This phase also provides a check of log-to-log depth control.

The interpretation program control parameters — such as R_w, GR and neutron shale limits, F_R constants, m, fluid density, etc. — are usually chosen in the wet intervals, and become part of the assumptions for the actual wellsite computer interpretation (Figure 6-11). These assumptions should always be examined in determining the validity of a wellsite computer interpretation.

The wellsite interpretation usually includes an analog presentation of water saturation, lithology corrected porosity, apparent grain density (computed lithology), shaleness indicator, and a gas detection flag (Figure 6-12).

REFERENCES

Dumanoir, J. L., Hall, J. D., and Jones, J. M., 1972, R_{xo}/R_t methods for wellsite interpretation; 13th Ann. Soc. Prof. Well Log Analysts Logging symposium, Tulsa, paper R.

REFERENCES FOR GENERAL READING

Best, D. L., Gardner, J. S., and Dumanoir, J. L., 1978, A computer-processed wellsite log computation: 19th Ann. Soc. Prof. Well Log Analysts Logging Symposium, El Paso, paper Z.

Bigelow, e. L., abt 1972, The "F" overlay technique and construction of the M.O.P.

Burgen, J. G., and Evans, H. B., 1975, Direct digital laser-logging: 50th Ann. Soc. Petr. Eng. Convention, Dallas, SPE-506.

Dumanoir, J. L., Tixier, M. P., and Martin, M., 1957, Interpretation of the induction-electrical log in fresh mud: Trans. Am. Inst. Min., Metallurg., Petr. Eng., **210**, 202.

Eaton, F. M., Elliott, J. W., Hurlston, F. D., Olsen, R. S., Vanderschell, D. J., and Warren, J. P., 1976, The cyper service unit — an integrated logging system: 51st Ann. Soc. Petr. Eng. Convention, New Orleans, SPE-6158.

Fertl, W. H., 1978, R_{wa} method: fast formation evaluation: Oil and Gas J., **76**, 37, 73-76.

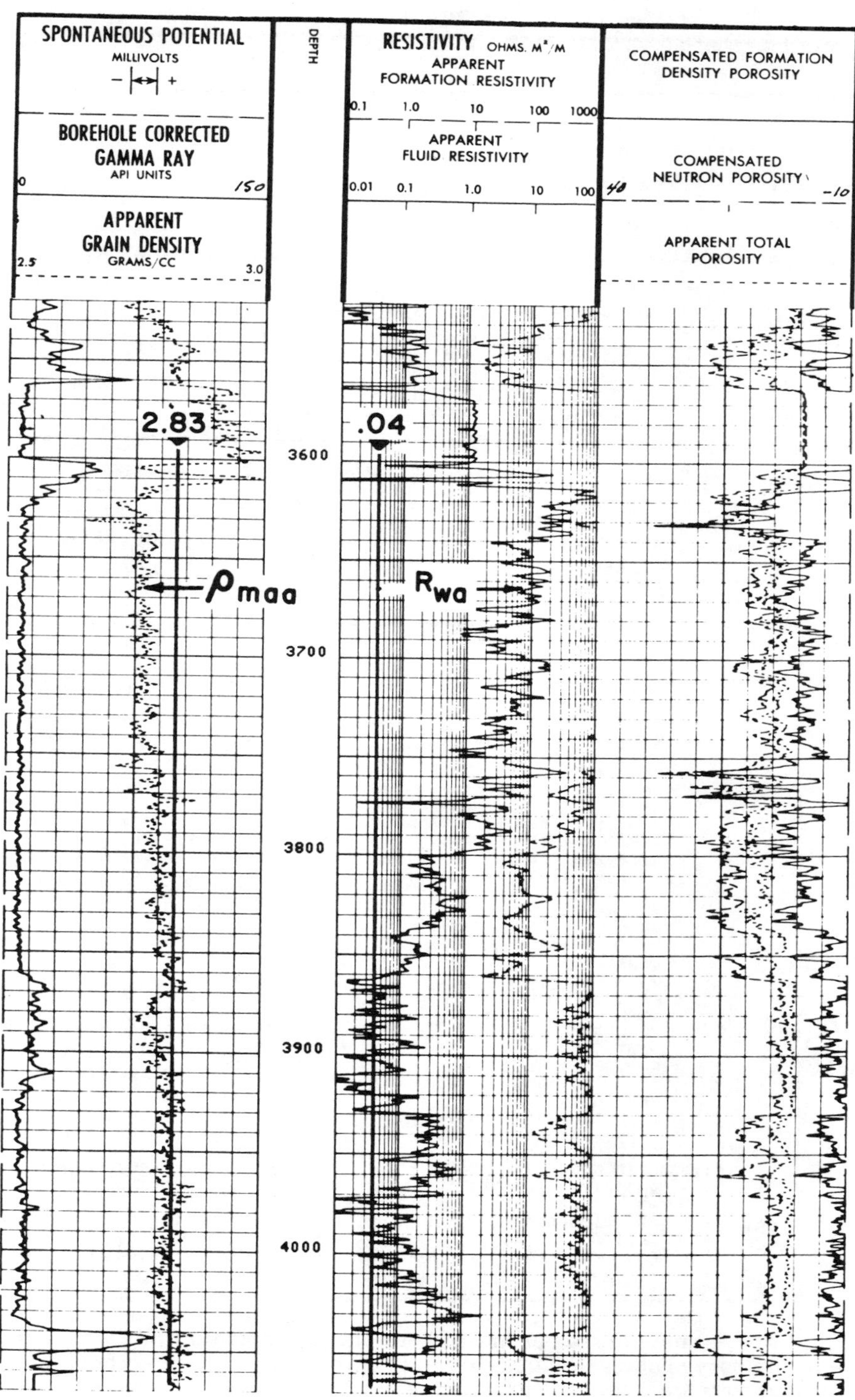

FIG. 6-10. Initial analysis of merged well logs which provides a depth check of the data and aids selection of interpretation program control parameters for the computation phase.

All interpretations are opinions based on inferences from electrical or other measurements and we cannot and do not guarantee the accuracy or correctness of any interpretations and we shall not except in the case of gross or willful negligence on our part be liable or responsible for any loss costs damages or expenses incurred or sustained by anyone resulting from any interpretation made by any of our officers agents or employees. These interpretations are also subject to our General Terms and Conditions as set out in our current Price Schedule. *A mark of Schlumberger

COMPUTATION PARAMETERS

DEPTH	R_w		GR		F = a/φ^m		NEUTRON		SHALE INDEX USED		PMAX	MD	DO	DO?	FLUID DENS	MATR	BS
	Free	Bound	Clean	Shale	a	m	Clean	Shale	GR SP BOTH	Neutron Used Y N	PHI MAX	Clean Matrix Density	Depth Offset Tape 1	Depth Offset Tape 2	FD	Neutron Matrix	Bit Size
4133	.04	.3	12	120	8	1.8	0.0	.3	GR	Y	.3	M13	—	—	1.3	LS	7.825
3873					1	2.3									1.2		

COMPUTATION CONSTANTS

FD = 1.2   WMHD = 10.6

Service Order Number

MUD FILTRATE DENSITY =

REMARKS: Computation used the following logs: LDTA, CNTA, SGTE, DSTD.

RxoF Present.

MDET = SALT

Calculations may be eroniuo in washed out zones.

PARAMETERS

NAME	VALUE	UNIT	NAME	VALUE	UNIT
BS	7.87500	IN	RW	0.040000	OHMM
KCON	1.00000		KEXP	1.00000	
FRT	ALLO		KES	E4	
UM33	8.99700		GRSS	GR	
UM31	13.7700		UM32	4.77900	
UM22	4.77900		UM23	8.99700	
UF	1.40000		UM21	13.7700	
RG32	2.64400		RG33	2.87700	
RG23	2.87700		RG31	2.71000	
RG21	2.71000		RG22	2.64400	
SIS	NONE		LDTA	PRES	
NMLA	ABSE		SDGC	M13	
GULM	999.000	GAPI	TPML	9.80000	
NLIM	0.010000		DLLM	0.350000	
LSWB	DISA		MDET	SALT	
GDSH	2.90000	G/C3	FRTC	MSFL	
DRUL	999.000	G/C3	GDCL	0.0	G/C3
TPDM	8.70000	NS/M	TPFM	7.20000	NS/M
FESX	DISA		TPSM	7.20000	NS/M
POUT	LIME		FEPT	NONE	
RWB	0.300000	OHMM	FCAL	PRES	
NPSH	0.300000		RWF	0.040000	OHMM
SPSH	0.0	MV	NPCL	0.0	
GRSH	120.000	GAPI	SPCL	-200.000	MV
PMAX	0.300000		GRCL	12.0000	GAPI
SINP	GR		NGSI	NEUT	
RXOF	PRES		WMUD	10.6000	LB/G
BHT	101.000	DEGF	SONI	ABSE	
HC	CALI		RTLF	DISA	
FD	1.30000	G/C3	MATR	LIME	
BARI	DISA		MDEN	2.71000	G/C3
FNUM	8.00000		FEXP	1.80000	
DO	0 0	F	BHS	OPEN	

FIG. 6-11. Assumptions input into the wellsite interpretation program were obtained from the preliminary analysis (Figure 6-10), to control the computer interpretation of Figure 6-12. The parameters are listed in the Computation Parameters section or more often with current logs in printout form (Parameter section) attached to the log.

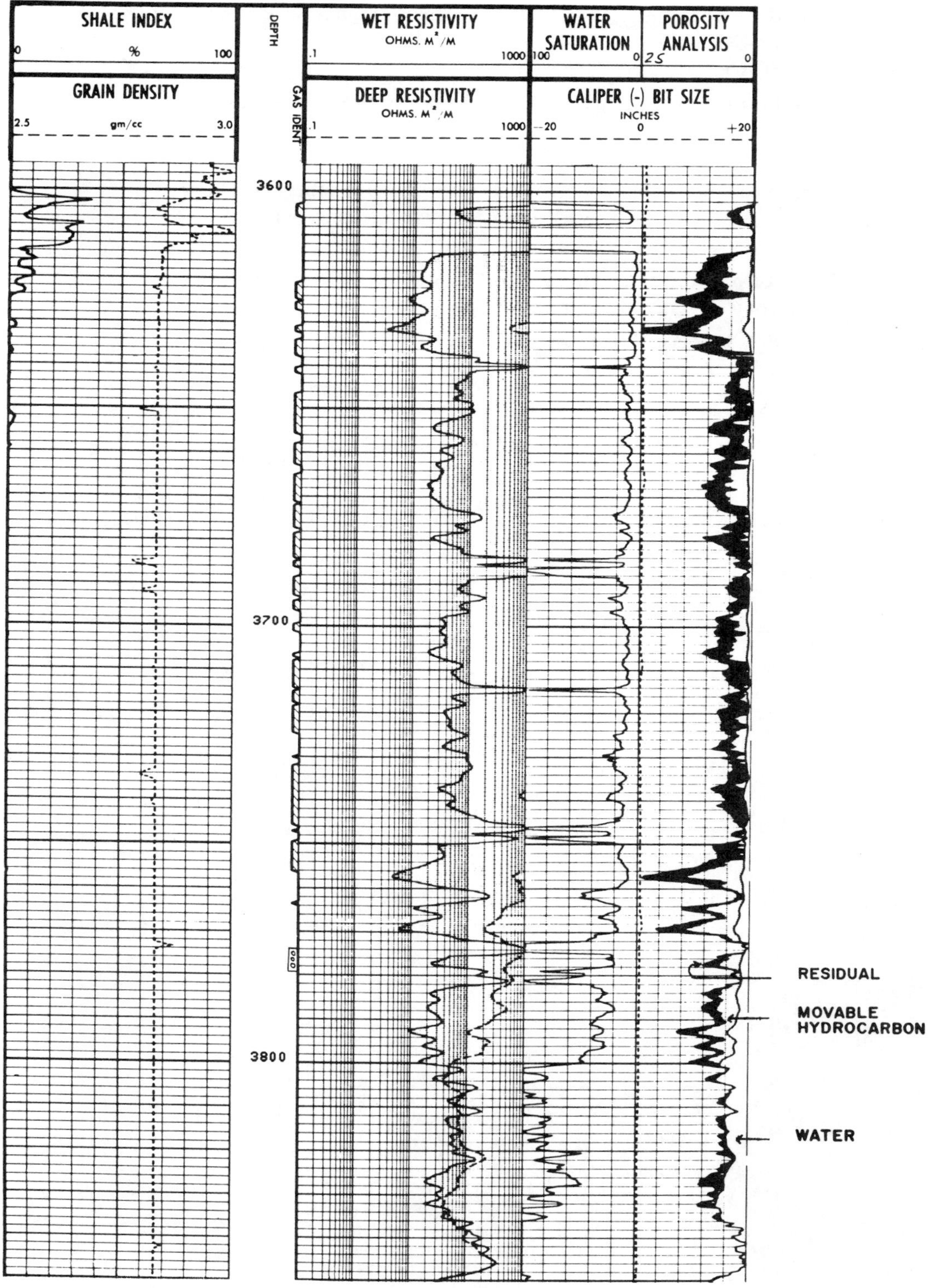

FIG. 6-12. Wellsite computer interpretation of a Michigan Niagaran reef section. In the porosity track (extreme right) the dark is residual hydrocarbon and the stipple is movable hydrocarbons under drilling mud invasion conditions.

Head, M. P., and Gearhart, M., 1977, Wellsite formation analysis using the DDL computer: 47th Ann. Calif. Regional Soc. Petr. Eng. Convention, Bakersfield, SPE-6541.

Heflin, J., 1974, Formation factor plot: Dresser Atlas technical memorandum, **5**, 2.

Studlick, J. R. J., and Gilchrist, W. A., 1981, A sensitivity study of Schlumberger's cyberlook computation and its comparison to other evaluation methods: 22nd Ann. Soc. Prof. Well Log Analysts Logging Symposium, Mexico City, Paper F.

Tixier, M. P., Alger, R. P., and Tanguy, D. R., 1960, New developments in induction and sonic logging: J. Petr. Tech, **12**.

Tixier, M. P., 1962, Modern log analysis: J. Petr. Tech., **14**.

Chapter 7

WELLSITE INTERPRETATION EXAMPLES

To show how Chapter 5 log interpretation principles work, Chapter 7 covers two wellsite interpretation examples. The very general interpretation examples used are designed to emphasize several points. They are not detailed petrographic studies but are similar to the preliminary interpretation required before deciding whether to set pipe on a well.

EXAMPLE 1: FRESH MUD

For the first example, three closely spaced, relatively thin intervals drilled with fresh (1 200 ppm NaCL) mud are examined. The log suite is a dual induction (Figure 7-1), and a simultaneous density-neutron log (Figure 7-2). A density derived F Log (Figure 7-3) is also used.

Verification of log data. — The first step before interpreting any log is to check both intralog curve tracking and interlog depth control (Figure 7-4). First, check the individual curves on each log for tracking. Each memorized curve should track against the primary curve depth measurement. The intralog depth control especially should be checked at TD, casing, washouts, and shale streaks. After these have been verified, the depths of the prime measurements on each log should be checked against the prime measurements on all the other logs. Often the logs will be on-depth with each other at the bottom of the well, where the depth tie is made by the engineer, but slowly become off-depth as they move uphole. If there are small discrepancies, the first-run base depth log should be determined, and subsequent logs annotated with new depth grid numbers. For large depth discrepancies the log should be returned to the service company for correction. If the logs were digitized and recorded on magnetic tape at the wellsite, to replay the tapes with new

(Text continued on page 116)

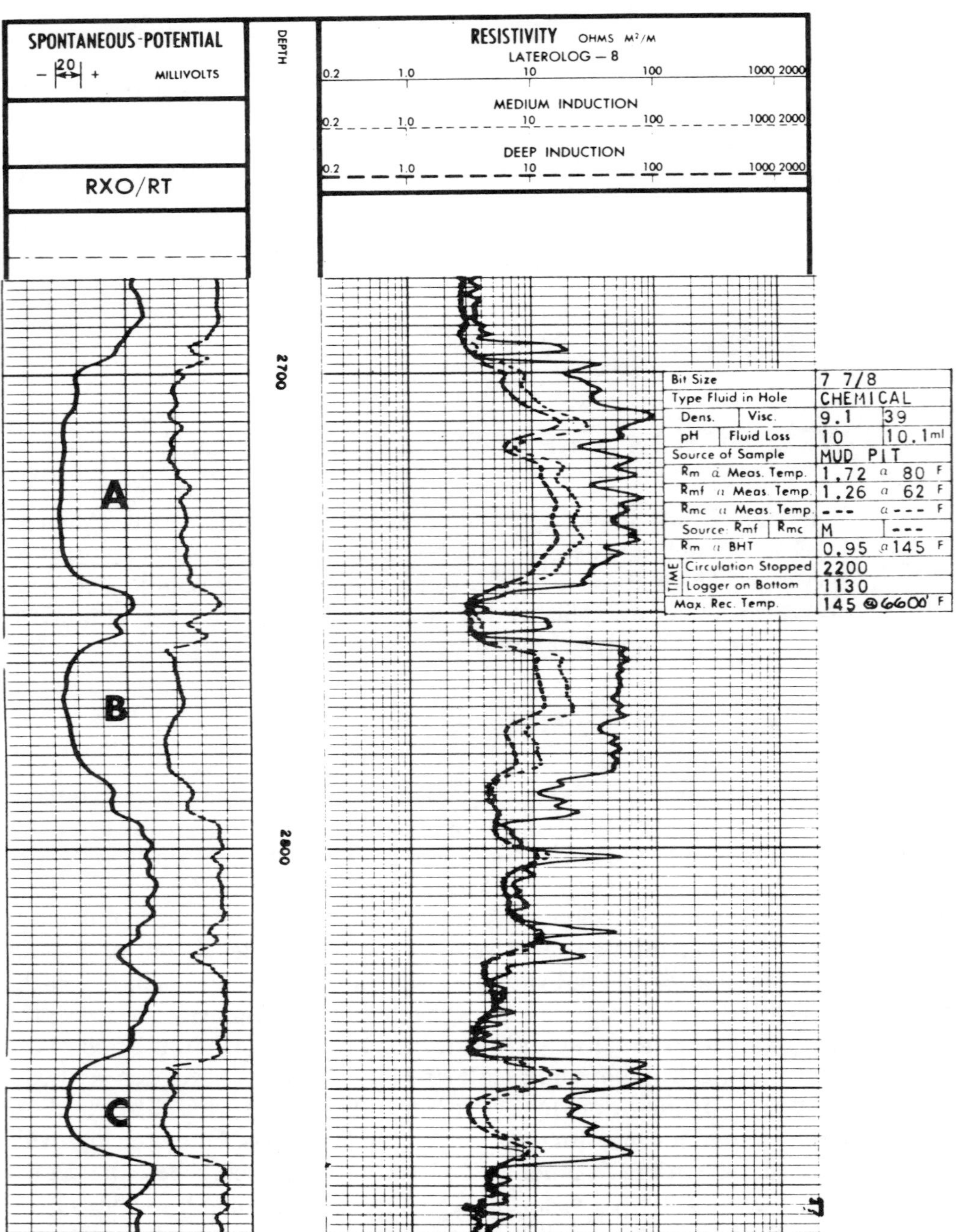

FIG. 7-1. Fresh mud example, a dual-induction log.

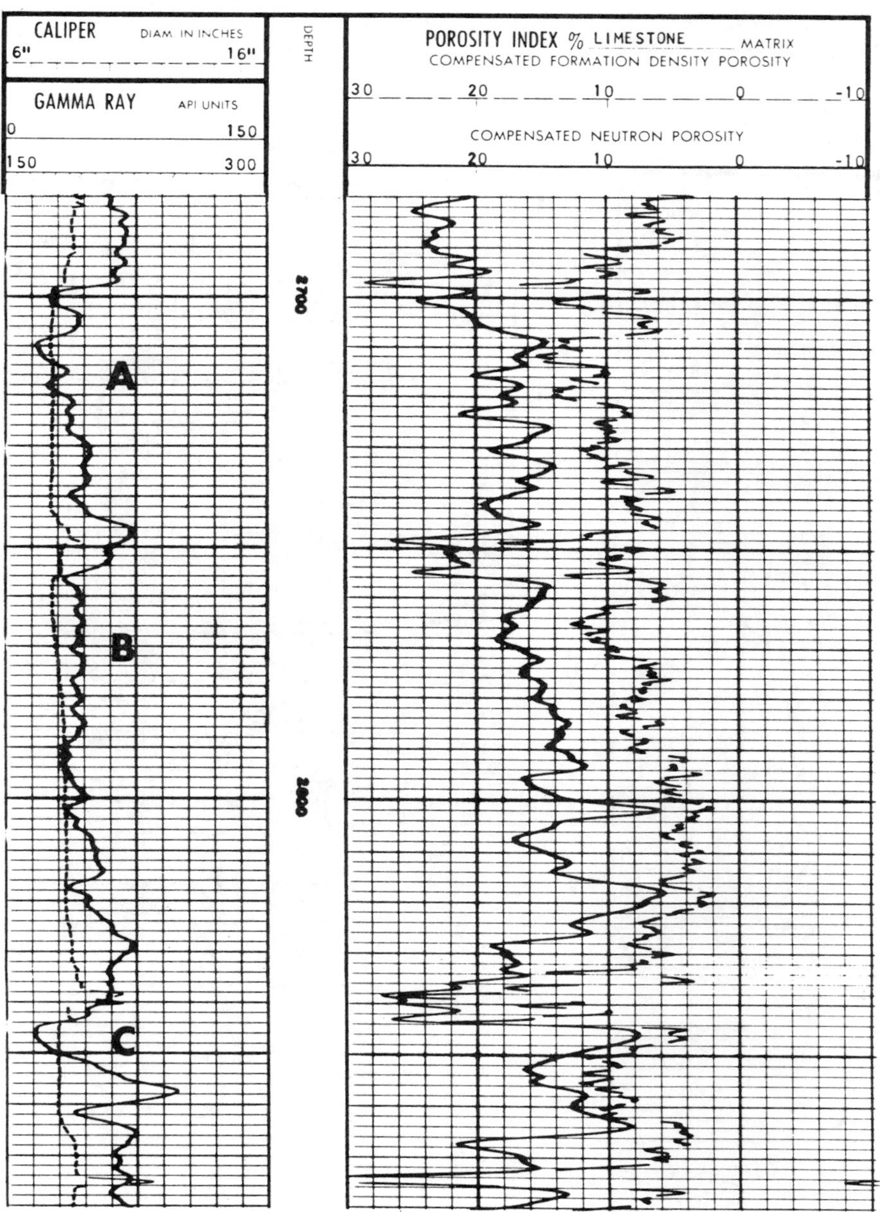

FIG. 7-2. Fresh mud example, a density-neutron log.

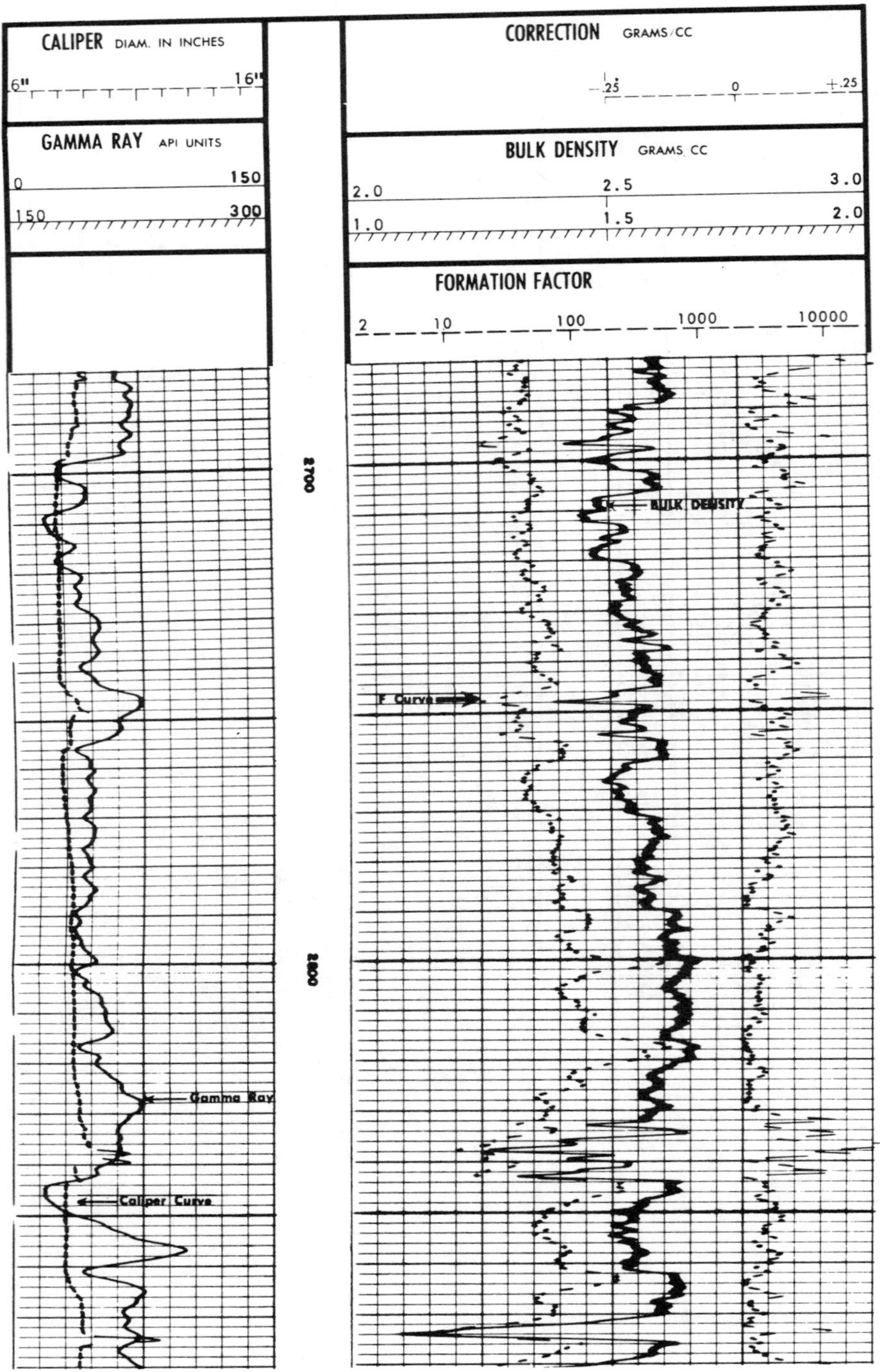

FIG. 7-3. Fresh mud example, a bulk density with *F*-log recording. The *F*-log assumes a matrix density of 2.71 g/cm³.

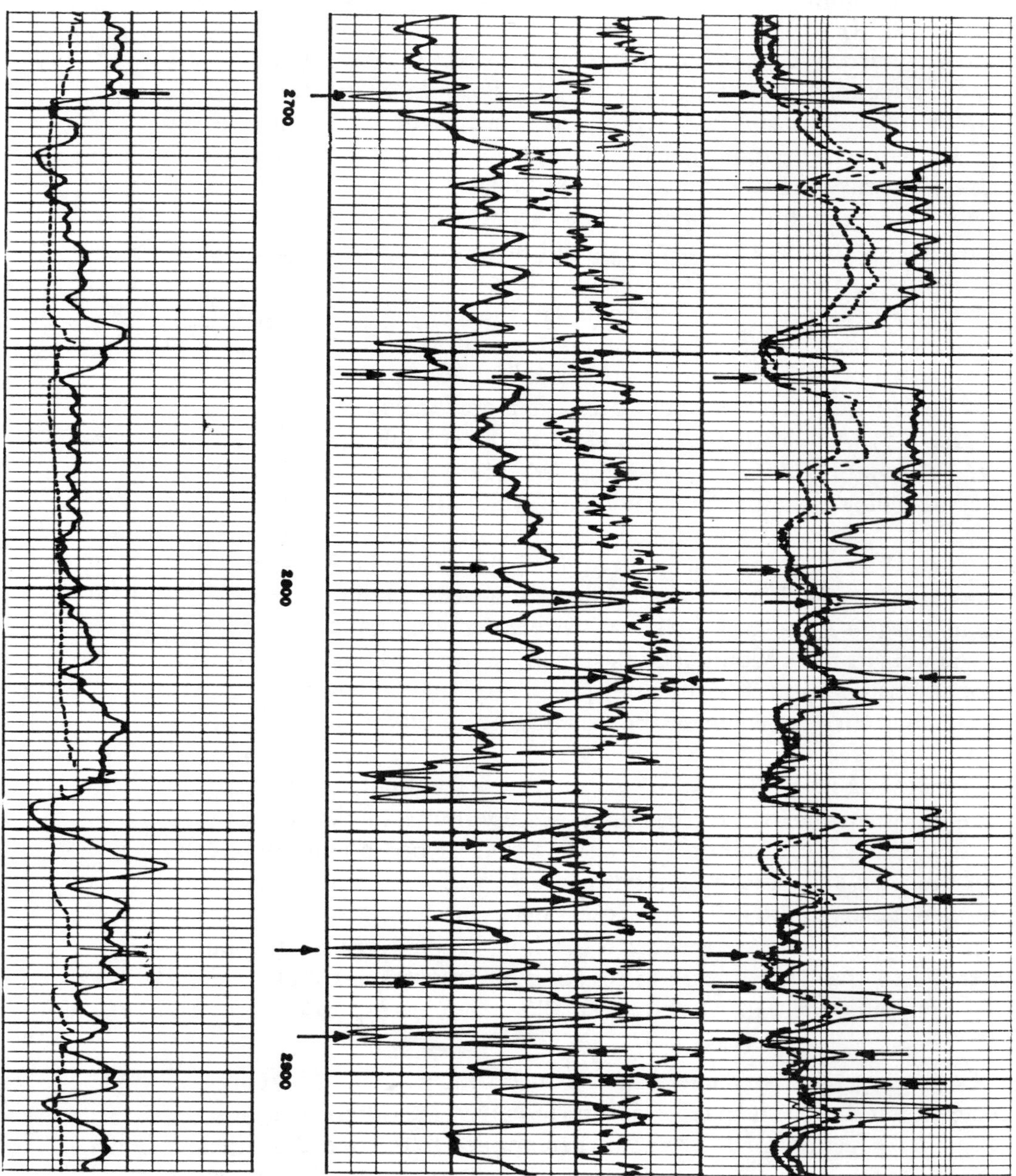

FIG. 7-4. Depth check between resistivity and porosity measurements. The gamma-ray curve memorization over this interval is inconclusive, but is proper on other sections of the log.

depth-offset information either at the wellsite or in their shop is a simple procedure.

If a tension curve were recorded, usually found on the right edge of Track III, it would indicate potential off-depth intervals due to tool sticking (Cooley, 1974).

After verifying that the log depths are correct or making annotated hand corrections, check the individual curves for correctness. First, examine the calibration tails. Correct calibrations do *not* ensure a good log, but conversely, improper calibration does ensure a bad log.

Next, the curves should be examined in "known" formations for proper response. Although this approach may seem pessimistic, good logs are easily proved and incorrect logs should be discovered before reams of calculation are generated from unsure data.

Interval Zoning. — After the quality of the logs is established, use the wellsite recorded indicators of Chapter 6 to cull wet intervals and select likely potential intervals. In this example, a *F*-log is recorded from the bulk density measurement and is overlain in the three intervals of interest (Figure 7-5). Adjust the curve to estimate a probable R_w value.

In this example the *F*-log was recorded for a limestone matrix. If the intervals being examined are sandstones, the *F*-log, as recorded, will be incorrect. Figure 7-6 shows how to revise the scale index to better present the *F*-log if a sandstone matrix is required. Although there is some error in rescaling, it is within reasonable limits for a wellsite indicator approach. Adjusting the *F*-log index will not work for dolomites.

The wellsite recorded indicators should establish which intervals need further investigation. However, all producing intervals in the area may be included for completeness and as a quality check on parameter selection in the area.

The desired intervals next are zoned for correctional application and computations, aimed at representing the largest intervals with one set of parameters. All logging tools have some edge or bed-boundary effects. The object is to zone the intervals in a way that also minimizes this problem. Select initial inflection points from the shallow resistivity curve because it has the best vertical resolution. Then compare the initial intervals to the deeper reading resistivity measurements and porosity measurements, breaking the zones smaller where necessary (Figures 7-7 and 7-8).

Correcting resistivity and porosity measurements. — The common correction technique is to crossplot the resistivity measurement ratios and obtain a correction factor which is then applied to the deepest measurement to obtain R_t (Figure 7-9 and Table 7-1). Remember that the corrections are good only if the downhole conditions are the same as those used to create the resistivity crossplot chart.

To enter the resistivity crossplot chart (Figure 7-9) the ratios of the medium deep and shallow deep must be determined. Two approaches are possible: (1) read each log value and compute the ratio, or (2) use the fact that the distance between the logarithmic recorded curves is the

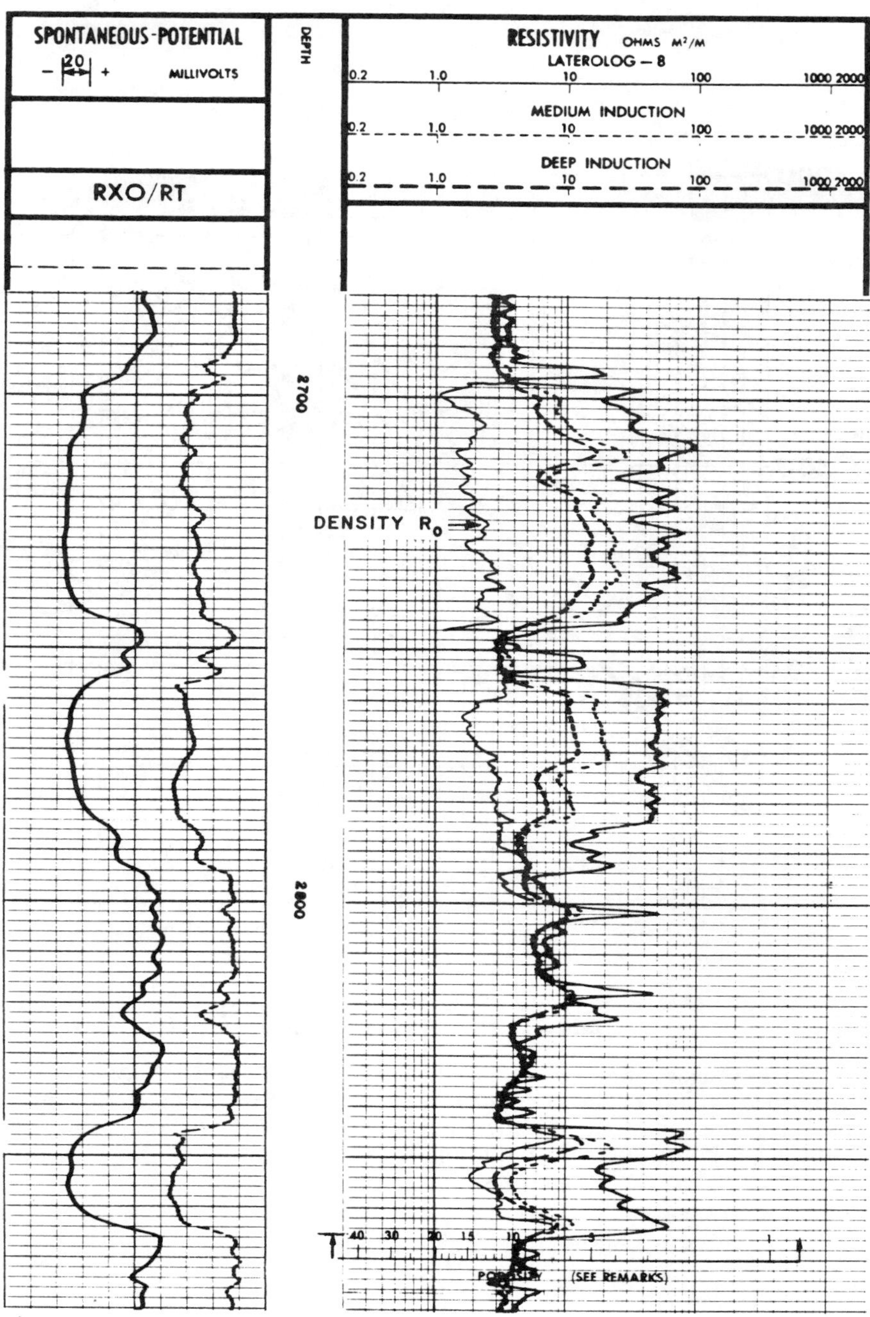

FIG. 7-5. Wellsite indicator check. The $F_R - R_o$ overlay technique. The upper two intervals are of interest, while the lower interval appears to be water bearing. An R_w of .045 Ω•m looks reasonable if the lower interval is in fact wet.

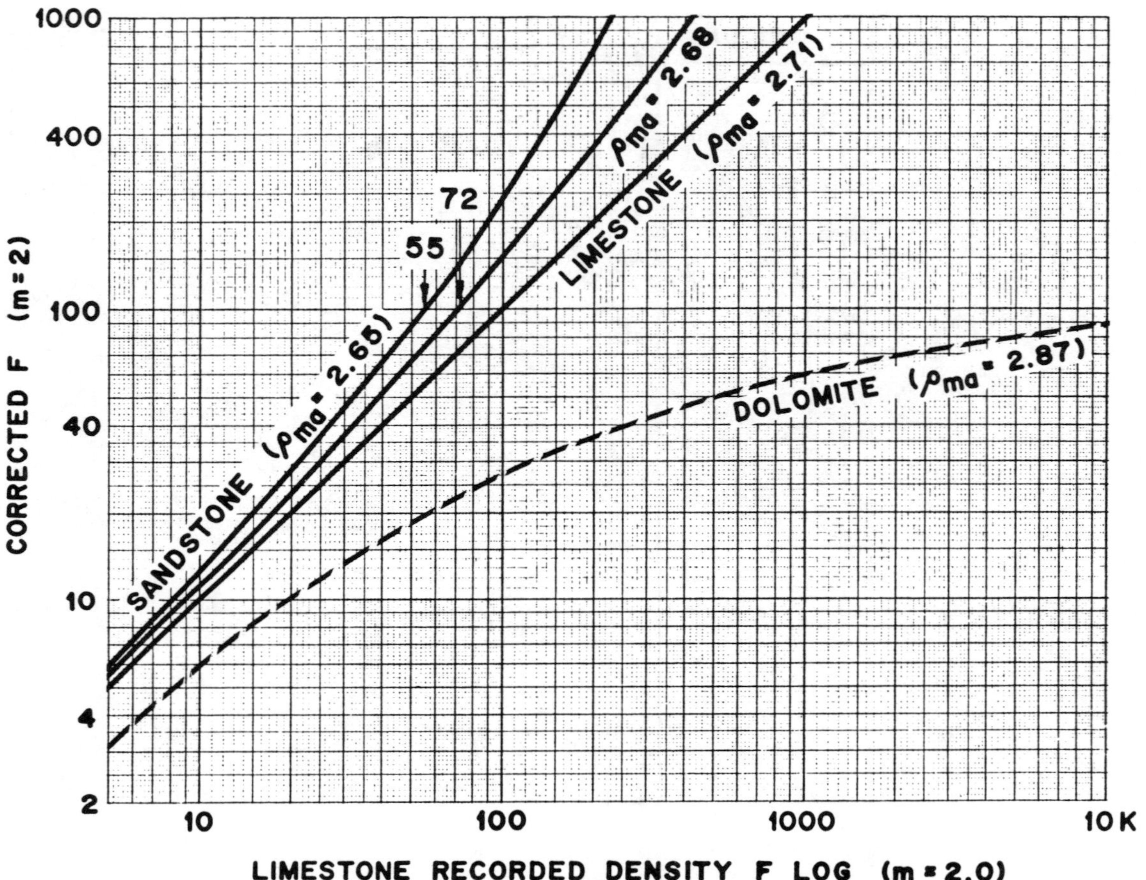

FIG. 7-6. To rescale a limestone matrix F-log, use an index of 55 for sandstone and 72 for limy sandstone in place of the normal 100 index for the R_o indicator. If the other lithology lines were parallel to the limestone line, the F-log recording would need only a new index. Nonparallelity indicates a porosity error of varying degrees. The error for dolomite is unacceptable.

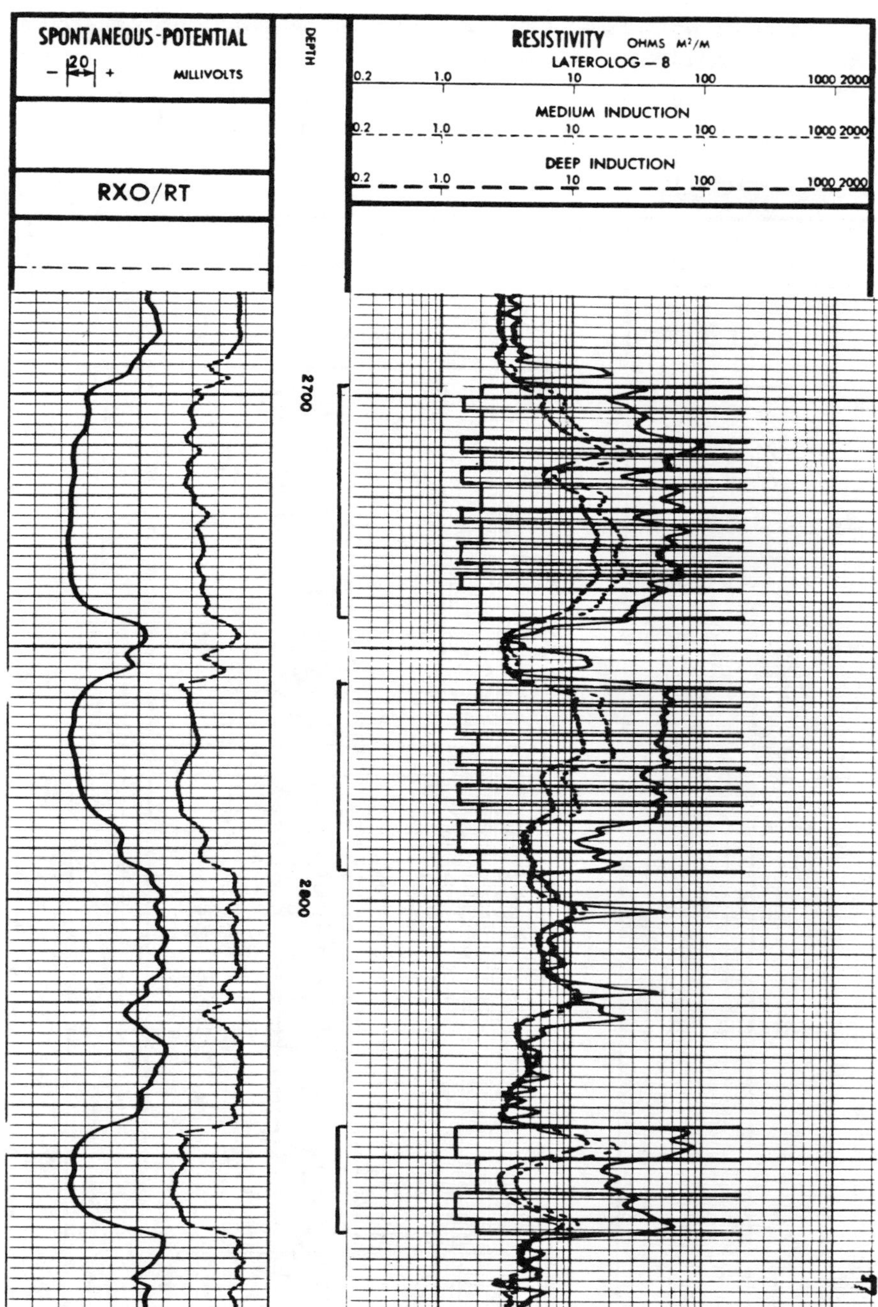

FIG. 7-7. Zoning the resistivity log measurements.

logarithm of their ratios. For the latter approach, either a conventional divider is used to compare the separation of the curves to the logarithmic grid scale or proportional dividers are used to transfer the curve separation ratios into the crossplot chart [for a detailed explanation see Blakeman (1962)]. The results are shown on Table 7-1.

At the same time the resistivity correction is obtained, an R_{xo}/R_t is determined from the crossplot (Figure 7-9) and entered in the data table

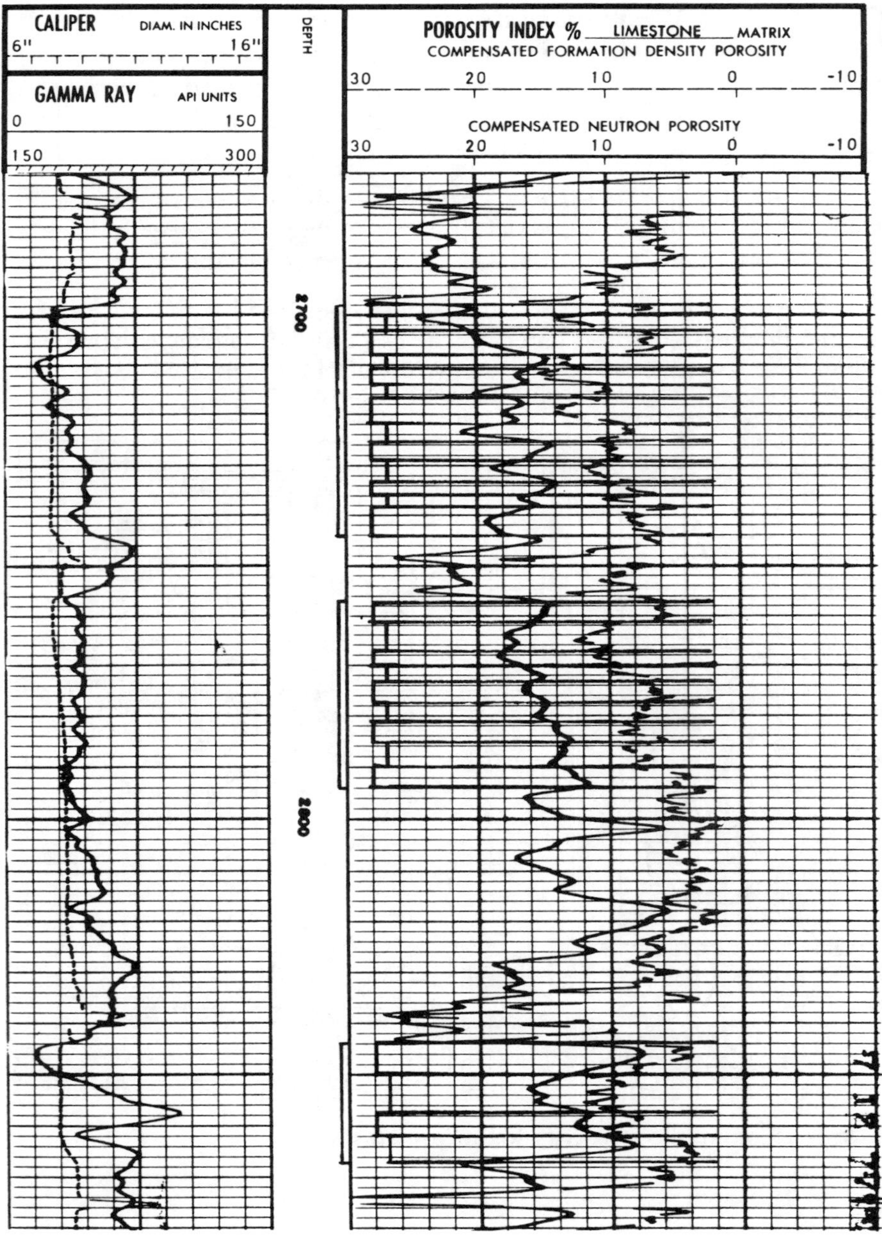

FIG. 7-8. Zoning the porosity log measurements.

(Table 7-1). This is combined with an R_{mf}/R_w ratio calculated from mud measurements and local knowledge to determine a ratio S_w (Table 7-2, left side) using Figure 5-24. The ratio S_{wr} (Table 7-2, left side) should be computed before an Archie equation S_{wa}, because there is a tendency to ignore S_{wr} if S_{wa} is calculated first. It is helpful to have both calculations in defining interpretation results and determining transition profile assumption accuracy.

The density-neutron porosity measurements are crossplotted to obtain a better porosity value and determine the lithology of the interval (Figure 7-10). While the crossplot charts are scaled in bulk density units, they can be entered easily with limestone density porosity by entering horizontally from the limestone matrix porosity line instead of converting porosity back to bulk density. The resultant porosity and apparent grain density are entered on Table 7-1.

With an appropriate R_w, the Archie equation water saturation S_{wa} [equations (5-1) and (5-2)], is determined for each interval. All assumptions (K_R, m, n, R_{mf}, R_w, matrix, etc.) are noted on the interpretation

Table 7-1. Determining R_t and lithology corrected porosity. The two resistivity ratios, medium deep and shallow deep, determine a R_t/R_{deep} correction to calculate R_t and at the same time determine a R_{xo}/R_t value (Figure 7-9). By crossplotting density and neutron porosity (shown in Figure 7-10), a corrected porosity and lithology are determined.

	Resistivity Corrections						Porosity Determination			
Depth	R_{med}/R_{deep}	$R_{shallow}/R_{deep}$	R_{deep}	R_t/R_{deep}	R_t	R_{xo}/R_t	ϕ_n	ϕ_d	ϕ_{xp}	Apparent grain density
	Interval A									
2698-2700	1.5	5.4	5	0.86	4.3	8.5	20	7	13½	2.84
2700-03	1.45	4.1	6	0.86	5.2	6	22	12	17½	2.82
03-08	1.45	4.1	8	0.87	6.9	6	20	7	13½	2.84
08-11	1.45	6.6	16	0.90	14.4	10	16	13	15	2.75
11-14	1.5	5	12	0.85	10.2	8	16½	14½	16	2.73
14-17	1.25	5	6	1.00	6	8	20	10	15	2.81
17-22	1.5	4.1	12	0.83	9.9	6	17½	13½	16	2.75
22-25	1.4	2.4	14	0.81	11.3	3½	21	8½	15	2.83
25-29	1.5	3.4	16	0.82	13.1	5	15	10	12½	2.76
29-33	1.5	3	15	0.79	11.8	4½	18½	11½	15	2.78
33-35	1.5	4	16	0.83	13.3	6	14½	9	12	2.76
35-38	1.5	3	14	0.79	11.1	4½	17	10	14	2.78
2738-44	1.55	3	11	0.72	7.9	5	19	8	14	2.82
	Interval B									
2757-61	1.6	4.8	10	0.88	8.8	7½	15	6	10	2.80
61-67	1.6	4.8	11	0.88	9.7	7½	17½	12	15	2.77
67-70	1.6	3.7	12	0.73	8.8	6	18	11	15	2.78
70-73	1.7	4.5	11	0.65	7.2	8	15½	10	13	2.76
73-77	1.5	5.6	6	0.87	5.2	9	16½	6½	11½	2.81
77-81	1.55	6.9	7	0.84	5.9	11	15	7½	11½	2.78
81-85	1.55	6.7	7	0.84	5.9	10½	14	8½	11	2.76
85-90	1.1	3	4½	1.00	4.5	6	14	8	11	2.78
2790-94	1.0	3.6	5	1.00	5	16	12½	4½	8½	2.79
	Interval C									
2844-50	1.7	5	12	0.69	8.3	8½	8	4½	7	2.75-clean
50-57	1.35	7	3	0.96	2.9	11	15½	10	13	2.77-clean
57-62	1.2	5	5	1.00	5	9	12½	8½	11	2.75-shaley
2862-67	1.3	6.5	8.5	0.98	8.3	11	8½	5	7	2.75-cleaner

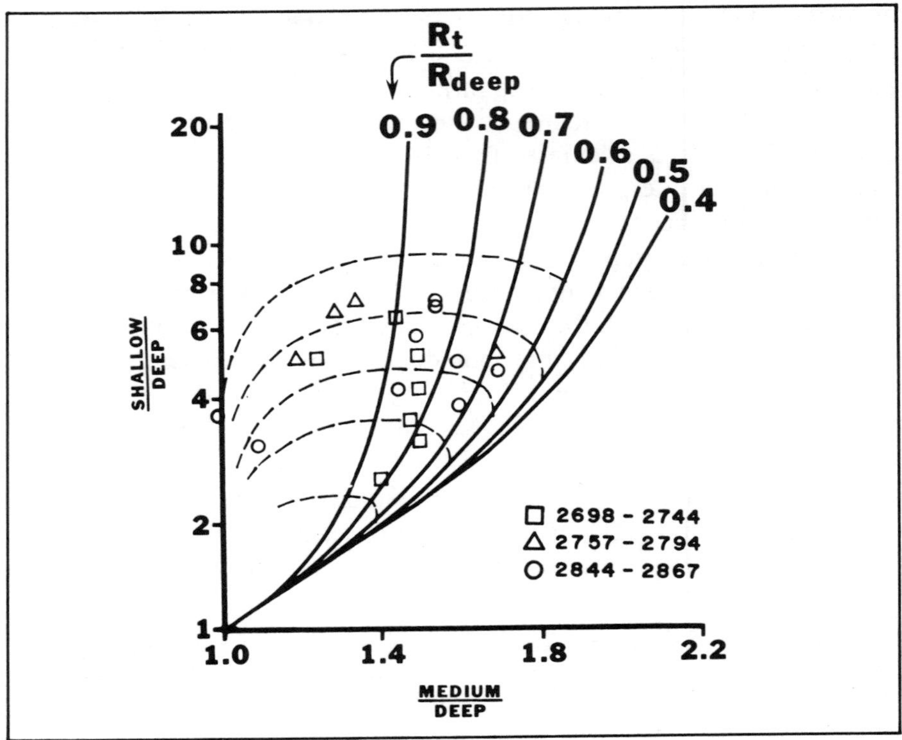

FIG. 7-9. Resistivity crossplot used to determine correction of deep measurement to R_t (solid lines) and a R_{xo}/R_t value (dashed lines) — all assuming a step, not a gradual, transition profile.

table (Table 7-2, right side) to make spotting of incorrect assumptions easier.

Interpreting results. — After the results are calculated for each prospective interval, they need to be interpreted to determine the appropriateness of the assumptions and the producibility of hydrocarbons. The first check compares the two water saturations to determine the validity of invasion profile assumptions used in the resistivity corrections, and the most likely water saturation. For clean formations

$S_{wa} > S_{wr}$ indicates shallow or a transition zone invasion, with S_{wa} being the best S_w choice,

$S_{wa} = S_{wr}$ confirms a step invasion profile and resistivity crossplot corrections are accurate, or

$S_{wa} < S_{wr}$ indicates an annulus invasion profile with the actual S_w less than S_{wa}.

In this example most Archie equation saturations are greater than ratio water saturations ($S_{wa} > S_{wr}$) which indicates a shallow invasion profile or a transition zone. The resistivity crossplot shows that most of the points of interval A (□) have moderate to deep invasion. Thus the water saturation difference indicates a transition invasion profile and hydrocarbons.

Table 7-2. Calculation of Archie's equation and ratio method water saturations using the corrected log values from Table 7-1. The upper interval A is probably hydrocarbon bearing, along with interval B above 2 770 ft. Interval B from 2 770 ft and all of interval C are probably water bearing which corresponds well with the density R_o curve indication of Figure 7-5.

	ARCHIE EQUATION				RATIO METHOD		
	R_w	ϕ_{xp}	R_t	S_w	R_{mf}/R_w	R_{xo}/R_t	S_w
	($K_R=1$, $m=2.0$, and $n=2.0$)				($R_{mf}=0.85$, $R_w=0.04$ Ω•m)		
	Interval A						
2698-2700	.04	13½	4.3	71	21.25	8½	56
2700-03	.04	17½	5.2	50	21.25	6	45
03-08	.04	13½	6.9	56	21.25	6	45
08-11	.04	15	14.4	35	21.25	10	62
11-14	.04	16	10.2	39	21.25	8	54
14-17	.04	15	6	54	21.25	8	54
17-22	.04	16	9.9	40	21.25	6	45
22-25	.04	15	11.3	40	21.25	3½	32
25-29	.04	12½	13.1	44	21.25	5	40
29-33	.04	15	11.8	39	21.25	4½	38
33-35	.04	12	13.3	46	21.25	6	45
35-38	.04	14	11.1	43	21.25	4½	38
2738-44	.04	14	7.9	51	21.25	5	40
	Interval B						
2757-61	.04	10	8.8	67	21.25	7½	52
61-67	.04	15	9.7	43	21.25	7½	52
67-70	.04	15	8.8	45	21.25	6	45
70-73	.04	13	7.2	57	21.25	8	54
73-77	.04	11½	5.2	76	21.25	9	58
77-81	.04	11½	5.9	72	21.25	11	66
81-85	.04	11	5.9	75	21.25	10½	64
85-90	.04	11	4.5	86	21.25	6	45
2790-94	.04	8½	5	105	21.25	16	84
	Interval C						
2844-50	.04	7	8.3	99	21.25	8½	56
50-57	.04	13	2.9	90	21.25	11	66
57-62	.04	11	5	81	21.25	9	58
2862-67	.04	7	8.3	99	21.25	11	66

Using a water saturation cutoff of 50 percent, all of the upper interval A is probably hydrocarbon bearing, as well as interval B above 2 770 ft. Interval B below 2 770 ft and all of interval C are probably water bearing. This agrees with the initial density R_o curve indication of Figure 7-5.

EXAMPLE 2: SATURATED SALT MUD

The second example, a thick carbonate Michigan Silurian-Niagaran reef section, has borehole conditions vastly different from the first example. The carbonate-evaporite section above the reef requires drilling of the reef section with saturated salt muds, typically 200 000+ ppm chlorides, requiring use of focused-resistivity devices. The R_{mf}/R_w ratio is near unity so a meaningful SP cannot be recorded. The R_t/R_w ratio exceeds 50 000 in hydrocarbon-bearing intervals, one of the highest known ratios. The log suite is a dual laterolog (Figure 7-11), a simultaneous density-neutron log (Figure 7-12), and a microlaterolog (Figure 7-13).

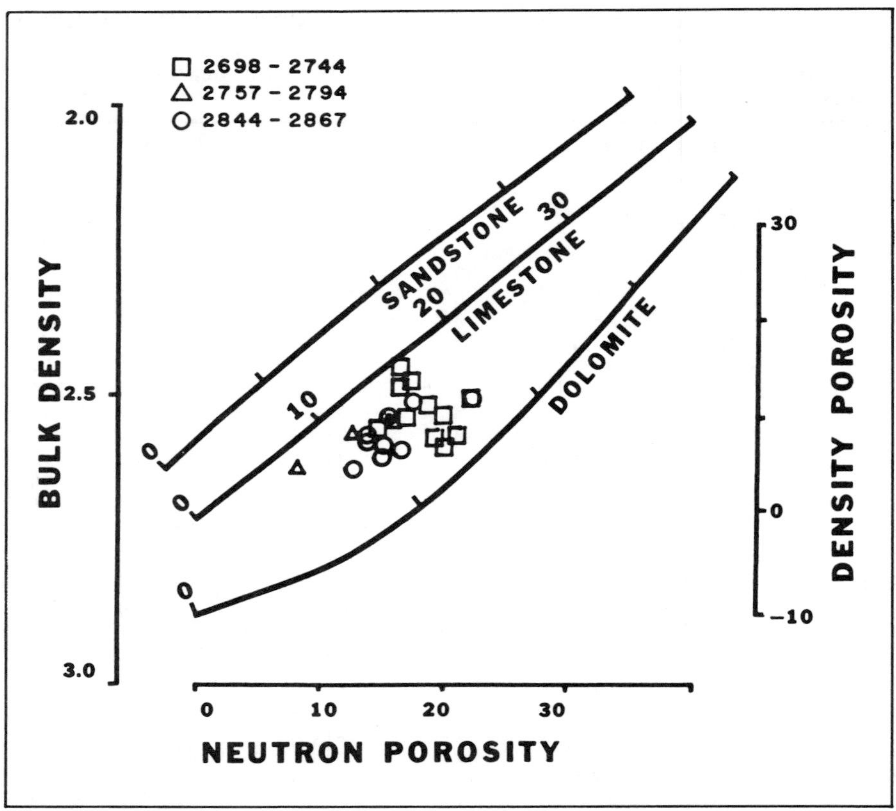

FIG. 7-10. Density-neutron crossplot used to determine lithology and porosity of the intervals.

Verification of data. — Initial checks include the log-to-log and intralog depths where the microlaterolog is found to differ by 2 ft from the other logs (Figure 7-14). Next, log quality is checked. In carbonate-evaporite sequences salt and anhydrite intervals are invaluable when checking measurement quality. Log depth and quality are correct.

Selection of interpretation interval. — Two hydrocarbon checks are available. First, the dual laterolog curves are checked for separation. Because the borehole conditions have R_{mf}/R_w approaching 1, separation is a hydrocarbon indicator. The top of the reef is at 3 574 ft and the bottom of the laterolog separation occurs at about 3 744 ft.

The density-neutron log (Figure 7-12) is examined for gas indications. This log is recorded with a Michigan convention: limestone neutron porosity and bulk density. With the chosen scales, the curves have the same deflections for a change in limestone porosity, but do not track in limestone. The separation is due predominately because the reef matrix is almost dolomite, although a small part is due to the format. The gas effect is determined by examining the log for a decrease in separation of the two curves. The decrease occurs at 3 618 ft with decreasing separa-

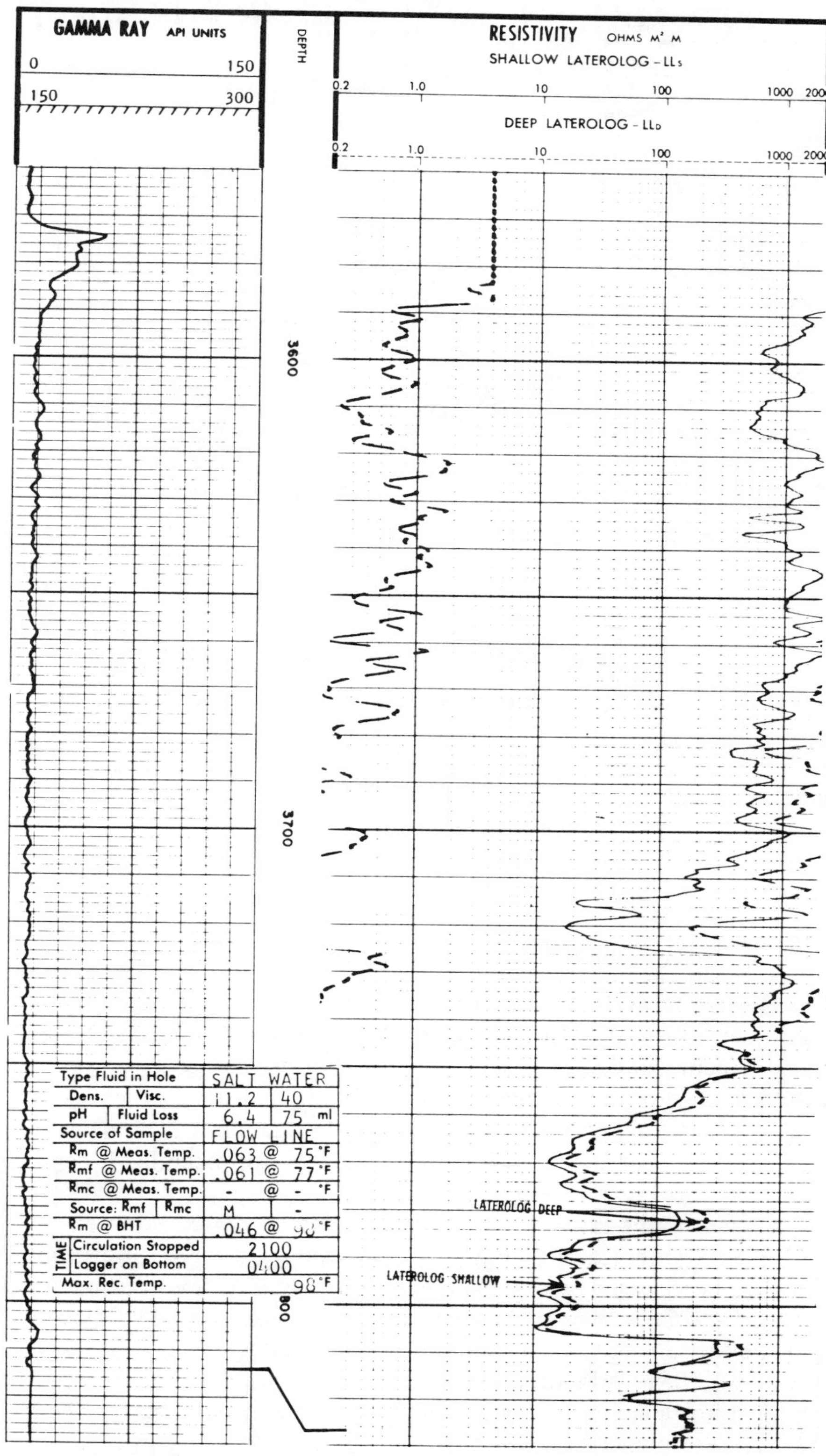

FIG. 7-11. Saturated salt mud example, a dual laterolog.

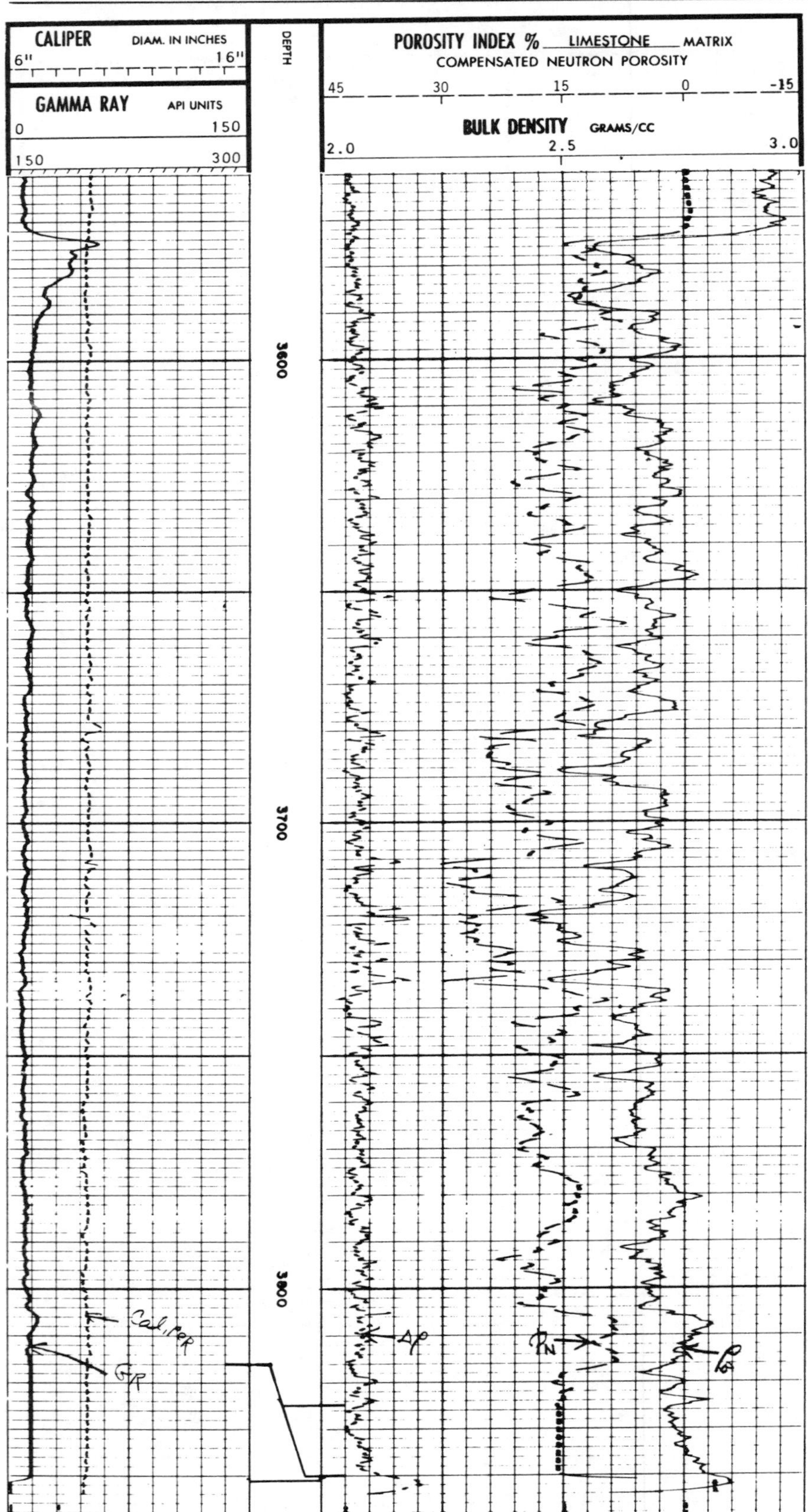

FIG 7-12. Saturated salt mud example, a density-neutron log.

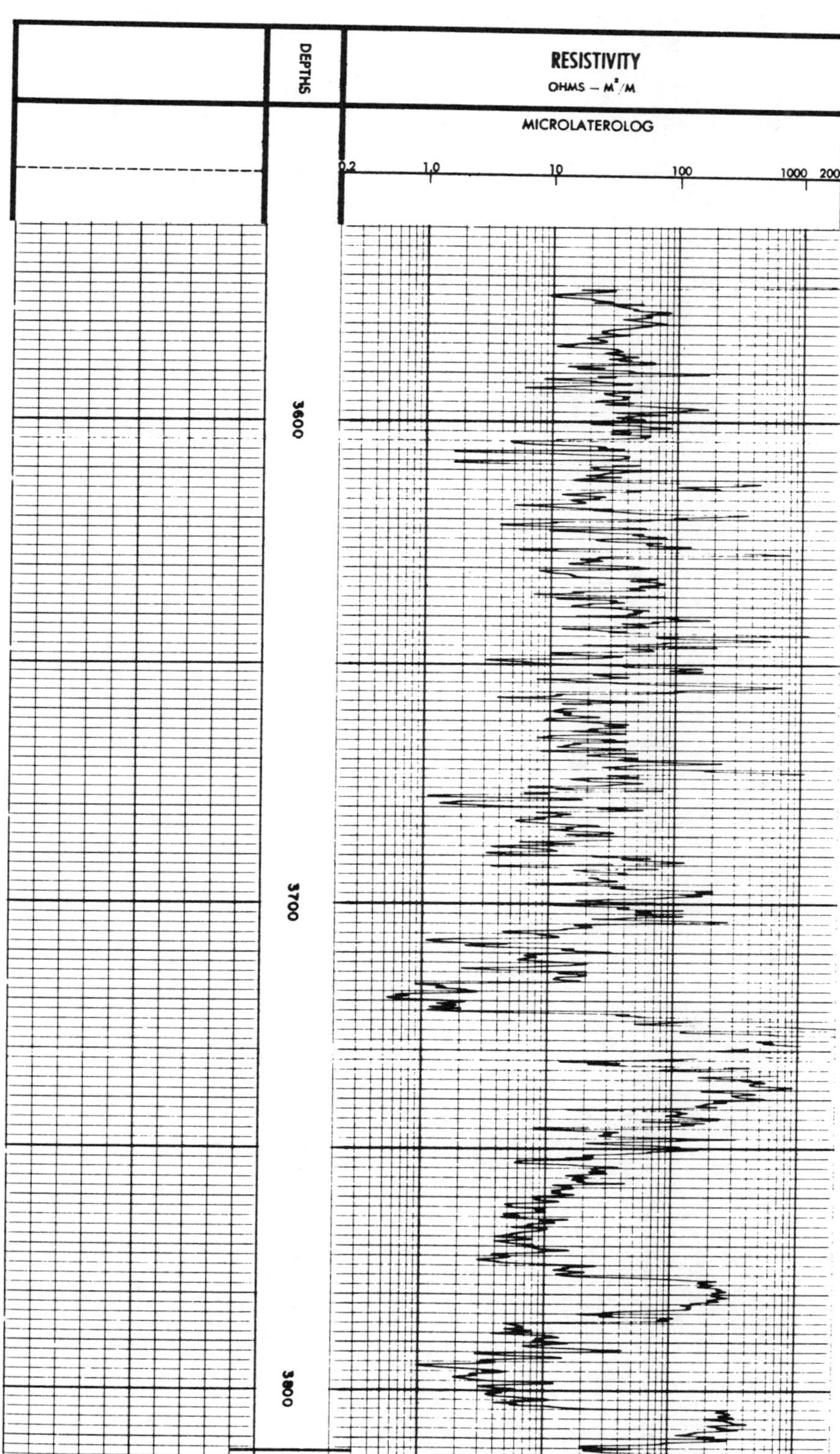

FIG 7-13. Saturated salt mud example, a micro-laterolog.

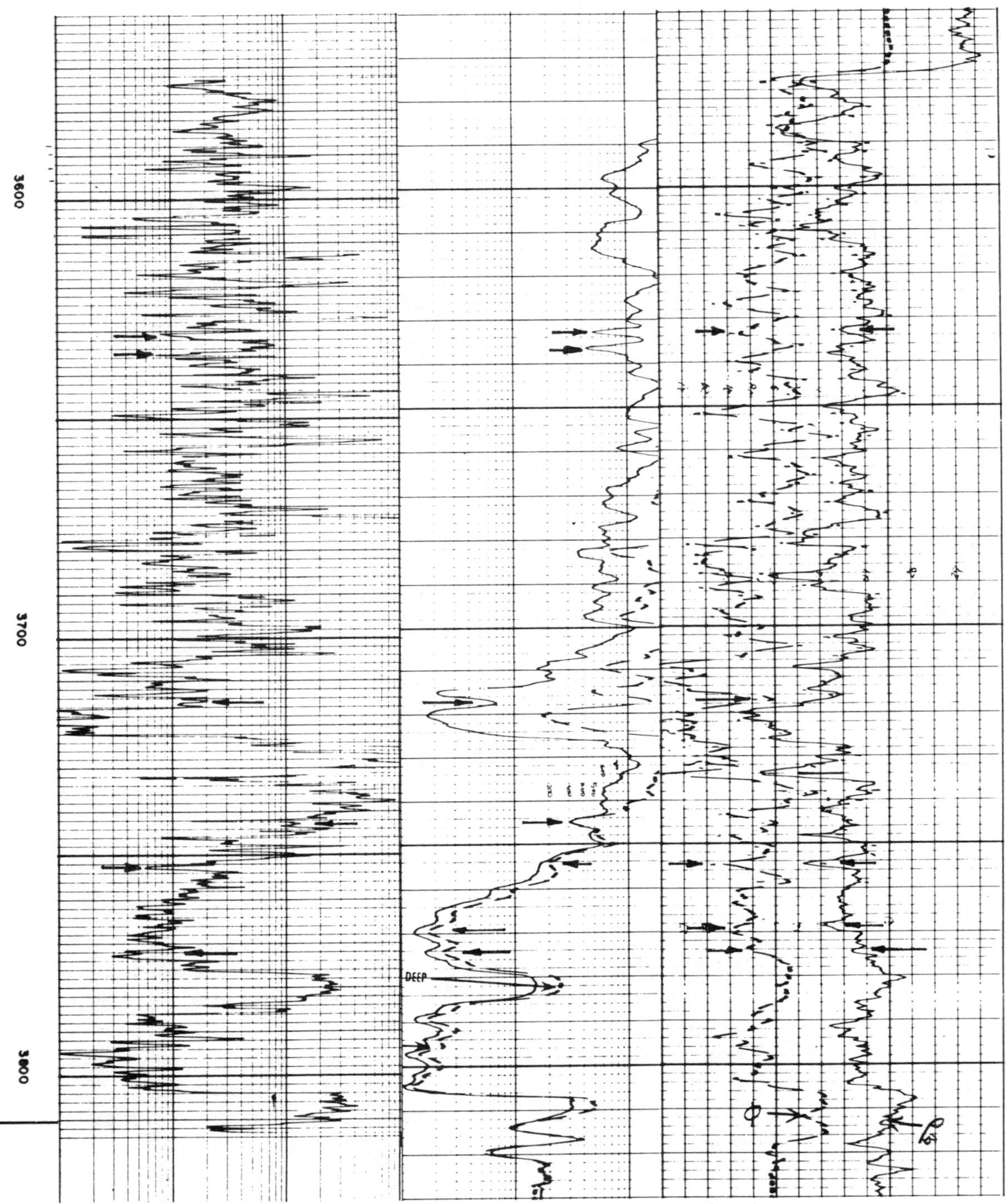

FIG. 7-14. Depth check of resistivity and porosity measurements. Note the microlaterolog, left track, is recorded 2 ft shallow as compared to the other measurements.

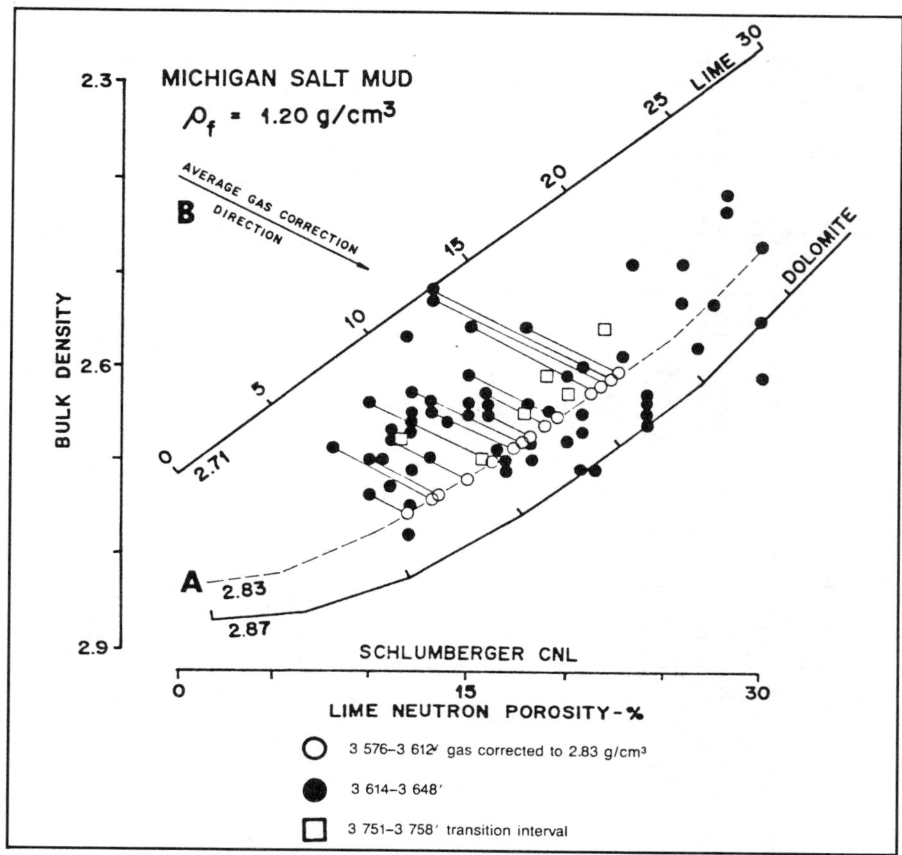

FIG. 7-15. Density-neutron crossplot for saturated salt mud conditions.

tion up to the top of the reef at 3 574 ft. Thus, the wellsite indicators yield a preliminary picture of:

3 574 — 3 618 ft gas bearing,

3 619 — 3 743 ft oil bearing,

and

3 744 — down water bearing.

Determining porosity measurements and formation resistivity factor. — The neutron porosity and bulk density are crossplotted (Figure 7-15) to obtain a better porosity value. This technique is all right in water and oil bearing intervals, but in the gas bearing interval the log values do not plot with a lithology consistent with the reef lithology (Figure 7-15, line A). Line A was determined by making a density-neutron overlay (Figure 7-16) and then plotting the values for the water and oil bearing intervals first, to get the lithology trend. Where the points begin to depart from the lower reef lithology trend (line A) indicates the gas-oil contact.

A correction is made, projecting along an "average gas correction" direction (Figure 7-15, item B) to the reef lithology line, and then the porosity is read and recorded on the data table (Table 7-3) with an * used to indicate a gas-corrected porosity value.

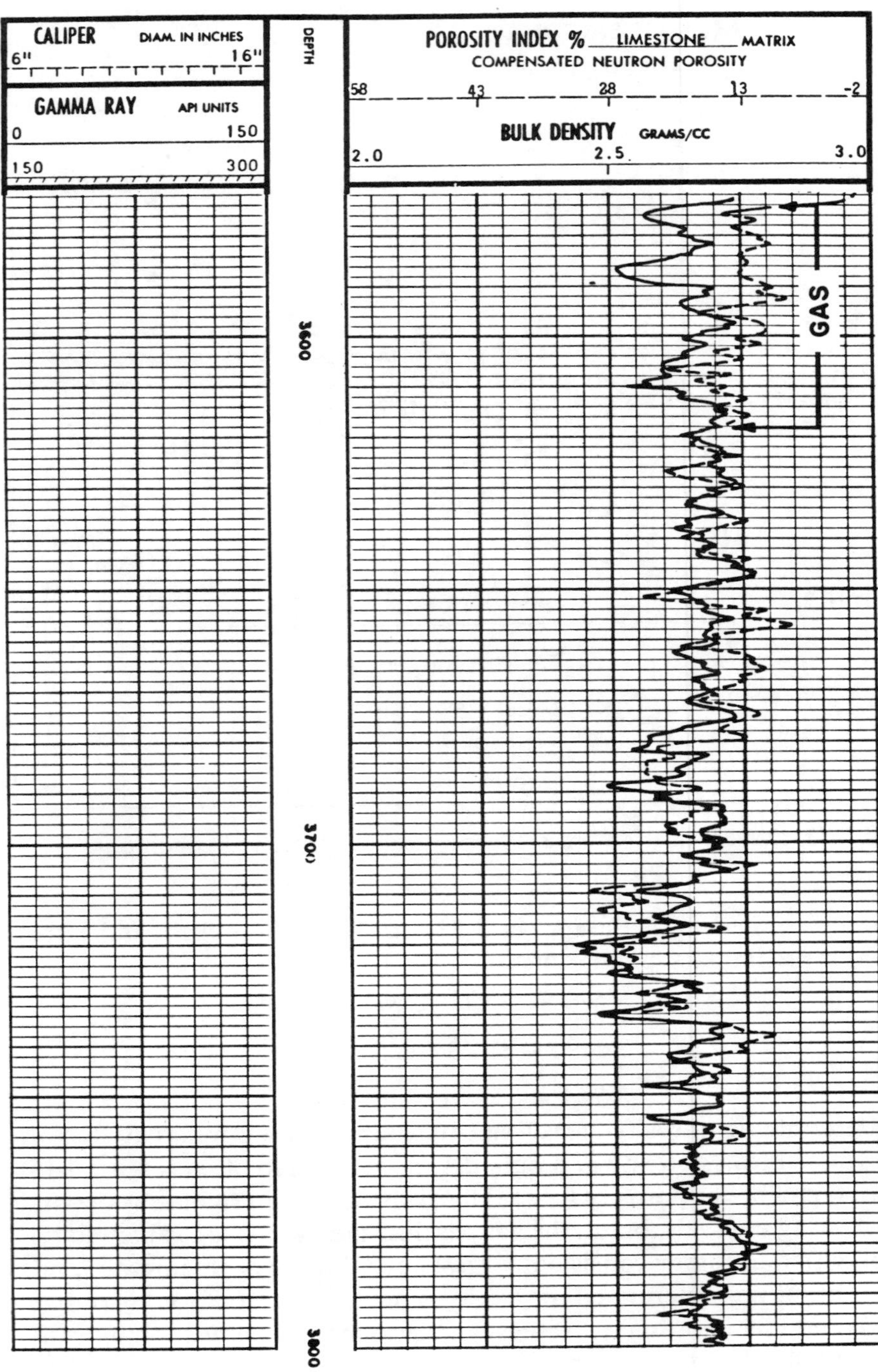

FIG. 7-16. Density-neutron gas detection overlay where the curve separation indicates probable gas-bearing formation from 3 574 to 3 618 ft. The neutron measurement is shifted to overlay the bulk density measurement when the grain density is 2.83 g/cm³ (Figure 7-15, line A).

Empirical studies indicate that, due to compaction of the reef and nonhomogeneous reef permeability, an m of at least 2.5 is required to transform ϕ to F_R (Figure 7-17). Further refinement (Labo, 1977), indicates the need for adjustments of the transform for low-porosity values which leads to the empirical Niagaran reef "variable m" relationship (Figure 7-17). With the advent of computer wellsite interpretations, this relationship is modified for easier use in the computer. The closest computer-adaptable fit to the variable m relationship is

$$F_R = 8\ \phi^{-1.8}. \tag{7-1}$$

Table 7-3. Data tabulation and porosity determination for the saturated salt mud example. Apparent grain density is for crossplot point location before gas correction. Crossplot porosity points flagged * have a gas correction (Figure 7-15, item B) to a 2.83 g/cm³ lithology.

Depth	ρ_b	ϕ_n	ϕ_{nd}	Grain density	F_R	R_o	Depth	ρ_b	ϕ_n	ϕ_{nd}	Grain density	F_R	R_o
3576	2.56	15	15.5*	2.75	200	8	3681	2.53	26	21.2	2.82	97	4
3582	2.70	10	8.0*	2.78	800	32	82	2.68	18	12	2.83	370	15
84	2.66	12	10.5*	2.77	477	19	84	2.65	24	17	2.86	165	7
86	2.52	13	16.5	2.71	178	7	86	2.63	24	17.5	2.85	152	6
88	2.53	13	16.0*	2.72	190	7	88	2.49	23	20.2	2.775	115	5
3590	2.70	10	8.0*	2.78	800	32	3690	2.60	18	14	2.79	260	10
92	2.69	8	8.0*	2.76	800	32	91	2.59	23	17.5	2.82	152	7
95	2.64	18	13.2	2.81	297	12	93	2.71	17	10.5	2.84	477	19
3598	2.74	10	8.5*	2.78	710	28	96	2.71	22	14	2.88	260	10
3600	2.64	13	10.5*	2.77	477	19	3698	2.67	21	14.7	2.84	230	9
02	2.68	11	9.5*	2.78	580	23	3700	2.70	17	11	2.83	430	17
03	2.64	15	12.5*	2.79	330	13	02	2.63	20	14.5	2.82	240	10
04	2.65	13	11.5*	2.78	400	16	05	2.71	12	7.7	2.80	850	34
06	2.60	21	16.0*	2.81	190	8	07	2.65	19	13.5	2.82	285	11
08	2.61	15	13.5*	2.77	285	11	3710	2.55	30	23.5	2.87	72	3
09	2.56	18	17.0*	2.77	165	7	12	2.64	24	17	2.86	165	7
3610	2.56	15	15.5*	2.75	200	8	13	2.61	30	22.5	2.90	80	3
11	2.63	16	13.0*	2.83	310	12	15	2.58	27	21	2.85	102	4
12	2.63	12	12.0*	2.78	370	15	17	2.64	15	11.5	2.79	400	16
14	2.70	15	9.7	2.82	566	23	3720	2.42	28	25	2.78	62	3
16	2.71	12	7.7	2.80	850	34	22	2.43	28	25	2.79	62	3
3620	2.65	19	13.5	2.82	285	14	23	2.49	26	22.2	2.80	82	3
24	2.70	13	8.7	2.80	688	28	24	2.53	28	22.2	2.84	80	3
27	2.71	21	13.2	2.87	273	11	26	2.49	26	22	2.80	85	3
3630	2.74	13	7.7	2.82	850	34	28	2.65	21	15	2.84	220	9
33	2.65	19	13.5	2.82	285	11	3730	2.66	24	16.5	2.87	178	7
36	2.65	12	9.5	2.77	588	24	31	2.60	21	16	2.82	190	8
3640	2.68	20	13.5	2.85	285	11	32	2.64	24	17	2.86	165	7
42	2.70	18	11.7	2.84	385	15	33	2.57	20	16.2	2.79	184	7
44	2.67	12	9	2.78	650	26	34	2.47	30	25.5	2.83	58	2
47	2.78	12	6	2.84	1300	52	38	2.70	10	6.7	2.78	1100	44
52	2.65	24	17	2.86	165	7	3740	2.66	14	10	2.70	525	21
54	2.67	11	7.5	2.78	900	36	41	2.64	13	10	2.77	525	21
55	2.67	11	8	2.77	800	32	43	2.60	21	16	2.82	190	8
56	2.74	13	7.7	2.83	850	34	45	2.65	15	11	2.79	430	17
57	2.69	7	5	2.75	1800	72	47	2.65	16	11.5	2.80	400	16
62	2.61	20	15.2	2.81	212	9	48	2.56	21	17	2.79	165	7
64	2.75	12	6.7	2.82	1100	44	3751	2.70	16	10.5	2.83	477	19
66	2.70	10.5	7	2.78	1000	40	53	2.70	16	10.5	2.83	477	19
68	2.65	13	10	2.78	525	21	55	2.56	22	18	2.80	140	6
3670	2.69	17	11.2	2.83	415	17	59	2.68	11.5	8.2	2.78	762	30
72	2.64	18	13.2	2.81	297	12	3762	2.65	18	13		310	12
75	2.73	11	6.5	2.80	1150	46	64	2.61	19	14.5		240	10
77	2.64	16	12	2.80	370	15	66	2.63	20	15		220	9
3679	2.57	12	11.2	2.73	415	17	3768	2.65	19	14		260	10

Because of the unusual $\phi - F_R$ transform, which analog recording equipment could not produce, it was necessary to handplot an R_o curve (Figure 7-18) from the crossplot porosity determinations.

With the proper F_R values, Archie equation water saturations are determined (Table 7-4) using

$$S_w^n = F_R\, R_w\, R_t^{-1}. \tag{5-1}$$

Obviously, the reef contains hydrocarbon. However, two more factors can be obtained from this log suite: (1) an indication of movability of the hydrocarbons and (2) determination of the transition interval between oil and water.

Table 7-4. Saturated salt mud example — data interpretation. $R_o = F_R\, R_w$, $R_w = .04\ \Omega\cdot m$, $S_w^2 = R_o\, R_{deep}$.

Depth	F_R	R_o	R_{deep}	S_ω	c	Depth	F_R	R_o	R_{deep}	S_ω	c
3576	200	8	2000	6.3	.0098	3681	97	4	2500	3.9	.0083
3582	800	32	2000	12.6	.0101	82	370	15	700	14.5	.0174
84	477	19	2000	9.7	.0102	84	165	7	1200	7.4	.0126
86	178	7	2000	5.9	.0098	86	152	6	1600	6.2	.0107
88	190	7	2000	6.1	.0098	88	115	5	3000	3.9	.0079
3590	800	32	2000	12.6	.0101	3690	260	10	1600	8.0	.0112
92	800	32	2000	12.6	.0101	91	152	7	1500	6.3	.0111
95	297	12	2000	7.7	.0102	93	477	19	1400	11.6	.0122
3598	710	28	2000	11.9	.0101	96	260	10	1400	8.6	.0120
3600	477	19	2000	6.9	.0073	3698	230	9	850	10.4	.0153
02	580	23	4000	7.6	.0072	3700	430	17	3000	7.0	.0077
03	330	13	4000	5.7	.0071	02	240	10	4000	4.8	.0071
04	400	16	4000	6.3	.0073	05	850	34	2500	11.6	.0090
06	190	8	4000	4.3	.0069	07	285	11	1600	8.4	.0113
08	285	11	4000	5.3	.0072	3710	72	3	600	6.9	.0162
09	165	7	2500	5.1	.0087	12	165	7	1700	6.2	.0105
3610	200	8	2500	5.6	.0087	13	80	3	1600	4.4	.0100
11	310	12	3000	6.4	.0083	15	102	4	500	9.0	.0189
12	370	15	3500	6.5	.0078	17	400	16	1600	10	.0115
14	566	23	4000	7.5	.0073	3720	62	3	180	11.7	.0293
16	850	34	4000	9.2	.0071	22	62	3	250	10	.0248
3620	285	14	4000	5.3	.0072	23	82	3	350	9.6	.0215
24	688	28	4000	8.3	.0072	24	80	3	600	7.3	.0164
27	273	11	4000	5.2	.0069	26	85	3	4000	2.9	.0064
3630	850	34	4000	9.2	.0071	28	220	9	4000	4.7	.0070
33	285	11	4000	5.3	.0072	3730	178	7	4000	4.2	.0069
36	588	24	4000	7.7	.0073	31	190	8	3000	5.0	.0080
3640	285	11	4000	5.3	.0072	32	165	7	3000	4.6	.0079
42	385	15	4000	6.2	.0073	33	184	7	2500	5.4	.0088
44	650	26	4000	8.0	.0072	34	58	2	1800	3.5	.0090
47	1300	52	4000	11.4	.0068	38	1100	44	1600	16.5	.0111
52	165	7	3000	4.7	.0079	3740	525	21	1200	13.2	.0132
54	900	36	4000	9.5	.0071	41	525	21	950	14.8	.0148
55	800	32	4000	8.9	.0071	43	190	8	950	8.9	.0143
56	850	34	4000	9.2	.0071	45	430	17	300	23.9	.0263
57	1800	72	4000	13.4	.0067	47	400	16	550	17.0	.0196
62	212	9	4000	4.6	.0070	48	165	7	525	11.2	.0190
64	1100	44	4000	10.5	.0070	3751	477	19	750	15.9	.0167
66	1000	40	4000	10	.0070	53	477	19	250	27.6	.0290
68	525	21	3000	8.3	.0083	55	140	6	200	16.7	.0301
3670	415	17	1900	9.3	.0105	59	762	30	130	48.4	.0399
72	297	12	1800	8.1	.0107	3762	310	12	80	39.4	.0511
75	1150	46	4000	7.2	.0069	64	240	10	35	52	.0759
77	370	15	4000	6.0	.0072	66	220	9	30	54	.0812
3679	415	17	2200	8.6	.0097	3768	260	10	22	68	.0962

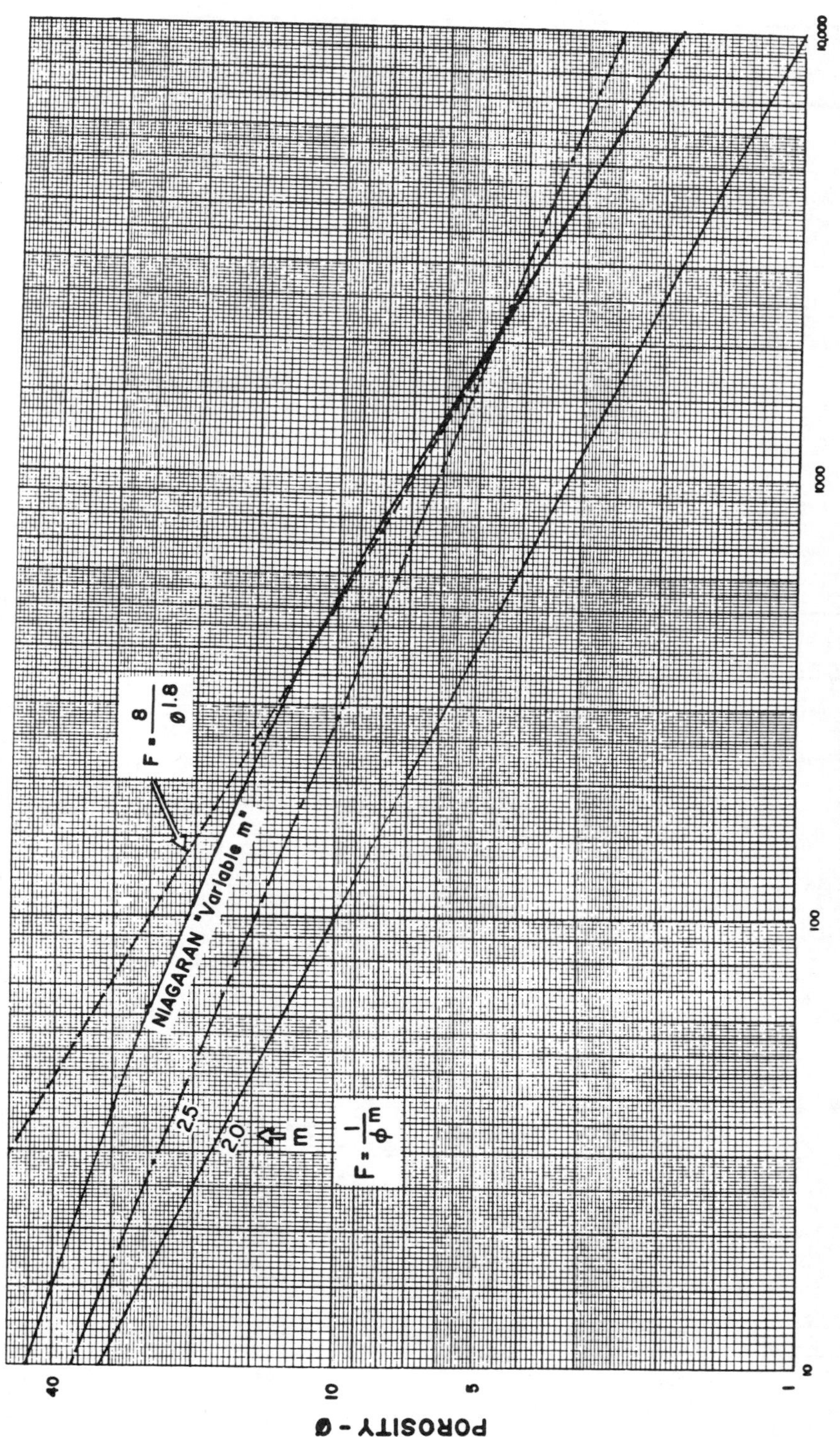

FIG. 7-17. Michigan Niagaran Reef ϕ-F_R relationships. The empirical "variable *m*" is the most correct, but for ease of use with wellsite computers a $K_R = 8$, $m = 1.8$ relationship is accurate below 13 percent porosity.

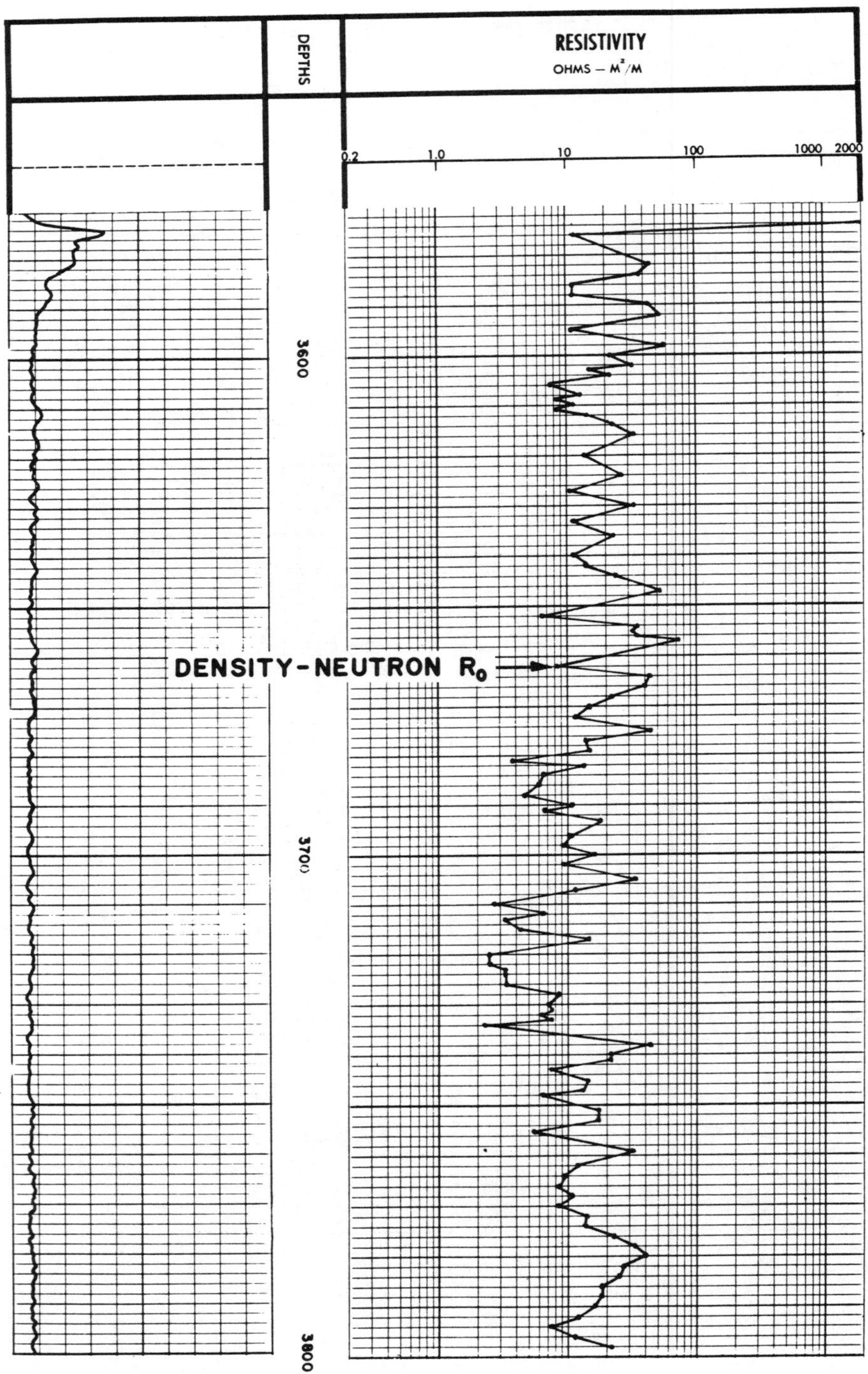

FIG. 7-18. A hand derived R_o curve, for a R_w of .04 Ω•m and the Niagaran "variable *m*" relationship.

Determining producible hydrocarbon-bearing intervals. — The movable-oil plot (MOP) technique, discussed in Chapter 6, is used to determine producible hydrocarbon-bearing intervals using R_o, R_{xo}, and R_{deep} measurements. The plot can be scaled in F units (Figure 6-6), porosity units (Figure 6-7), or resistivity units (Figure 7-19). In the resistivity units an F of 100 is located on the R_o and R_{deep} logs and on the depth-adjusted microlaterolog using the appropriate fluid resistivity. The results are the same with any construction technique used.

The separation between R_{xo} and R_{deep} indicates movable flushed hydrocarbons, and the R_o and R_{xo} separation indicates residual nonflushed hydrocarbons, *if* the producing conditions are simulated accurately by the drilling conditions.

The interface between hydrocarbon-bearing and wet intervals is not sharp, especially in low porosity formations such as this carbonate example, because of capillary action. The transition interval is an interval where water saturation increases with increasing depth (Figure 7-20). Above the transition interval the hydrocarbons are at irreducible water saturation (S_{irr}). Although some water is in the pore spaces, it is tightly bound to the formation matrix with minimum water saturation. Because hydrocarbons can be produced from this interval water free, it is important to know if the formation is at S_{irr} and where the bottom of the S_{irr} interval or top of the transition interval is located.

Because there is a porosity effect on both, they have been shown empirically to be determined by crossplotting porosity and water saturation (Figure 7-21), which is also known as a c plot (Morris and Biggs, 1967) where

$$c = \phi\, S_w. \tag{7-2}$$

An interval at irreducible water saturation will plot on a hyperbolic constant c line, if there is no change in lithology. Where the points began to plot at consistently increasing c values, with increasing depth, is the top of the transition interval; at 3 753 ft in this example. This interpretation technique is very sensitive to good input data.

Interpreting the results. — Interpretation indicates this obviously is a hydrocarbon-bearing reef, with the density-neutron separation showing a gas cap from 3 574 to 3 618 ft (Figure 7-16). The oil-bearing interval is from 3 619 to 3 798 ft, although the base of water free production is at 3 753 ft (Figure 7-21). The movable-oil plot shows considerable non-movable hydrocarbons are from 3 725 to 3 759 ft (Figure 7-19) and therefore the producible oil interval is from 3 619 to 3 724 ft.

REFERENCES

Blakeman, E. R., 1962, A method of analyzing electrical logs recorded on a logarithmic scale: J. Petr. Tech., **14**, 8, 844-850.

Cooley, B. B., 1974, The delta tension curve for better log quality: Trans. 15th Ann. Soc. Prof. Well Log Analysts Logging Symposium, McAllen, Paper F.

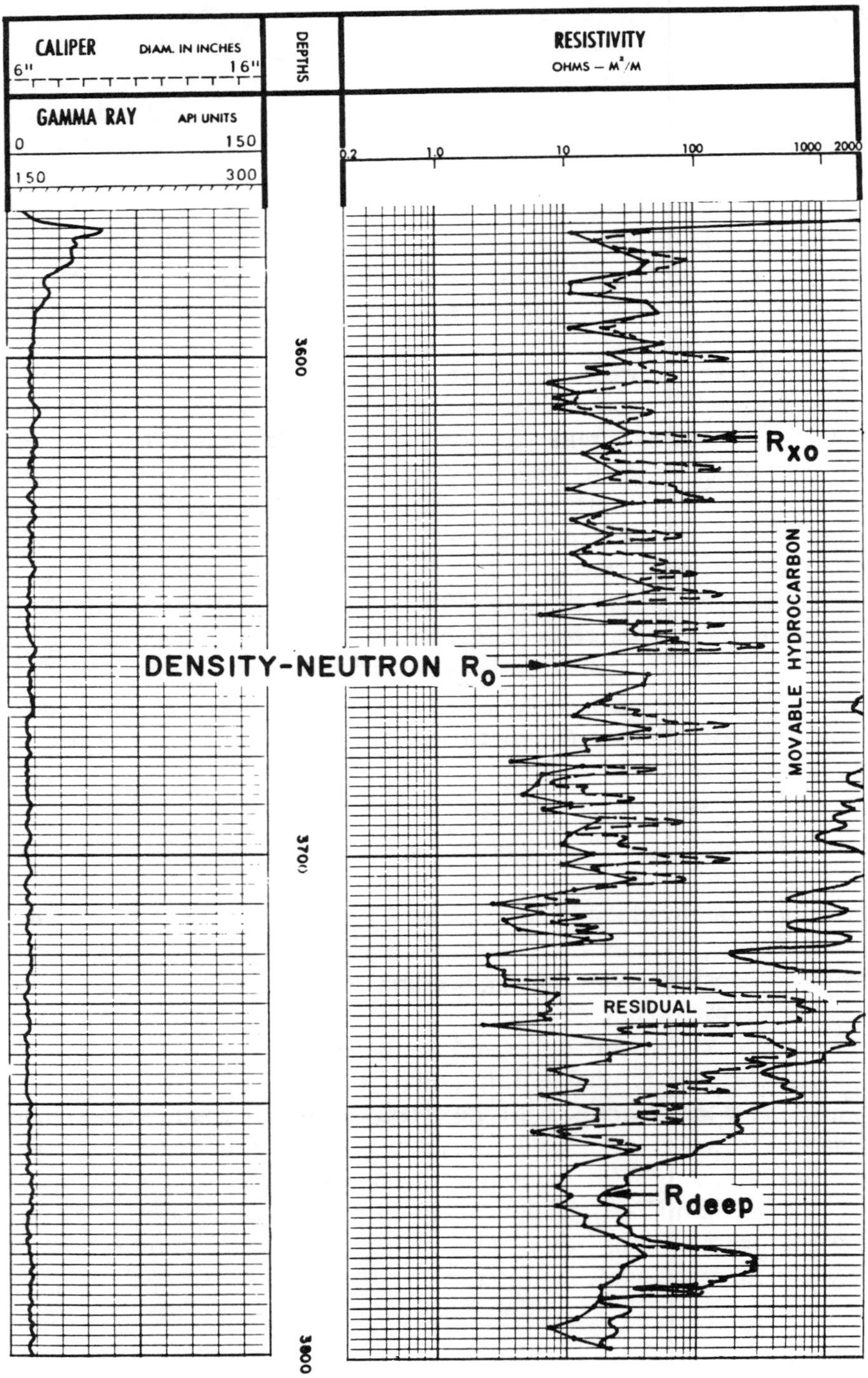

FIG. 7-19. Movable-oil plot (MOP). The separation between R_{xo} and R_t indicates movable hydrocarbon under mud invasion conditions, and that between R_o and R_{xo} is residual or nonmovable hydrocarbon.

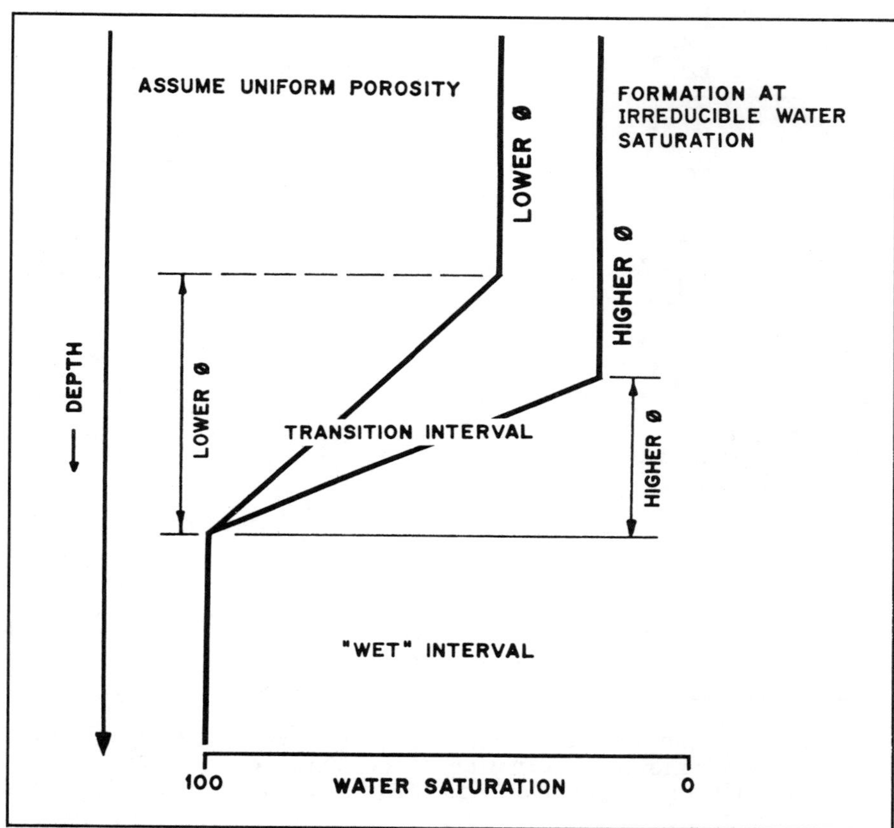

FIG. 7-20. Transition interval showing that capillary pressure causes a thicker transition interval for lower porosity formations. Also note that irreducible water saturation, S_{irr} is higher for lower porosity formations. With varying porosity, the transition interval and S_{irr} become more difficult to determine.

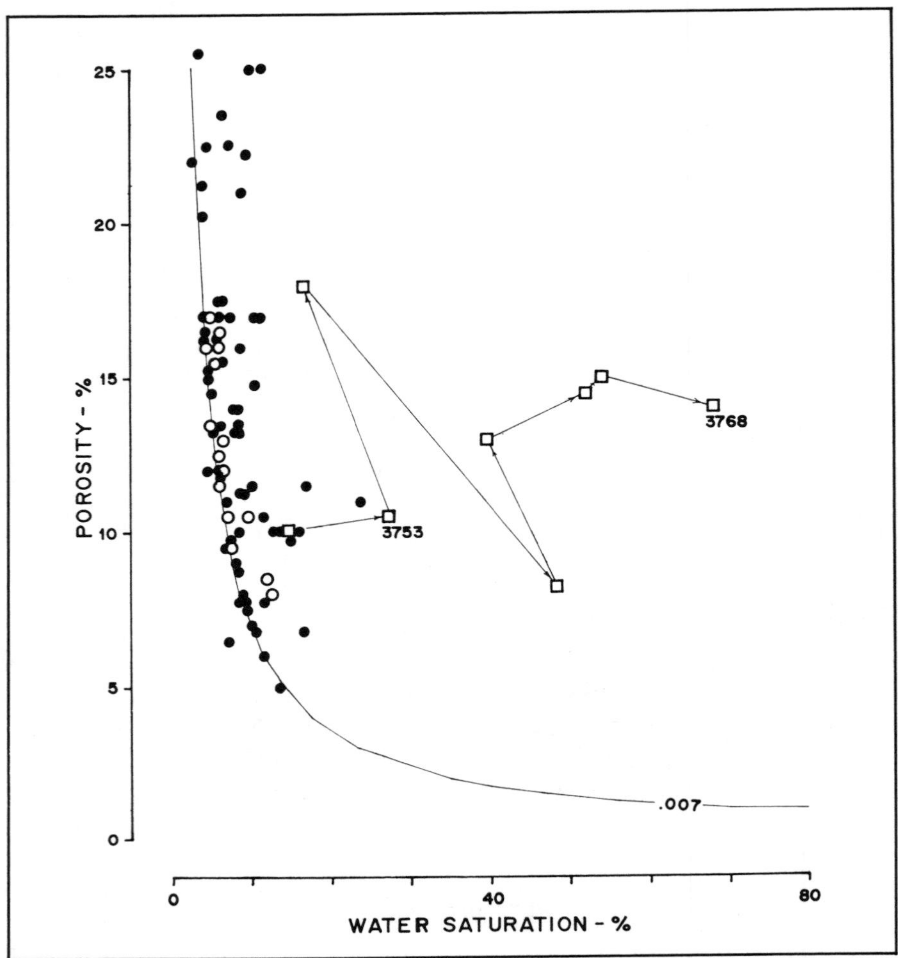

FIG. 7-21. A "c" plot of water saturation and porosity to determine the top of the transition interval. This is shown by a departure of the points from a constant c line and an increasing S_w with depth. In this example 3 753 ft is the top of the transition interval.

Labo, J., 1977, Interpretation of Silurian-Niagaran reefs in the Michigan basin: Trans. 18th Ann. Soc. Prof. Well Log Analysts Logging Symposium, Houston, Paper I.

Morris, R. L., and Biggs, W. P., 1967, Using log derived values of water saturation and porosity: Trans. 8th Ann. Soc. Prof. Well Log Analysts Logging Symposium, Denver, Paper X.

Tixier, M. P., 1958, Porosity balance verifies water saturation determination from logs: J. Petr. Tech., **10**, 7, 161-169.

Chapter 8

DENSITY LOG PRINCIPLES

The gamma-gamma density log measures formation electron density and, through tool calibration, converts the measurement to limestone bulk density. Lane-Wells Company marketed the first commercial density tool in the mid 1950s. The density log, originally developed to aid in surface gravity data interpretation, has evolved as an independent log analysis device which provides another indirect means to measure porosity. The second-generation tool employs a two-detector system that compensates for near-borehole problems, while the third-generation tool includes a measurement that aids in determining formation chemical composition or lithology and allows a better porosity determination from the bulk density measurement.

INTERACTION OF GAMMA-RAYS WITH THE FORMATION

The gamma-gamma density tool uses a chemical source to bombard the formation with gamma rays (Figure 8-1). A detector at a fixed distance from the source measures the remnant gamma-ray cloud. The interaction of gamma rays with the formation atoms can take one of three forms depending on the energy level of the incident gamma-ray photons.

1. At low energy levels, below about 100 keV or 0.1 MeV, the gamma-ray photon, incident on an atom, will produce a photoelectron (Figure 8-2). The interaction depends on the number of electrons-atom or the atomic number (Z). The atomic number reflects the chemical composition of the atom and thus the photoelectron effect is a lithology indicator.

2. At higher energy levels, from 0.075 to 2 MeV, the interaction is predominately by Compton scattering (Figure 8-3), the ejection of a Compton recoil electron and an "incident" gamma ray of slightly lower

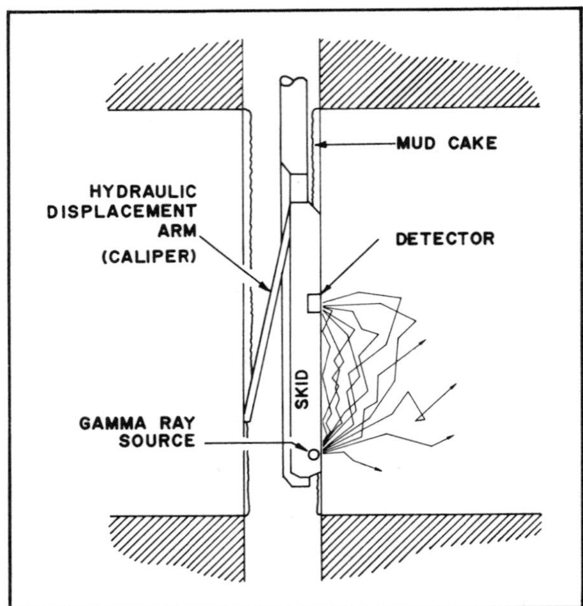

FIG. 8-1. The downhole density system where the source emits gamma rays which interact with the formation primarily through Compton scattering. The remaining diffused gamma rays are measured at the detector. The hydraulic back-up arm forces the skid against the borehole wall allowing the top of the skid to plow away most of the mud cake.

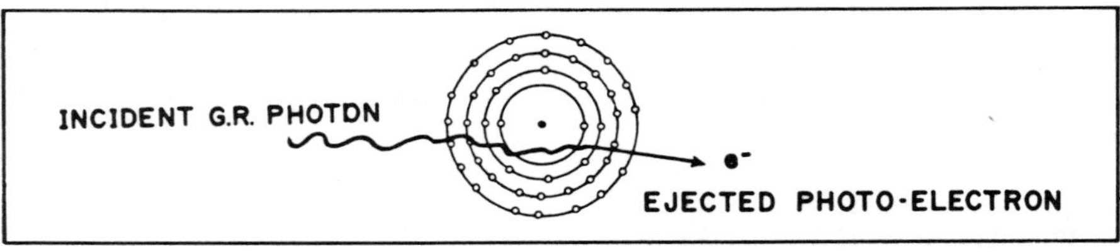

FIG. 8-2. The photoelectron effect. The incident gamma ray is captured by the atom and a photoelectron is ejected.

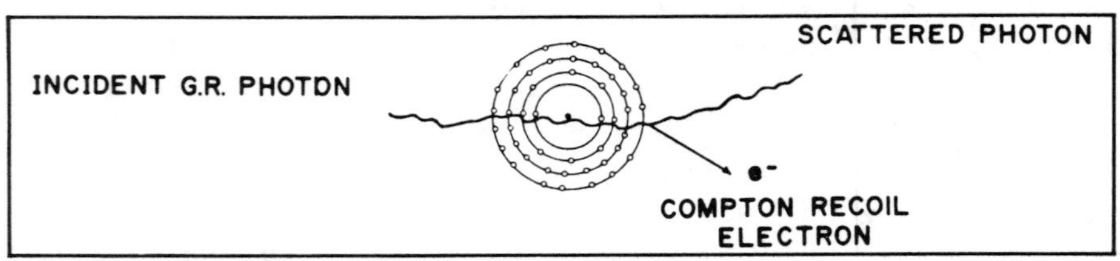

FIG. 8-3. Compton scattering. The incident gamma ray ejects a Compton recoil electron and in the process looses some energy. The likelihood of energy reduction is a function of the formation electron density.

energy. Incident gamma-ray attrition is directly proportional to the number of electrons per unit volume or the formation electron density.

3. Above 2 MeV the interaction is by pair production. However, this is uncommon as the conventional gamma-ray logging sources have energy levels considerably less than 2 MeV.

Because the Schlumberger density tool uses a 0.66 MeV Cesium 137 source, Compton scattering is the predominate mode. As the source gamma rays defuse through the formation they lose energy through Compton scattering. After they have lost sufficient energy to be captured, which depends on the electron density of the formation material, the detector uses a discrimination level to further minimize the photoelectron effect. At a fixed distance from the source, when a low-electron density material is present, there is a high-detector count rate of *remaining* backscattered gamma rays. With a high-electron density material there is a low-detector count rate, and a greater statistical count rate variation.

Bulk density (ρ_b) is related to electron density by

$$\rho_b = \rho_e \, N \, (Z/A) \tag{8-1}$$

where,

ρ_e is electron density,
N is Avogadro's number (6.02×10^{23}),
Z is material atomic number,
and A is material atomic weight.

Because electron density is measurable, bulk density can be calculated if the Z/A ratio of the material is known. Fortunately, the Z/A ratios for elements commonly encountered in borehole logging are fairly consistent (Table 8-1). For chemical compounds, Z/A is replaced with

$$\frac{\Sigma \, Z's}{\text{mol} \bullet \text{wts.}}$$

For common reservoir rocks this ratio is also fairly constant.

Quartz (sandstone) 0.4993
Calcite (limestone) 0.4996
Dolomite 0.4988

Table 8-1. Common formation elements encountered in hydrocarbon exploration.

	Z	A	Z/A
H	1	1.007	0.992 1
C	6	12.011	0.499 5
N	7	14.006	0.499 8
O	8	15.999	0.500 0
Na	11	22.989	0.478 5
Mg	12	24.312	0.493 4
Al	13	26.981	0.481 8
Si	14	28.085	0.498 4
Cl	17	35.453	0.479 4
K	19	39.098	0.485 9
Ca	20	40.08	0.499 0

For simplicity the Z/A ratio is assumed to be 0.5000 and the tool is calibrated to read the correct bulk density in water-filled limestone. This calibration yields nearly correct bulk densities in the other prime reservoir rocks, sandstone and dolomite. However, it does result in up to 4 percent error for nonreservoir minerals (Figure 8-4). The main divergent materials are gypsum and salt (halite).

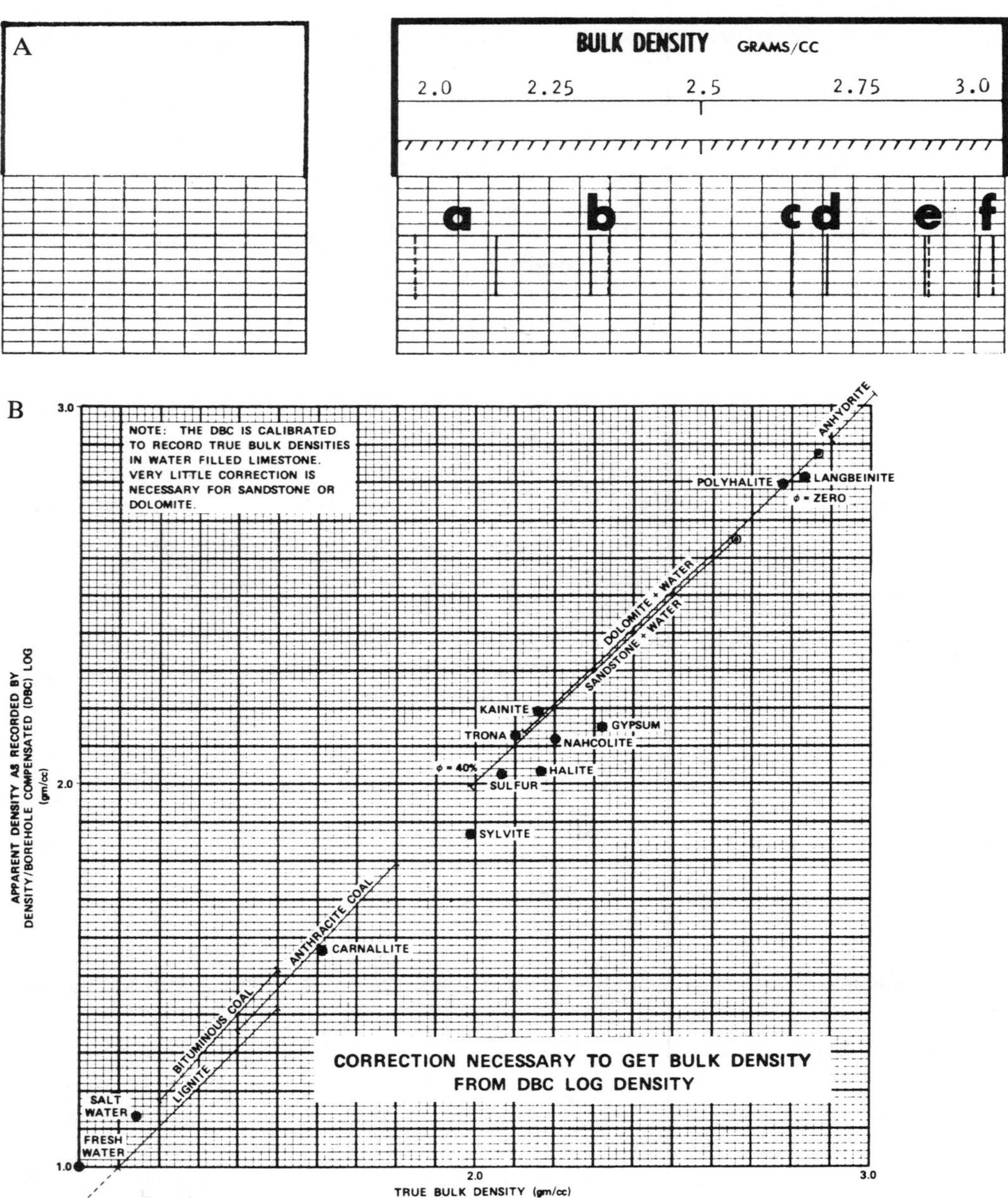

FIG. 8-4. A comparison between the measured density ρ_a and the true bulk density. Z/A was assumed to be exactly 0.5000 and the tool was calibrated to read the correct water-filled limestone density. (A) The magnitude of these errors as seen on a log grid. (B) (Courtesy of Birdwell) includes additional nonhydrocarbon bearing minerals.

SINGLE DETECTOR TOOL

The first-generation formation density tool used a single detector and presented a linear count rate measurement (Figure 8-5), with an added nonlinear bulk density scale. With this tool accurately measuring density in rugose boreholes and in the presence of mudcake was difficult (Figure 8-6). The sonde was pressed against the borehole wall with considerable force which tended to wipe away or to minimize the mudcake. Mudcake is manually corrected for if the mudcake thickness and density are known (Figure 8-7).

Hole rugosity in the extreme case (Figure 8-5, shown by the caliper breaks at 10 041 and 10 046 ft) can make the density measurement useless. In more subtle cases, where the measurement is not completely torn up, reliability of the log measurements is doubtful and they cannot be corrected manually. Figure 8-8 shows conversion of Schlumberger FDL count rates to density and porosity. To overcome the problems of mudcake and rugose borehole the second-generation compensated density tool was developed.

The Schlumberger density calibration system used a two-step indirect method. The Cesium logging source was placed in the skid and the tool was calibrated in an aluminum block secondary density standard at the shop. The source was then removed and a two-position field calibration jig was clamped on the skid. In the far-source position the count rate was adjusted to duplicate the reading in the aluminum block, 400 standard SWS density counts representing a 2.60 g/cm³ density. The near-source position was adjusted for 1 000 more standard counts for a check of tool operation (Figure 8-9).

The Welex Density Log used an aluminum block calibrator at the shop and a magnesium field calibrator block clamped over the skid and source. The field calibration block provides a 1 885 Welex standard density unit check point (Figure 8-10), for a 1.73 or 1.74 g/cm³ field calibration point.

TWO DETECTOR COMPENSATED TOOL

The compensated formation density tool adds a second detector closer to the source that is strongly influenced by the near-borehole effects that caused problems with the single-detector system. With no mudcake or other near-borehole problems, the two detector count rates plot (Figure 8-11, point A) on a log-log straight line that can be scaled in density units. If mudcake is present, the crossplot point falls away from the density line (point B). Empirical studies (Wahl, 1964) indicated a direction of correction to determine a corrected bulk density. Both the corrected bulk density and the applied correction (Figure 8-12) are presented on the compensated density log (Figure 8-13).

The compensated density tool automatically corrects for mudcake, as is shown by comparing the caliper mudcake indication (Figure 8-14) and the applied density corrections. *(Text continued on page 152)*

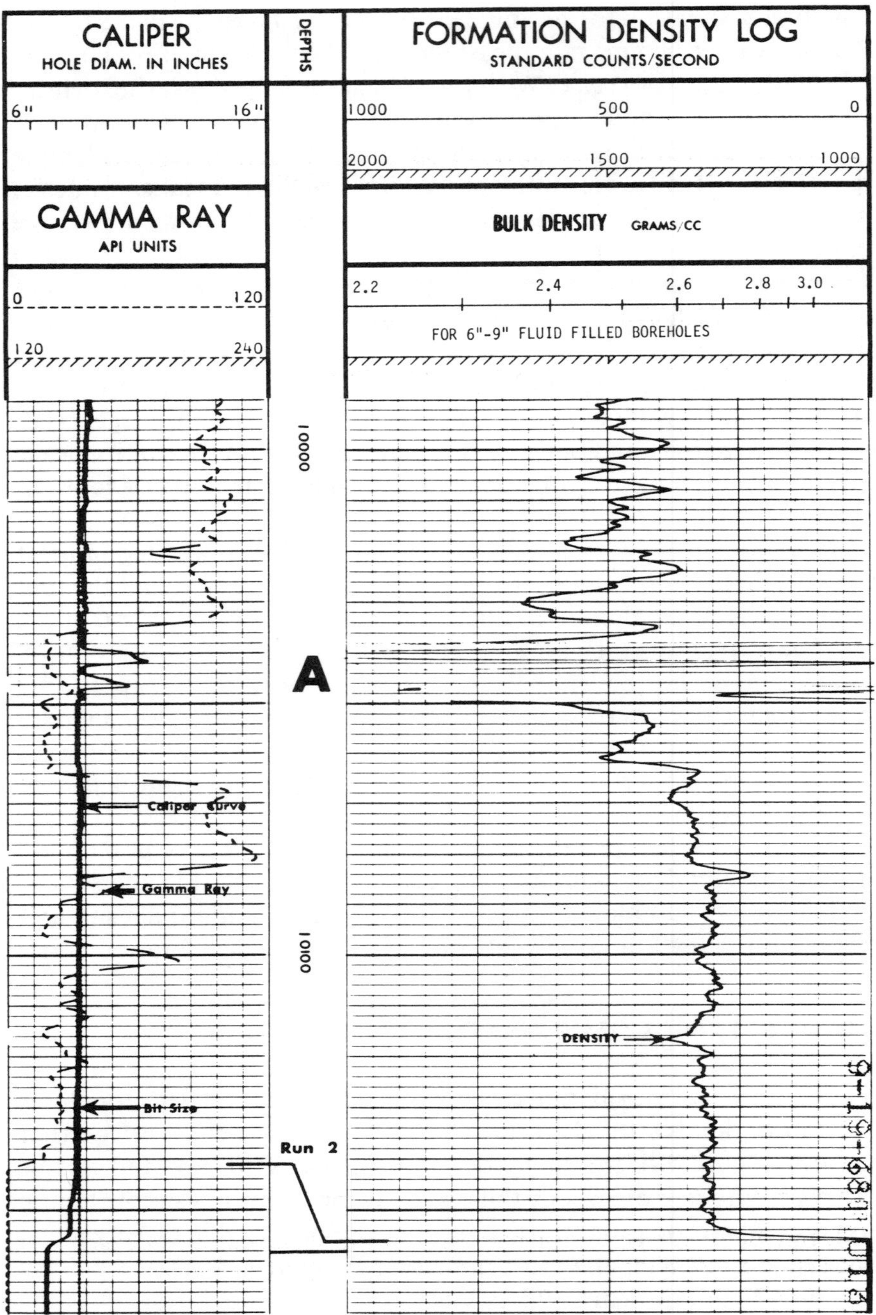

FIG. 8-5. The single-detector Schlumberger formation density log (FDL). Note the system's inability to handle rugose hole effects (A). This log is readily identified by the lack of a compensation curve and scaling in linear count rates and nonlinear bulk density.

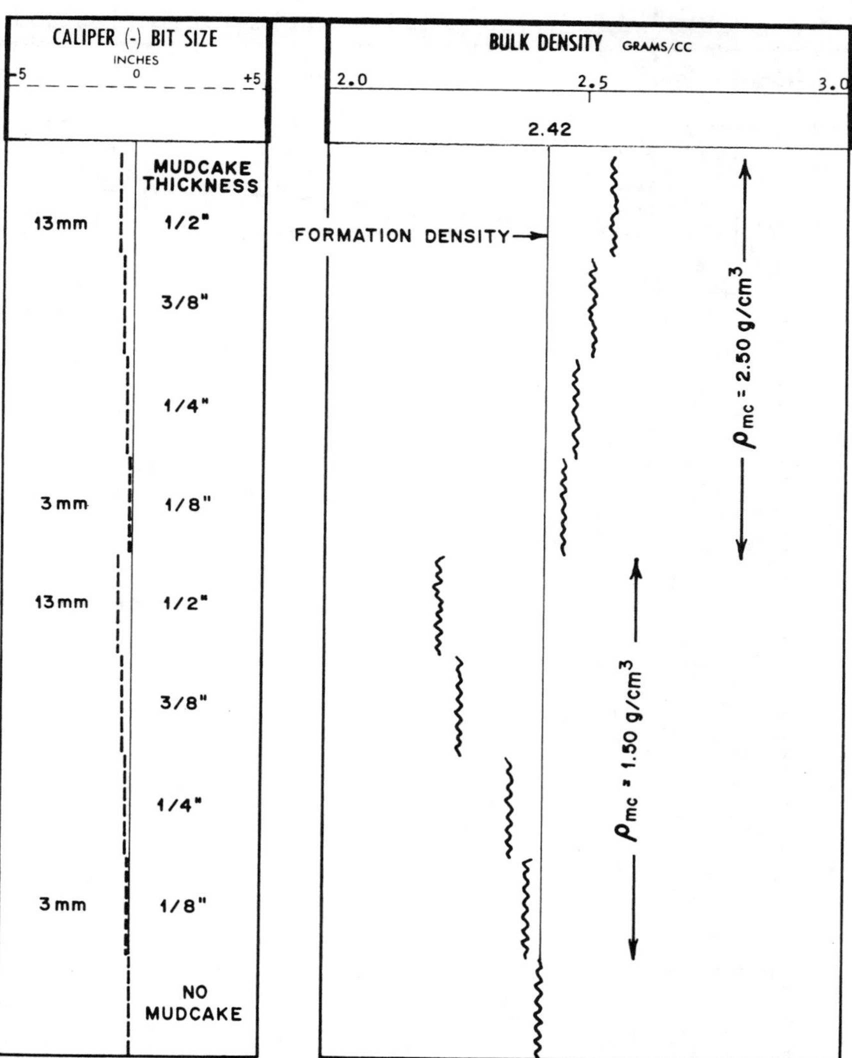

FIG. 8-6. Magnitude of the effects of mudcake on the single-detector density tool as adapted from Wahl et al. (1964) Figure 9, © 1964 SPE-AIME.

A

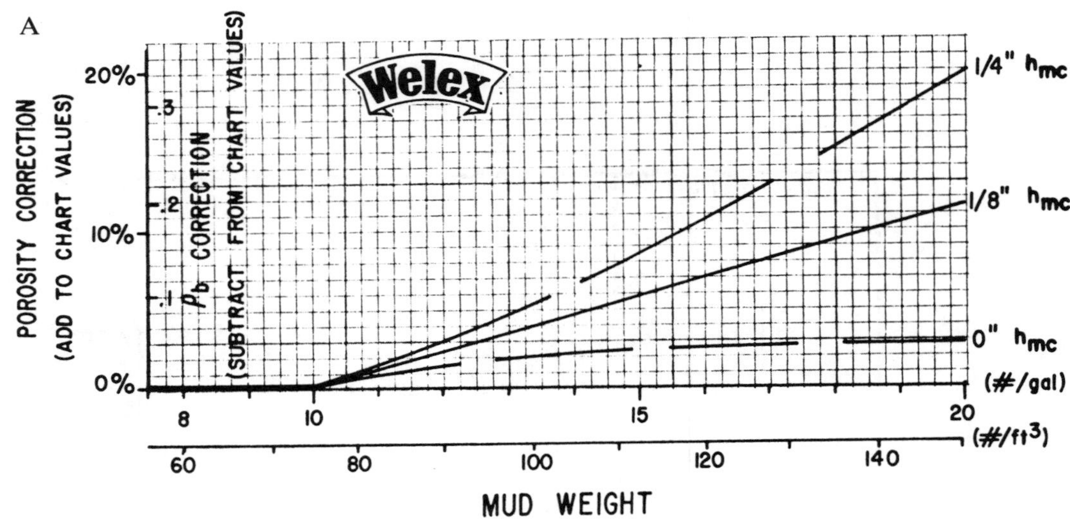

B

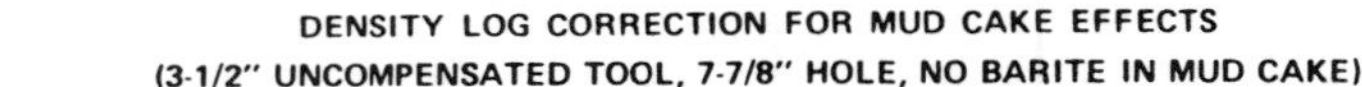

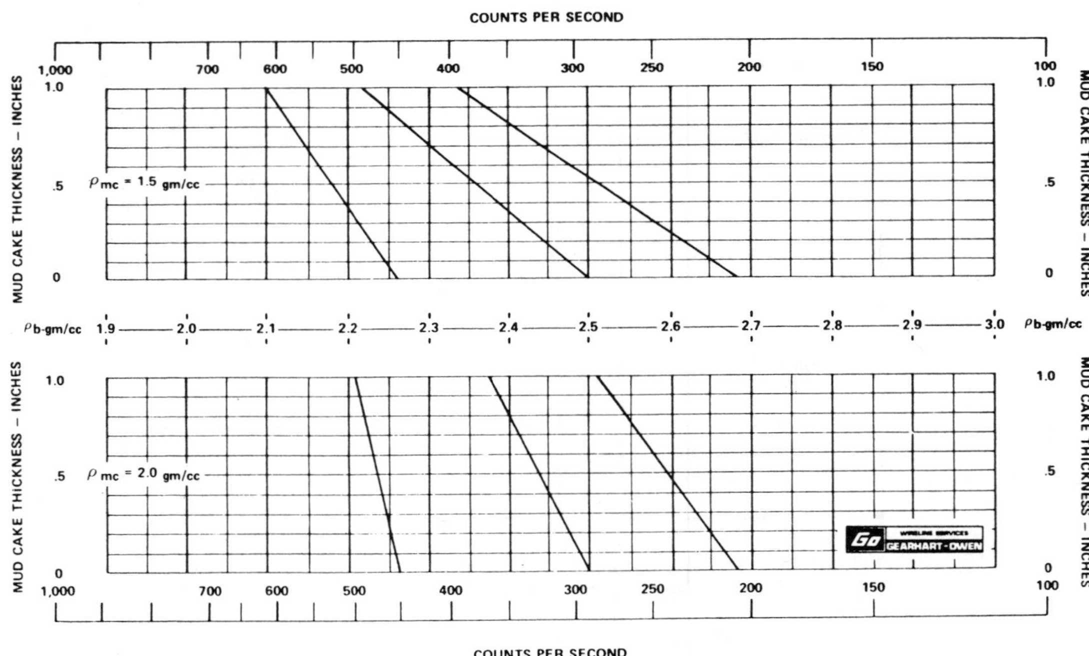

FIG. 8-7. Manual corrections used on older, mostly obsolete single detector density tools. They are presented primarily to show the magnitude of the effects of mudcake thickness and mud weight or mudcake density on single detector measurement of formation density. These corrections are not needed for compensated density tools. Courtesy Welex, a Halliburton Company, and Gearhart.

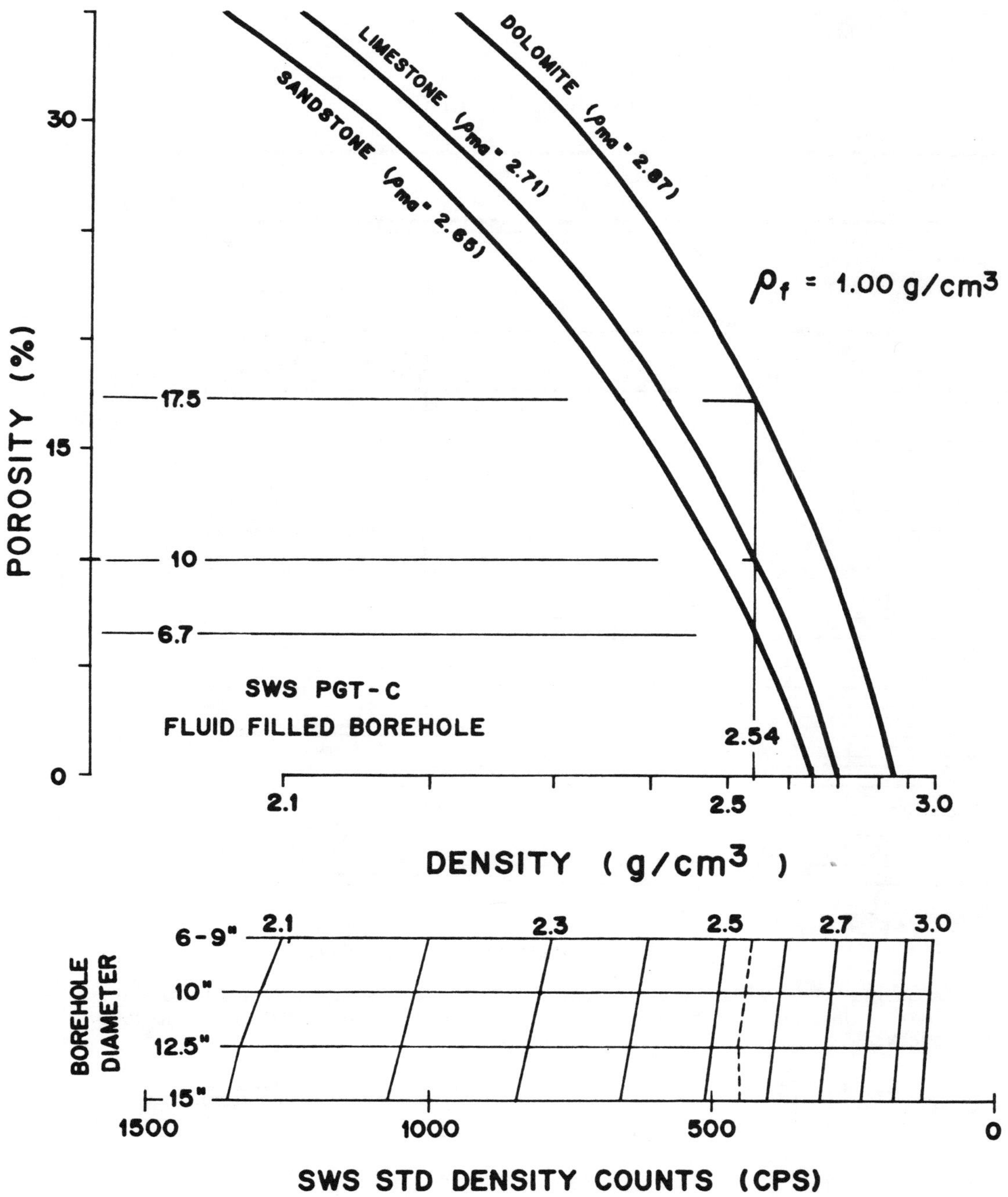

FIG. 8-8. Obtaining bulk density and porosity from Schlumberger's single-detector FDL measured count rate and borehole diameter.

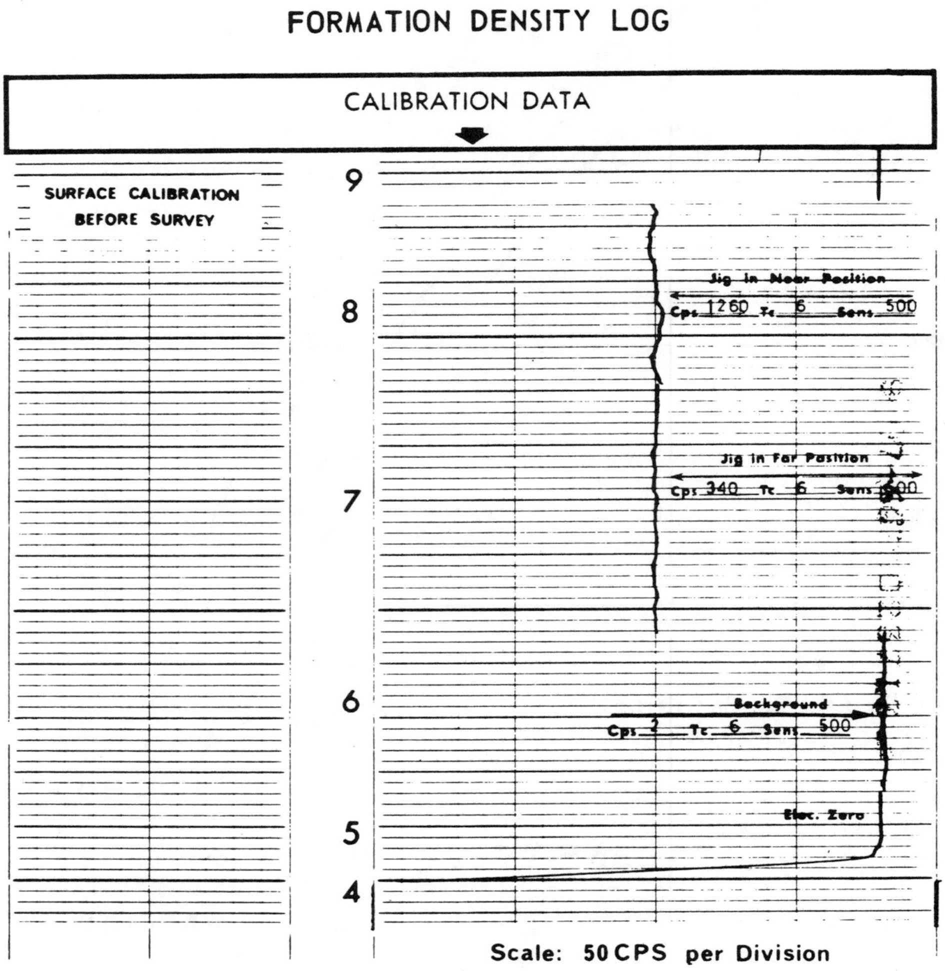

FORMATION DENSITY LOG

CALIBRATION DATA

SURFACE CALIBRATION BEFORE SURVEY

9

Jig In Near Position

8 Cps 1260 Tc 6 Sens 500

Jig In Far Position

7 Cps 340 Tc 6 Sens 500

Background

6 Cps 2 Tc 6 Sens 500

Elec. Zero

5

4

Scale: 50 CPS per Division

FIG. 8-9. Schlumberger's FDL field calibration using a two position field calibration jig: setting up galvo's 1 through 4; calibrate zero, 5 and 9; background radioactivity, 6; and calibration and check, 7 and 8.

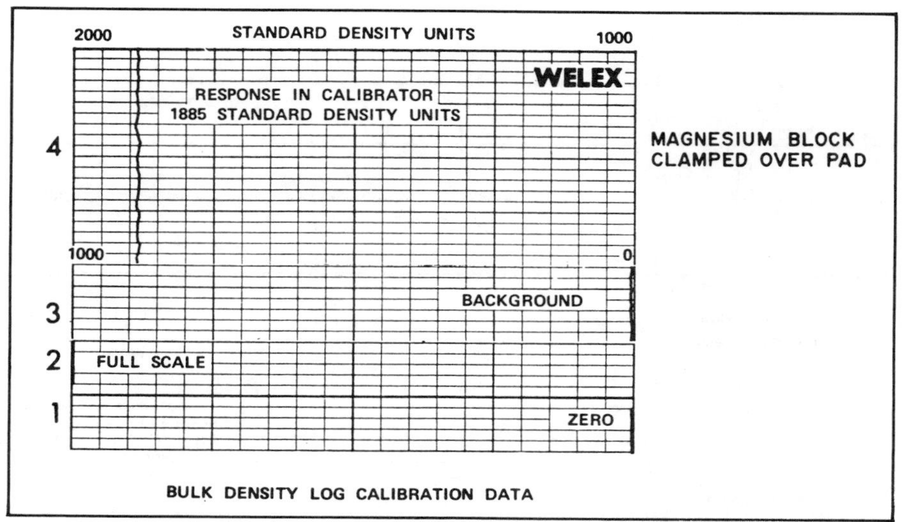

FIG. 8-10. Welex's Density Log single-point field calibration: recorder checks, 1 and 2; background radioactivity, 3; and calibration, 4.

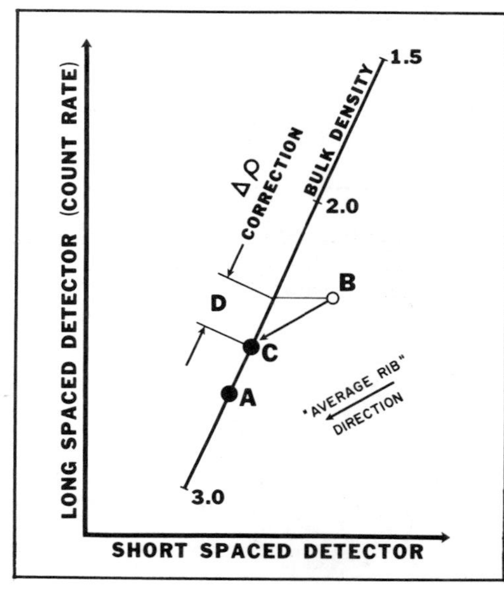

FIG. 8-11. The compensated density system uses the count rates from two different spaced detectors. If no mudcake or rugose borehole effects are present, the count rates crossplot on the "spine," at A for example, directly determining formation density. However, mudcake or hole effects cause the count rate crossplot to fall off the line (B). The panel uses an empirically determined "average rig," B to C direction, to make a correction back to the spine, to C. The compensated density (C) and the correction (D) are both presented on the log, Figure 8-13.

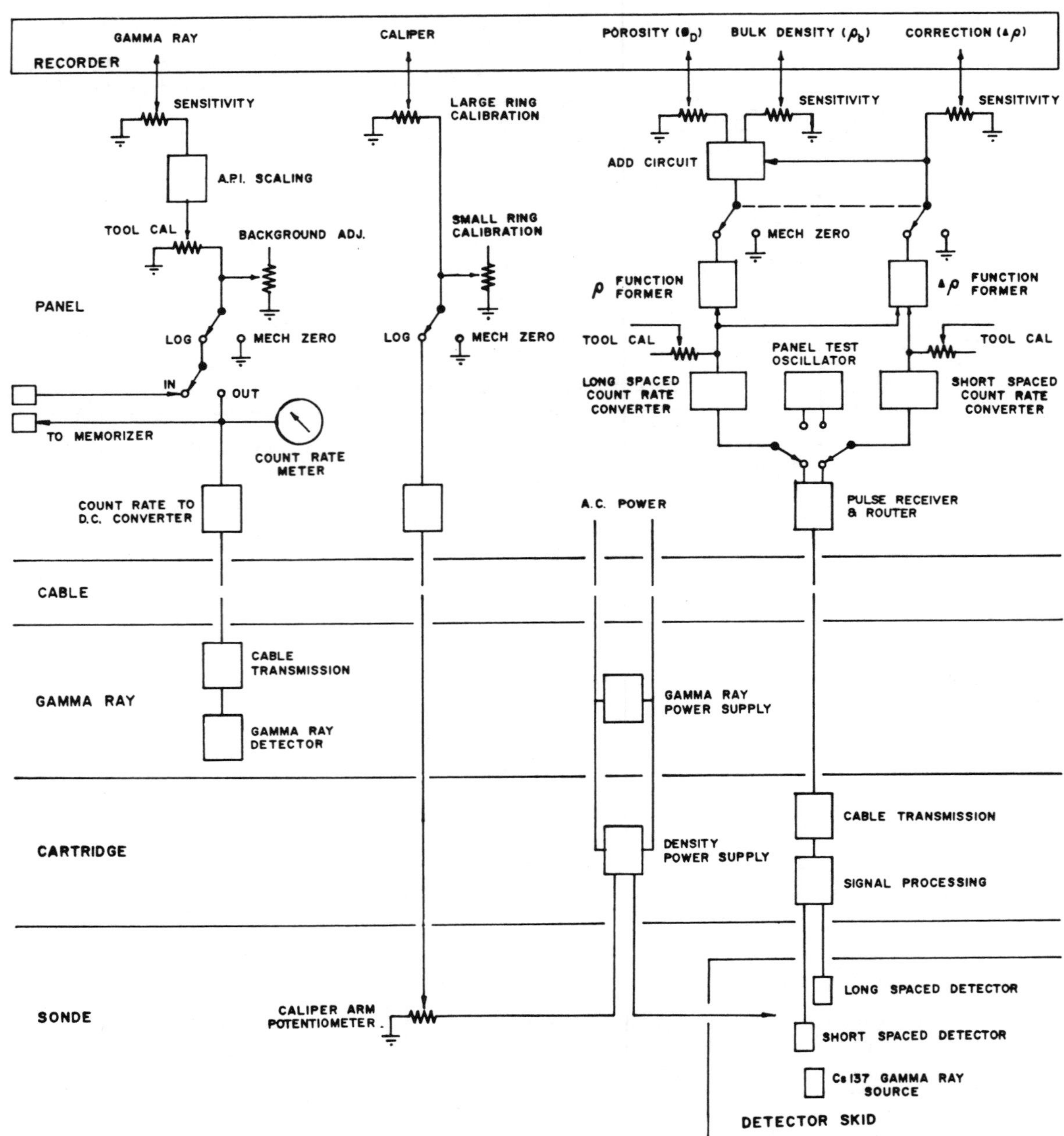

FIG. 8-12. A block diagram of Schlumberger's analog compensated density panel (Adapted from Waller, 1975) and downhole logging tool.

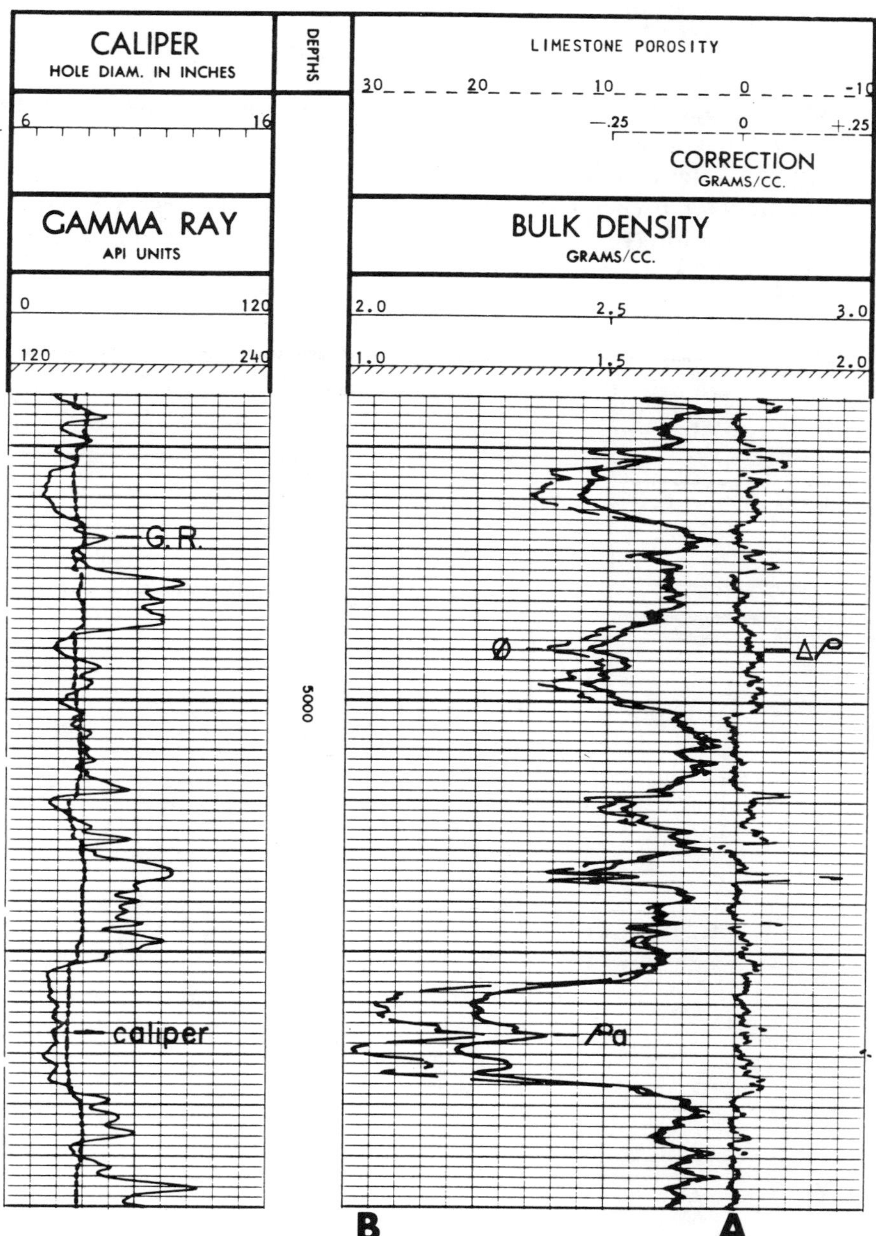

FIG. 8-13. The basic compensated density log presentation: bulk density scaled 2.0 to 3.0 g/cm³ along with the compensation (Δρ) presented with its zero at the center of track III, at A, or near the edge of track II, at B. This latter is common in hard rock presentations. A density-derived porosity curve may be presented, or alternatively, presented on a separate film with the neutron-derived porosity curve.

The compensated density tool also sets Z/A in equation (8-1) to 0.5000 and is calibrated to read the correct apparent bulk density in a water-filled calcite borehole. The result is

$$\rho_a = 1.0704\ \rho_e - 0.1883. \tag{8-2}$$

The errors in sandstone, limestone, and dolomite are minimized and the errors in other rocks tolerated. Fortunately, the other errors that *are* important to the geophysicist are small.

Schlumberger uses a two-point shop calibration and a single-point wellsite calibration (Figures 4-1 and 4-2) which was discussed in Chapter

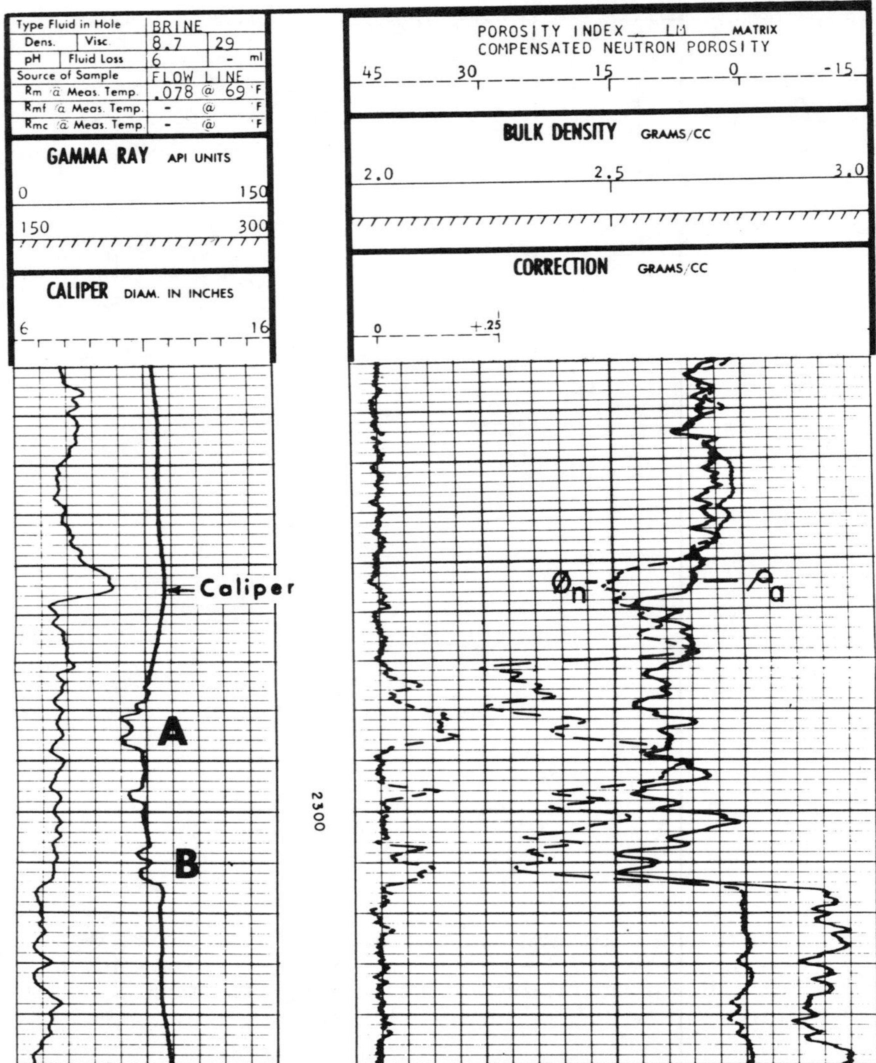

FIG. 8-14. Compensated density correction. The presence of mudcake shown by the caliper deflections at A and B are reflected in the matching compensation corrections just to the right of the depth track.

4. The Gearhart/GO Wireline calibration (Figure 4-6) uses a two-point direct digital wellsite calibration method. Dresser Atlas (Figure 8-15) and Welex (Figure 8-16) also use a two-point wellsite calibration.

MEASUREMENT QUALITY CHECKS

After verifying the calibrations of the density log, the log should be checked in known formations for correct log readings, i.e., halite/salt = 2.03 g/cm³, and anhydrite = 2.98 g/cm³. Known tight formations should also be checked for matrix value.

Logs with complete failure of either of the two detectors are so obvious that they are not released. However, there is always the possibility of either improper downhole detector operation or intermittent failure.

Improper detector operation is most noticeable in unusual density correction curve action. The correction curve normally lies consistently on one side of the zero line, depending on the mud properties. With normal-weighted muds the correction would be positive and for barite-weighted heavy muds the correction is usually negative. In tight impermeable formations with smooth boreholes, as indicated by smooth caliper readings, the correction curve reads near zero.

The log should be checked in many locations, because the detectors sometimes drift into and out of proper operation under downhole conditions. Use the 2 inch log or half-scale prints to look for correction curve drift, slow drift with depth or fixed correction offsetting intervals (Figure 8-17).

Another check, though somewhat less reliable, is a check of the density correction operation inside casing (Figure 8-18). The density correction curve normally saturates about −0.25 to −0.30 g/cm³. If the correction curve has a suspected positive offset, then the casing saturation level should be checked to see if it confirms the correction offset bias. In Figure 8-17 the density is too high in the salts and anhydrites just below the casing. The density correction is offset by two large divisions (0.10 g/cm³), which is about the error in the density measurements. The correction inside the casing also saturates about two large divisions too low, making the correction action suspect. In this example the log was rerun with the same tool, with the correction turned off, and the density measurements were correct in the salts and anhydrites. Unfortunately most computer run logs have curve cutoffs that prevent seeing the correction saturation in casing.

The density tool does not normally read over 3.00 g/cm³. Very low or zero count rates are expected except for tool failures. Spikes over 3.00 g/cm³ should be suspect and other measurements carefully examined for possible intermittent tool failure.

Low density readings should also be carefully examined for stuck tool effects (Figure 8-19) or washouts (Figure 8-22). A significant increase in wireline tension (Figure 8-19, item A, beginning at 14 264 ft) indicates the tool is sticking to the borehole wall. In this example, when the tool finally broke free, at 14 223 ft, the sonde lost borehole contact and pro-

(Text continued on page 159)

A

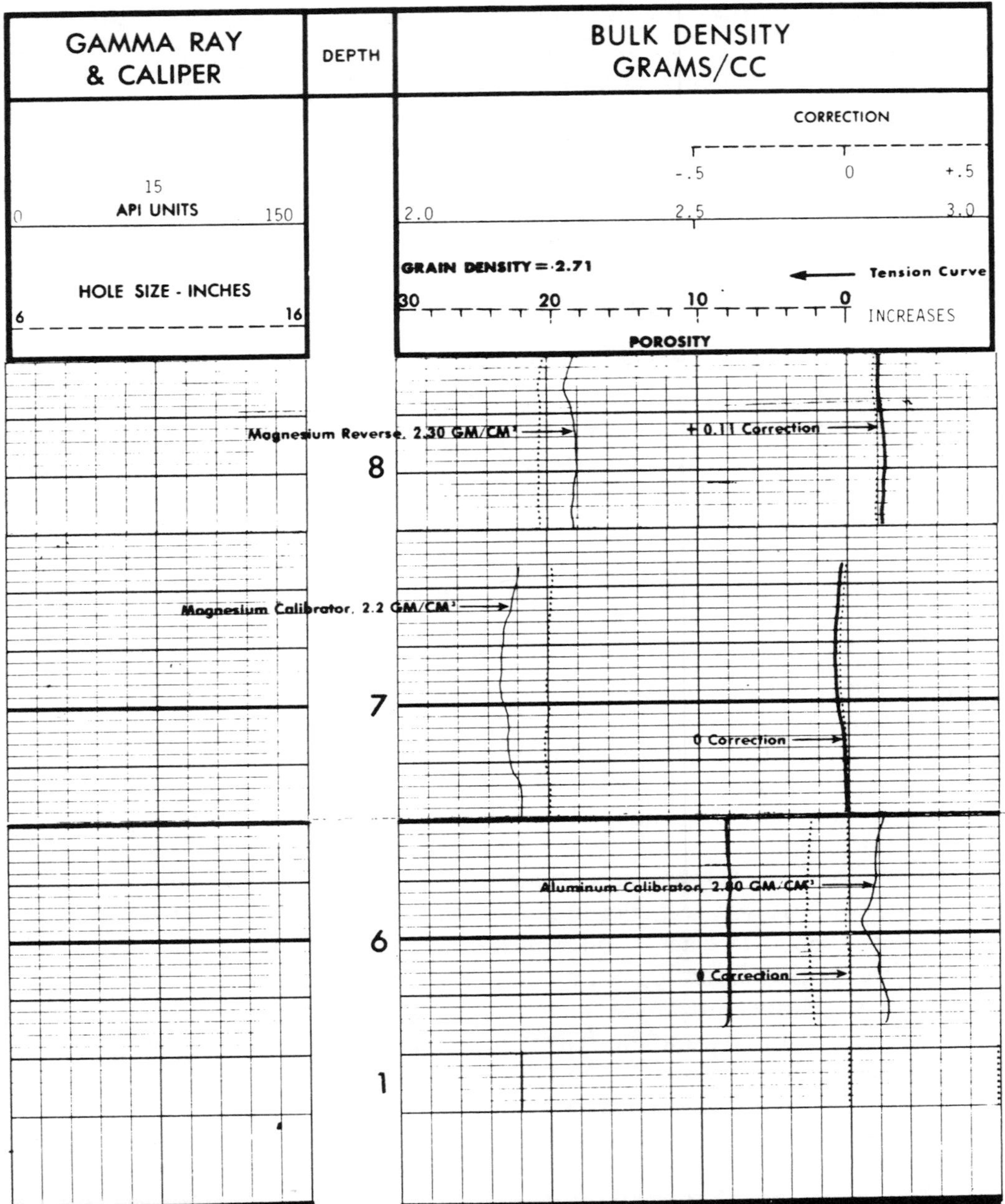

STEP 8 reading will vary with tool type:

Record of computer readout of instrument response to magnesium calibration block reversed and applied to the Densilog instrument. This calibration of the 4-3/4" O.D. (Series 2207 and 2210) Densilog instrument and the 3" O.D. (Series 2208 and 2211) Densilog instrument is different due to design differences. The bulk density galvo for the Series 2207 and 2210 reads 2.40 gm/cc and the correction galvo reads +.15 gm/cc. The bulk density galvo for the Series 2208 and 2211 reads 2.30 gm/cc and the correction galvo reads +.11 gm/cc.

FIG. 8-15. (A) Dresser Atlas' CDL calibration using two field calibration blocks; recorder/panel adjustments, 1-3; caliper calibration, 4 and 5; and tool calibration, 6 through 8. Also note the smaller compensation scale as compared with Schlumberger's density presentation, Figure 8-13. (B) Opposite above. Their digital density calibration printout is shown to the right (Dresser Atlas, 1986).

B

```
        LOG NAME CDL     ASSET NO. 60600     UNIT NO. 6360
  CALIBRATION ENTERED ON 01/19/85 AT 23:41:03           LARGE TOOL
                          SOURCE NUMBER 637

CURVE   COUNTS/SEC.      SCALING FACTORS    RATIO    ENG. VALUE  UNITS
        BLOCK  VALUE     ADD.     MULT.     MG/AL
----    ---------      ------------------ --------  ----------  ----

(1)     DATE 01/05/85        CALIBRATION          TIME 12:23:43

LSD     MG     323.9
SSD            448.2
DEN                                                    2.200   G/CC
CORR                                                   0.000   G/CC
LSD     AL     103.0     0.000    1.0000   3.1459
SSD            292.8     0.000    1.0000   1.5306
DEN                                                    2.800   G/CC
CORR                                                   0.000   G/CC
LSD     MGRV   292.4
SSD            500.0
DEN                                                    2.480   G/CC
CORR                                                   0.226   G/CC

(2)     DATE 01/05/85   PRIMARY VERIFICATION      TIME 12:38:01

LSD     AL     104.4     0.000    1.0000
SSD            292.9     0.000    1.0000
DEN                                                    2.785   G/CC
CORR                                                  -0.008   G/CC

(3)     DATE 01/19/85  BEFORE LOG VERIFICATION    TIME 18:59:17

LSD     AL     105.9     0.000    1.0000
SSD            298.4     0.000    1.0000
DEN                                                    2.799   G/CC
CORR                                                   0.013   G/CC

(4)     DATE 01/20/85   AFTER LOG VERIFICATION    TIME 05:29:58

LSD     AL     107.5     0.000    1.0000
SSD            298.2     0.000    1.0000
DEN                                                    2.781   G/CC
CORR                                                   0.003   G/CC
```

FIG. 8-15.

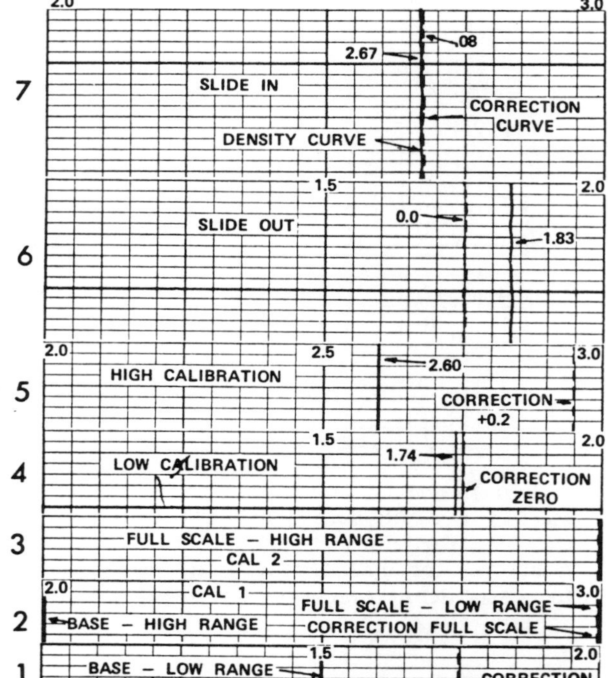

FIG. 8-16. Welex's CDL calibration using a two-point field calibrator: recorder adjustments, 1-3; panel checks, 4 and 5; and tool calibration, 6 and 7.

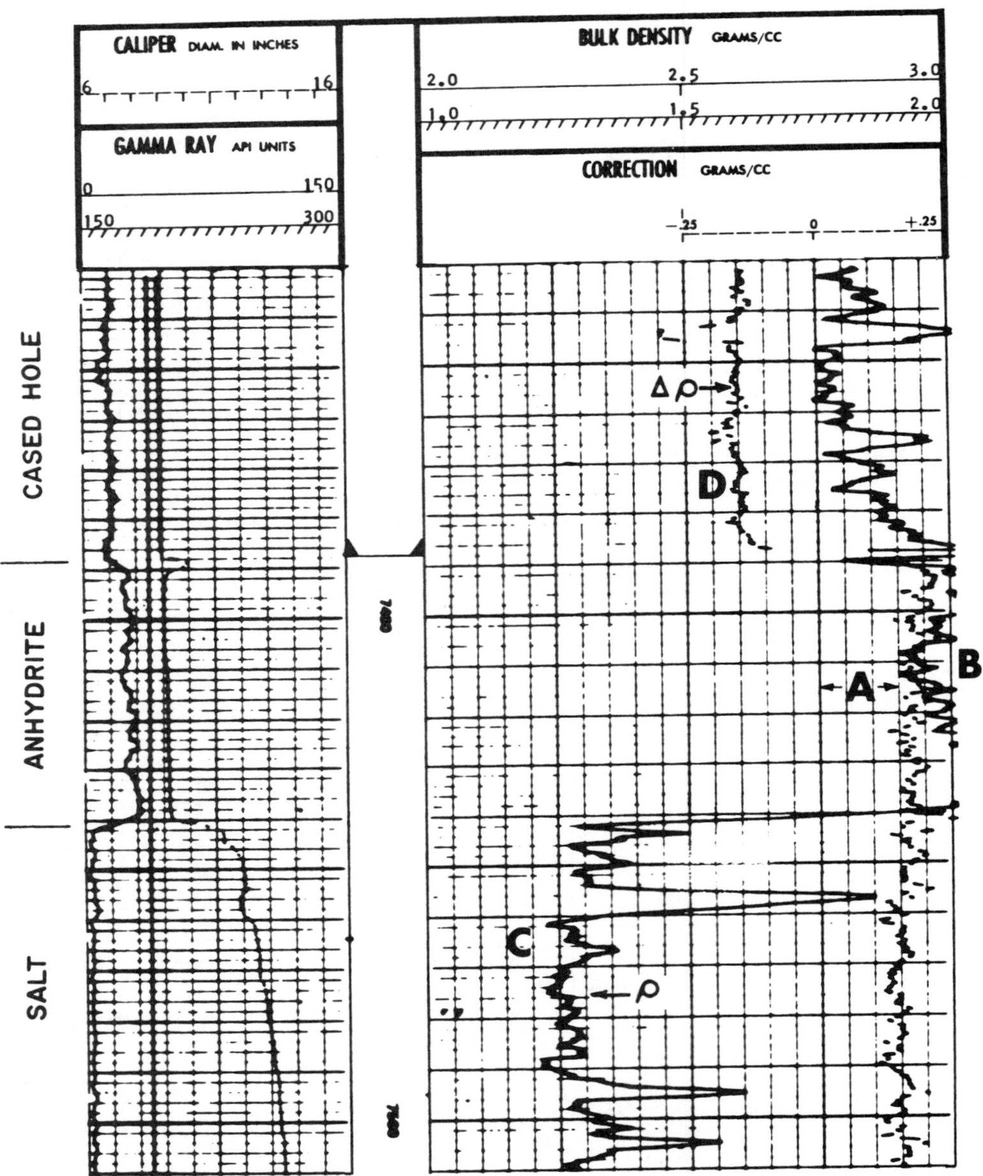

FIG. 8-17. Offset of the correction curve. Note that the Δρ correction curve does not return to its zero (A). The bulk density readings for the anhydrite interval (B) and salt (C) are too dense. Also the correction does not saturate at −.30 g/cm³ (D) inside casing. All of these indicate an incorrect log. The log was rerun with the correction turned off (see Connolly, 1974, figure 20) and the density values became reasonable. Thus the short spaced detector was not operating properly.

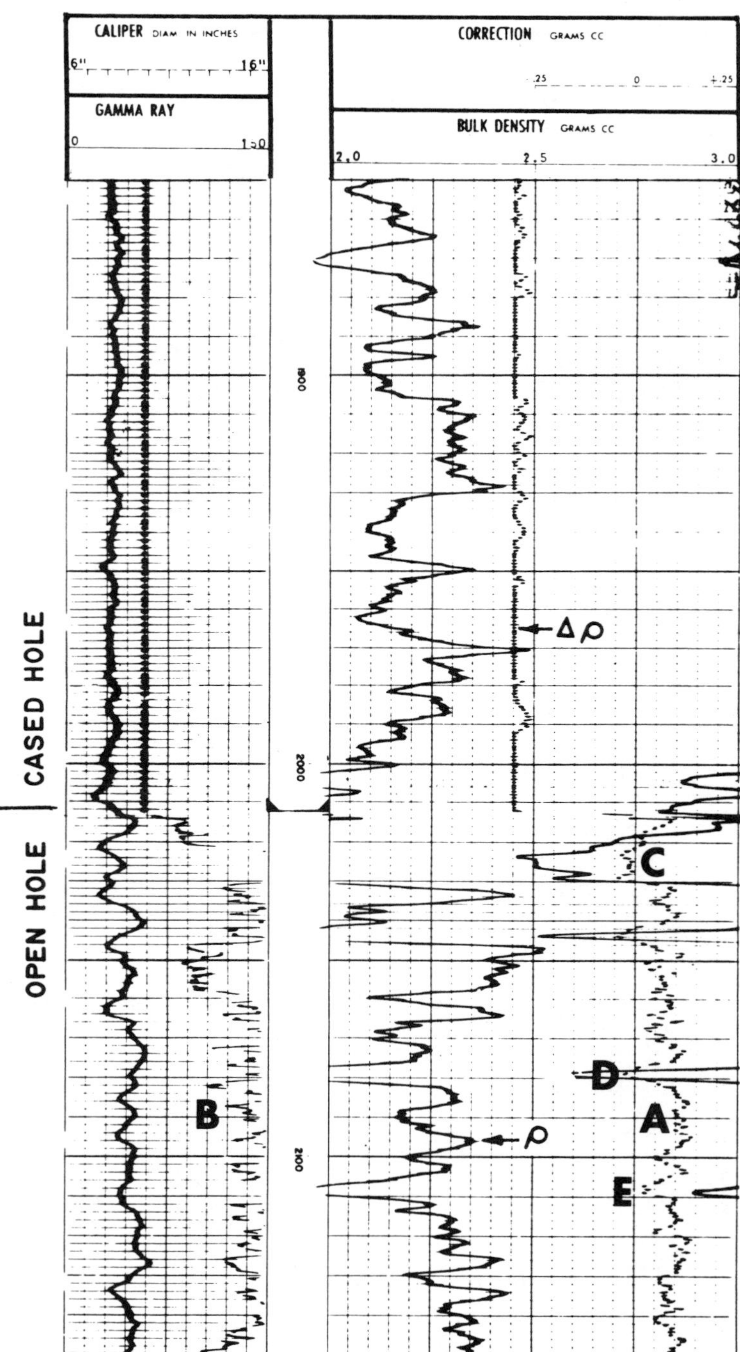

FIG. 8-18. Proper compensation action upon entering cased hole with saturation near $-.30\ g/cm^3$. Also note that the compensation is large at A in the open-hole section due to rugose borehole (erratic caliper readings, B), but the compensation does return to near zero upon occasion (C, D, and E).

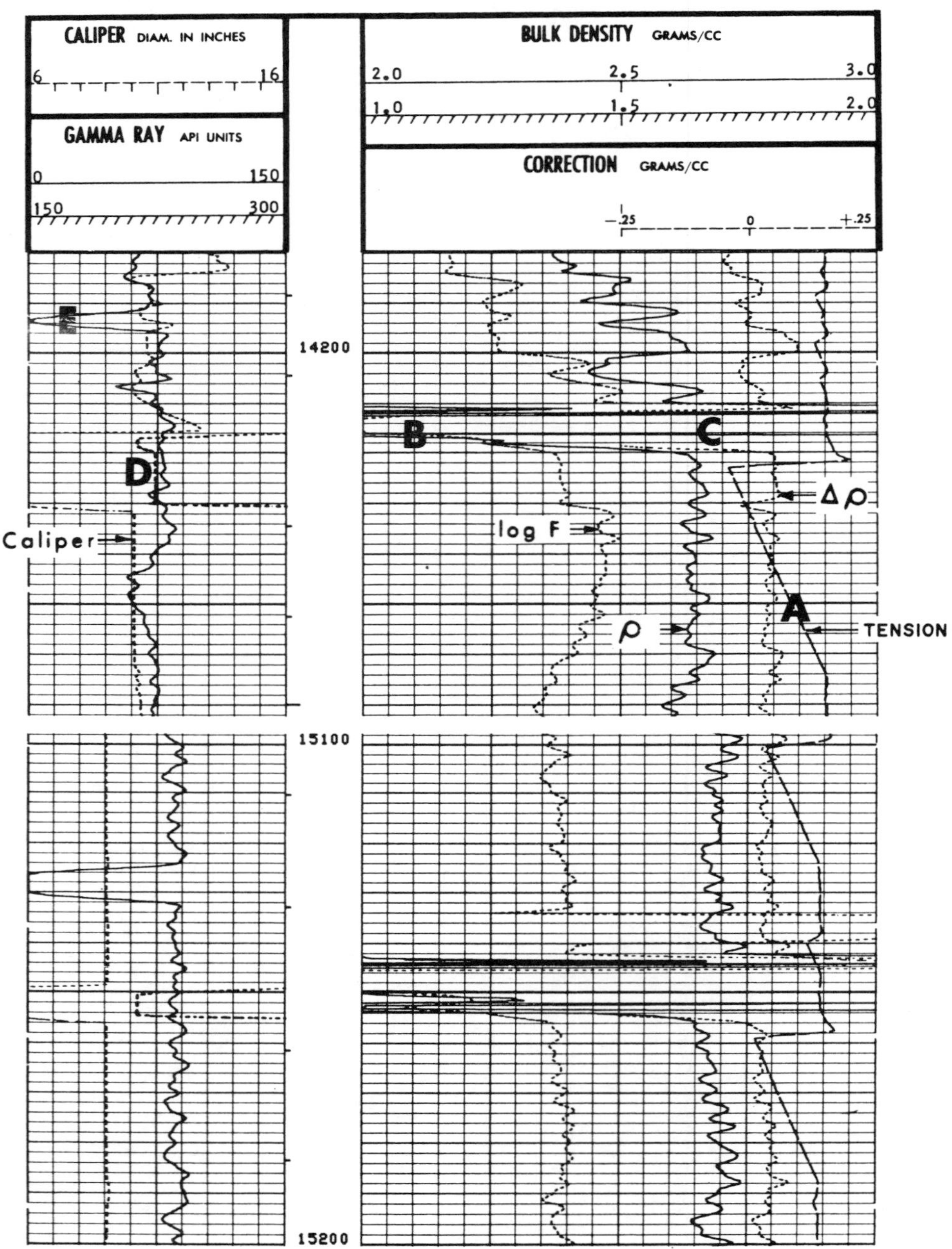

FIG. 8-19. Incorrect density reading due to tool sticking: the increase in wireline tension (A) indicates the tool is sticking intermittently to the borehole wall. The incorrect low-bulk density (B) is suspicious because of drastic negative correction (C) after the sonde becomes free, and the improbable 1 inch diameter caliper measurement (D). The gamma ray also quits (E) at the same stuck depth but is presented higher in depth. This last is not present in other gamma-ray runs.

duced the incorrect low bulk density (item B). This is also shown by the suspicious negative correction (item C) and the improbable 1 inch borehole diameter as measured by the caliper (item D). The computer interpretation of this interval of the density log (Figure 8-20) incorrectly portrays this as porous (item F), with a slight hydrocarbon content (item G), although it is really shale. However, not all low-density measurements are anomalous. Coal stringers cause measurements in the 1.32 to 1.80 g/cm³ range.

The density measurement is effected by the presence of light hydrocarbons in the formation pores. Low pressure gas in the pore spaces can reduce fluid density to near zero which then reduces the apparent density below its real value. The apparent formation density is a measurement of matrix rock and density of the fluids in the pores. If pore fluid density is reduced to near zero by low pressure gas, the density tool records this as increased porosity or decreased density and the light hydrocarbon effect requires a manual correction (Figure 8-21).

If the borehole enlarges until the source-detector skid loses contact with the borehole, it measures almost the drilling mud density (Figure 8-22). The correction or compensation measurement reads only a slight correction if the mud is the same density near the tool as in the washout. The normal Schlumberger caliper back-up arm keeps the skid against the borehole wall for holes up to 16 inches but special longer arms may be used for larger boreholes. Because the hole did not washout without reason, the other logs and the sample log should be checked for clues to the missing interval lithology.

MANUAL CORRECTIONS

The Schlumberger density skid geometry was designed to operate in a 6-9 inch diameter borehole. For boreholes larger than 9 inches a slight manual correction is required (Figure 8-23, upper). Older tools need borehole size and mud weight corrections.

The correction system does have limitations. The analog panel system requires introduction of an average linear direction of correction, or "average rib," although it is not accurate for all conditions. Empirical studies (Figure 8-24) show a diversion between the needed correction and the linear correction applied for values above 0.10 g/cm³, which is two large divisions for most log presentations. Most service companies use the same log scale for bulk density and correction but Dresser Atlas' Compensated Densilog™ [1] and Welex's Compensated Density Log are exceptions.

Measured density in rugose boreholes with badly caved shale sections, shown in Figure 8-25 by erratic caliper readings and large density correction, needs correction. An approximate correction using hole shape and

---

[1]Dresser Atlas trademark

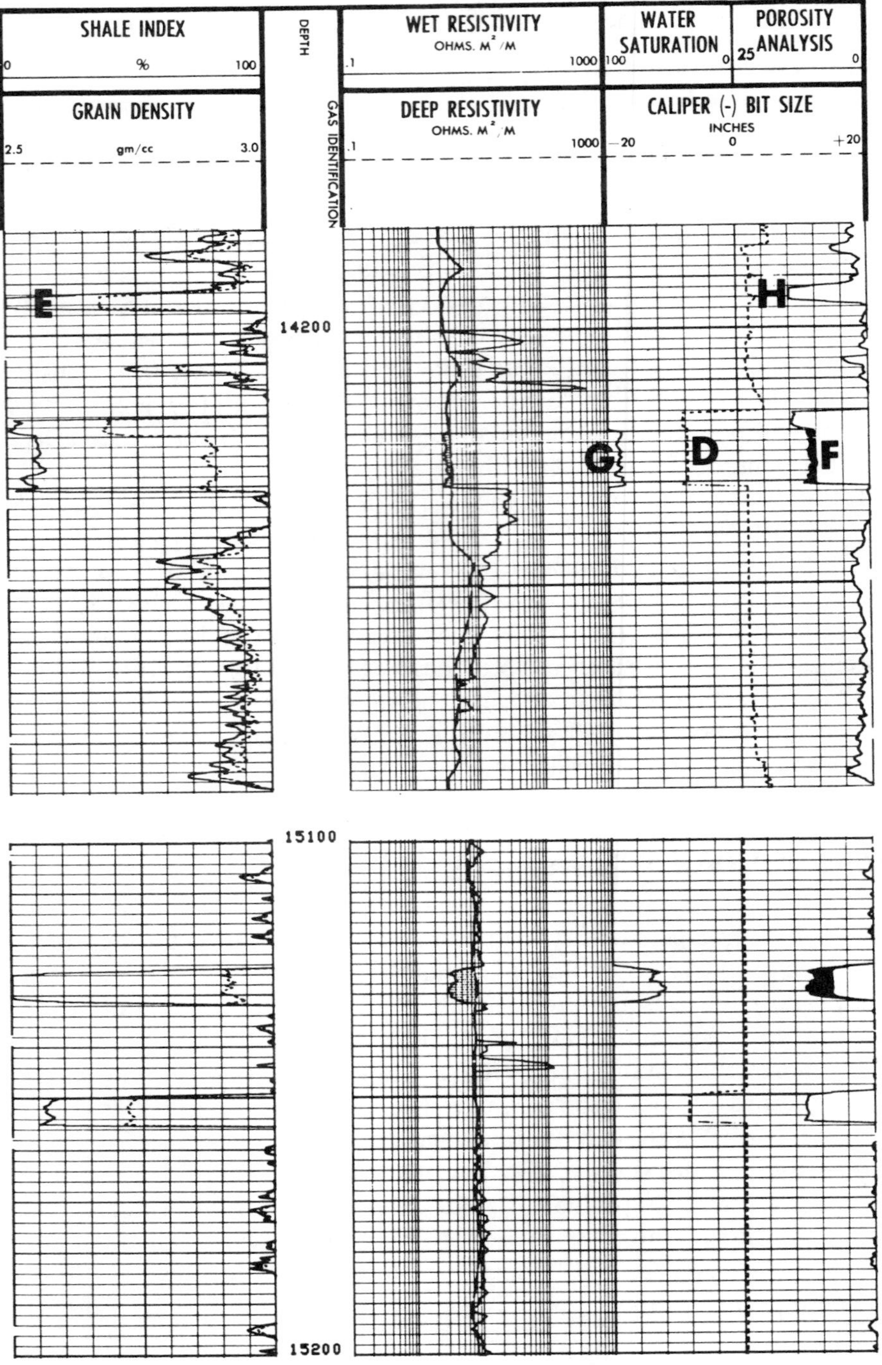

FIG. 8-20. Wellsite computer interpretation of Figure 8-19: the stuck density sonde measurement is incorrectly portrayed as porous (F) with a slight hydrocarbon content (G). The gamma ray quiting (E) is portrayed as a clean, porous interval.

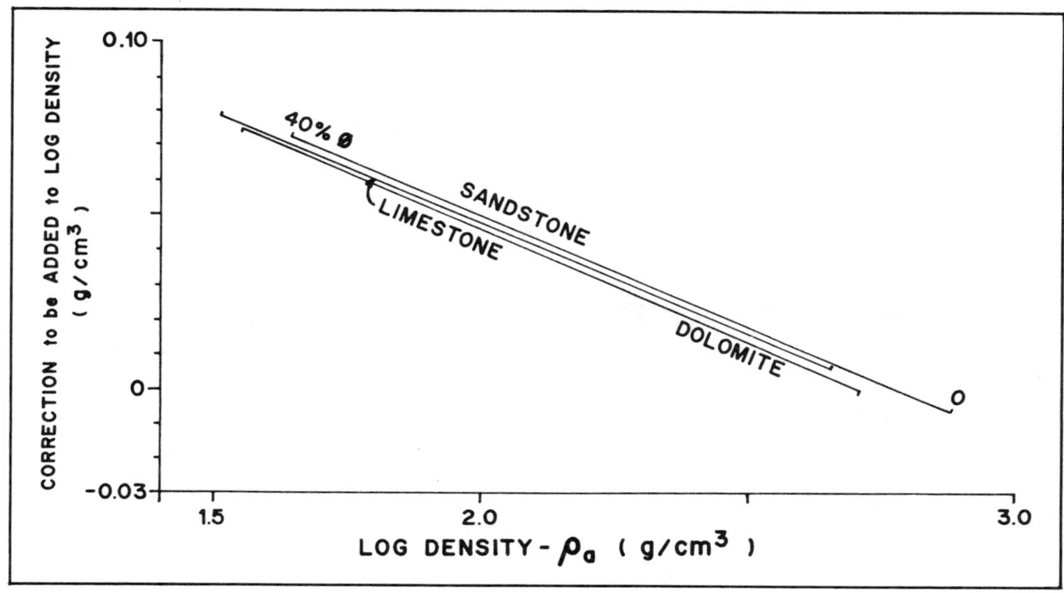

FIG. 8-21. Density corrections due to low pressure gas or air in pore spaces. Adapted from Tittman (1964).

to some extent matrix is

$$\rho_{corr} = \overline{\rho}_a + [(d\ \mathrm{CAL}^{-1})\ (\rho_a - \overline{\rho}_a)\ (GR_{max} - GR)\ (GR_{max} - GR_{min})] \tag{8-3}$$

where,

ρ_a is measured density,
$\overline{\rho}_a$ is average density of interval,
GR is measured gamma ray,
GR_{max} is maximum gamma-ray reading (usually in shale),
GR_{min} is minimum gamma-ray reading,
d is borehole diameter,

and

CAL is the bit size.

A considerably more elegant correction is possible if a complete computer evaluation is available which uses equation (5-10)

$$\rho_{corr} = \rho_{ma}\ (1 - \phi - V_{sh}) + \rho_{sh} V_{sh} + \rho_w \phi S_w + \rho_h \phi (1 - S_w) \tag{8-4}$$

where all the values on the right side of the equation are obtained from the computer analysis of the logs.

Most of the density tool deviations from true density, Z/A effects, borehole diameter, "average rib" compensation, and light hydrocarbons, have minor corrections. However washouts, loss of borehole contact, tool sticking, intermittent tool electronic failure, improper detector output, and very rugose borehole are significant.

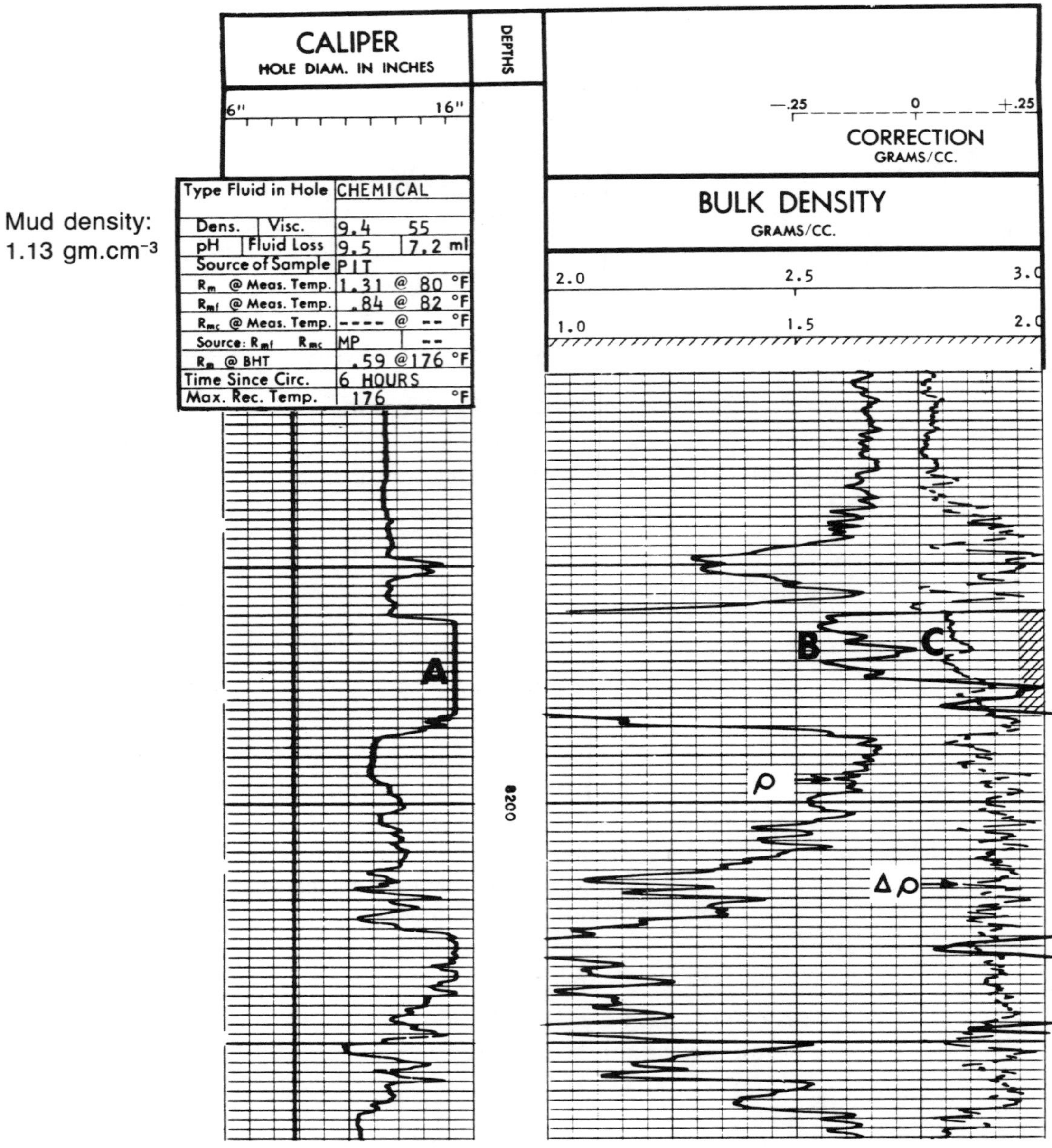

FIG. 8-22. Density sonde standoff in large borehole: the caliper flat topping (A) indicates that the skid is probably not in contact with the borehole wall and the tool is measuring the mud density (B). Because the mud is relatively uniform in density there is little correction (C).

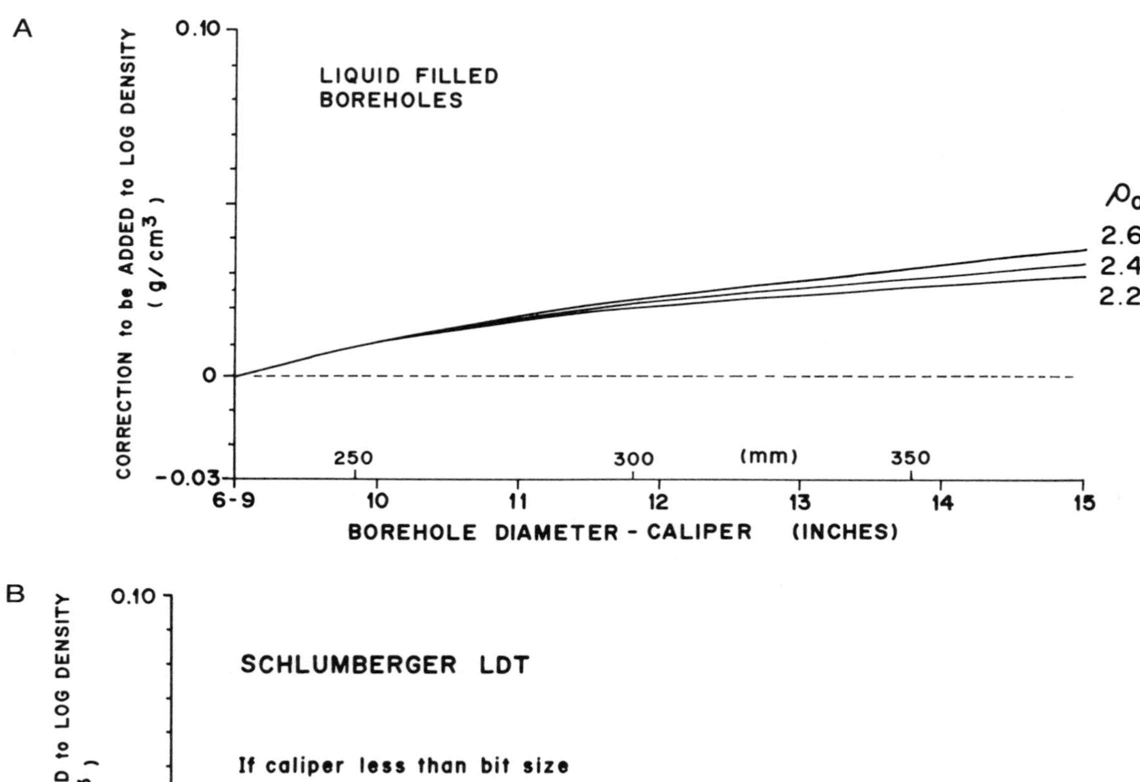

FIG. 8-23. Borehole size corrections for Schlumbergers' FDC (A) and LDT (B) density tools. The FDC data is from Wahl et al (1964) Figure 11, © 1964 SPE AIME. The corrections would be slightly greater for air- or gas-filled boreholes as shown in Wahl (1964) Figure 12. The LDT data is adapted, for comparison with the FDC data, from Schlumberger (1984) Figure Por-15.

A

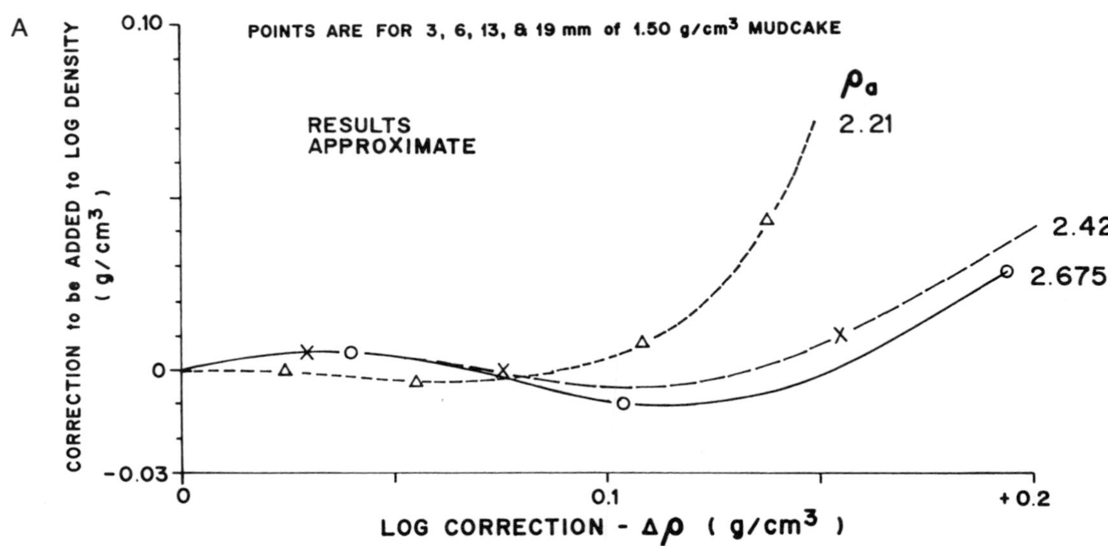

B

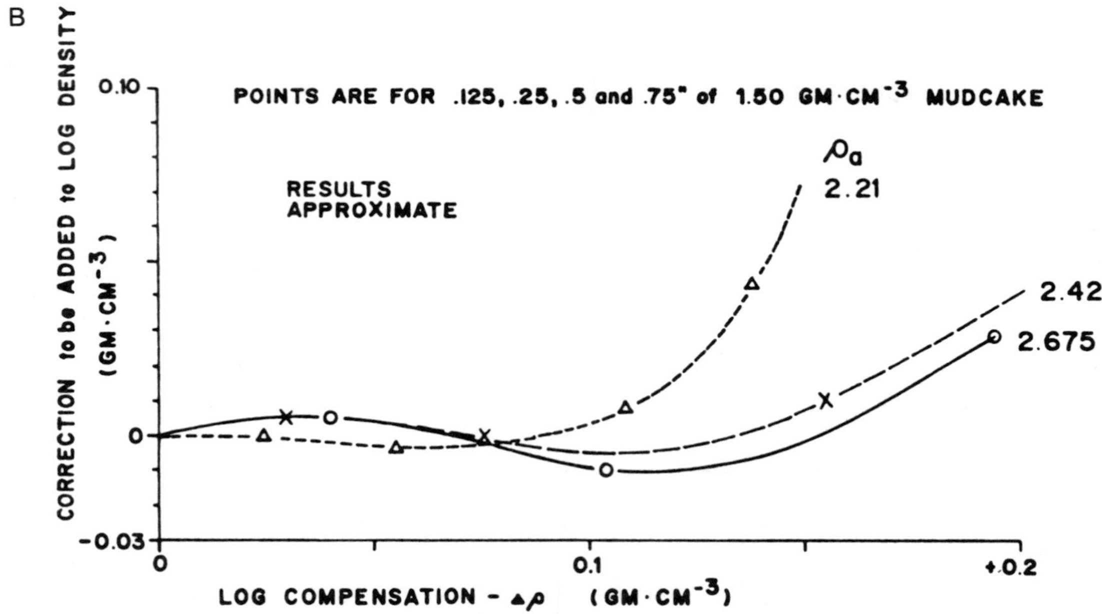

FIG. 8-24. Incomplete $\Delta\rho$ computation due to introduction of an "average rip" correction. There is good agreement up to 0.10 g/cm³ correction and then divergence for all three formation densities. Adapted from Wahl et al (1964) Figure 8, © 1964 SPE-AIME.

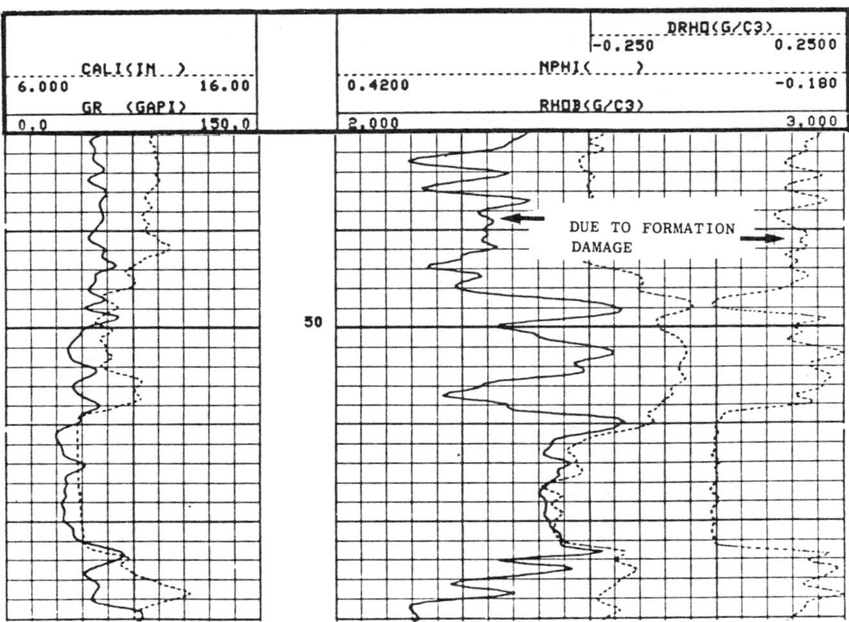

FIG. 8-25. Very rugose borehole effects are shown by the erratic caliper and large correction.

LOGGING THROUGH CASING

Figure 8-26, showing density log measurements both before and after casing was run, indicates the potential and problems of through-casing density measurements. The presence of the casing and cement causes the correction to be driven considerably negative, resulting in too low density by about 1.10 g/cm^3 in this example. But the magnitude of the deflection of the streak at 6 450 ft is approximately the same, 0.60 g/cm^3, for cased- and open-hole measurements. A linear shift of the cased-hole density measurements could bring the cased-hole data closer to the proper density.

Fertl and Wichmann (1977), directly compared open- and cased-hole Compensated Densilogs and found agreement within ± 3 sandstone porosity units, or 0.05 g/cm^3 (Figure 8-27). They accounted for the casing by placing a comparable piece of casing between the detectors and the calibration jig during the surface calibration sequence.

Figure 8-26 also shows the magnitude of statistical difference in cased-hole measurements. Dennis Schieman, with Schlumberger in Michigan, ran several uncalibrated density logs through casing. His main problem was repeatability of the log measurements which varied from very close match to up to three large divisions separation, or about 0.15 g/cm^3.

The main variable, cement thickness, is the probable cause of most of the variations between measurements. The density detector only radially samples 3 percent of the borehole. If the casing is not centered accurately in the borehole, the amount of cement between tool and formation varies considerably depending on tool orientation. This variation causes a

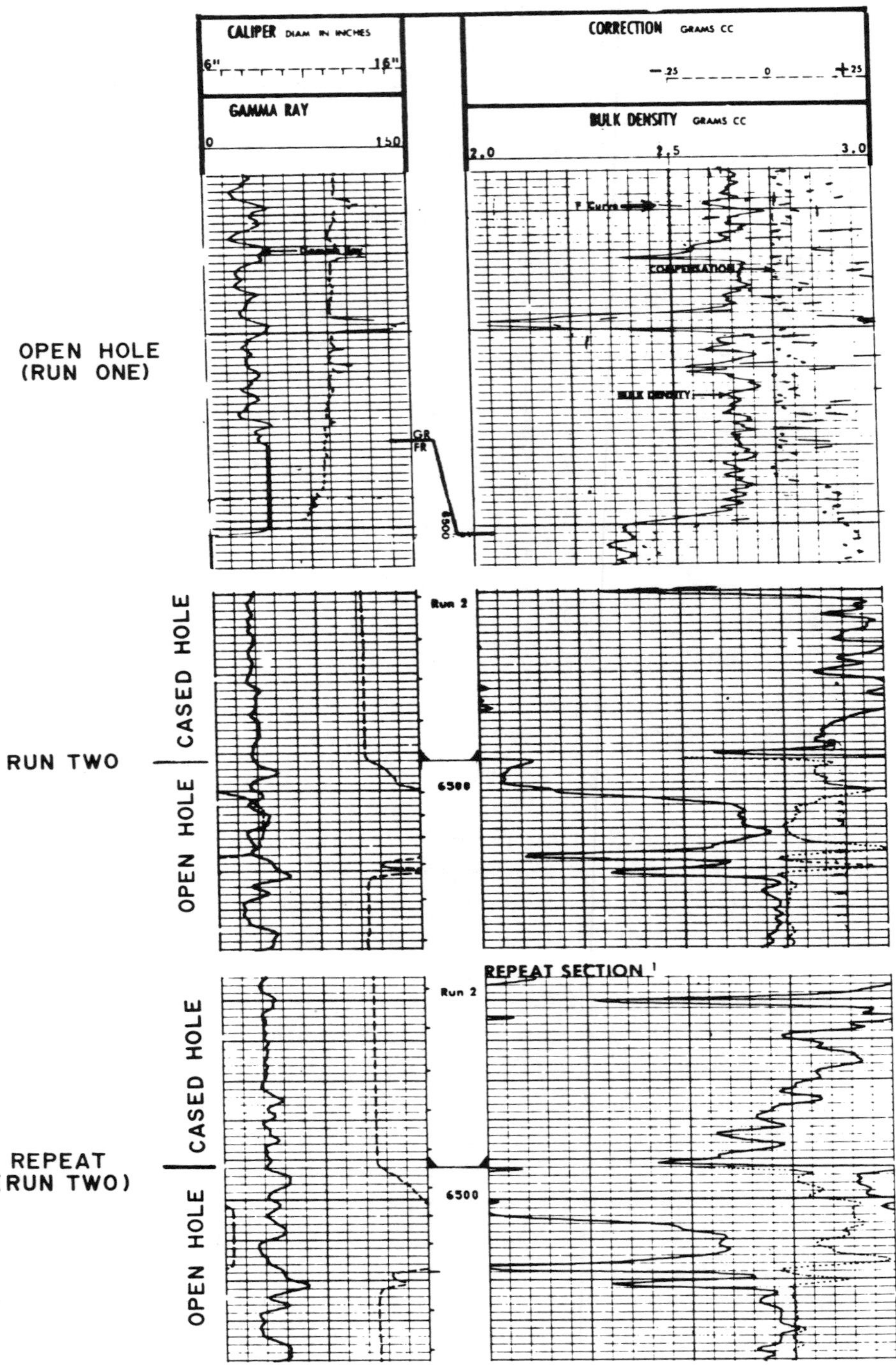

FIG. 8-26. Compensated density log run in open hole (top) and cased hole (center and bottom). There is an offset difference in the measurements but the magnitude of the deflections are comparable — especially the peak at 6 450 ft. Note the statistical difference of the two cased-hole passes.

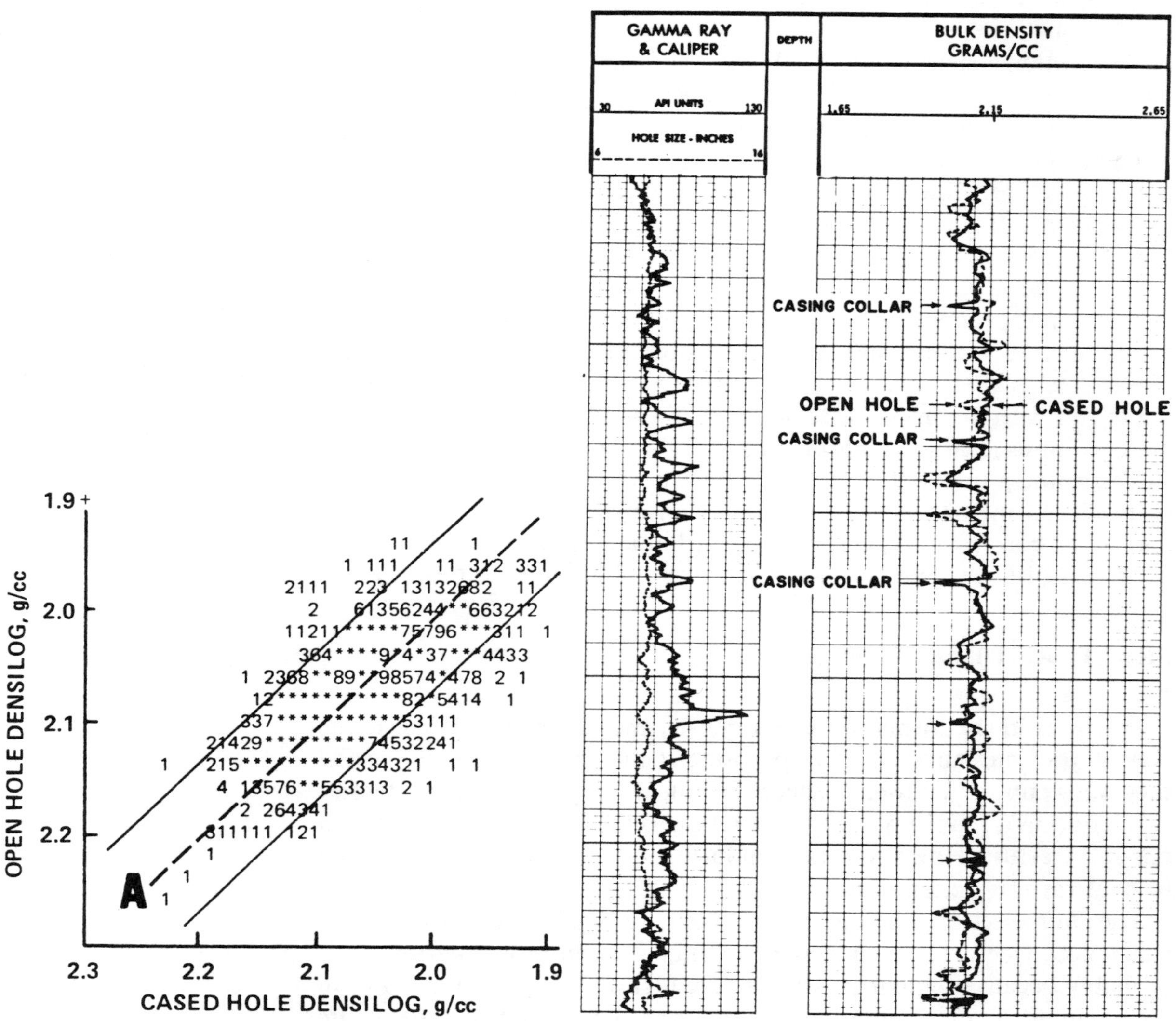

FIG. 8-27. Comparison of open-hole and cased-hole compensated density measurements. The cased-hole density log was calibrated with a comparable piece of casing between the skid and calibration jig. Note that while readings are close, very few are in exact agreement, line A. (Fertl and Wichmann, 1977). Courtesy Dresser Atlas, Dresser Industries, Inc., and Oil and Gas Journal.

problem of greater magnitude than is dealt with when correcting for mudcake thickness. The detector skid won't likely make the exact same trace up the casing on each cased-hole logging pass. By contrast, in the open-hole condition where boreholes are often slightly elliptical the skid tends to have one of two preferred orientations during the open-hole logging pass.

An additional consideration is the skid face curvature. In casing smaller than 5-½ inches inside diameter, the gap between the skid face and casing wall greatly affects the measurement.

Thus, while it is possible to make a through-casing density measurement, the results are always subject to considerable error. The necessary requirements for good through-casing measurements are an in-gauge borehole, no rugose borehole problems, and a good cement job on a well-centered casing string.

Short of an accurate through-casing compensation calibration method, there are three possible methods to determine a linear shift of the density measurements after the log is run with open-hole calibration. If the density log is run with a compensated neutron log — the density data can be matched in selected intervals manually or by using a computer crossplot to determine a linear shift. A cased-hole borehole gravity meter could be used in a method similar to check shot adjustment of the acoustic log. A last correction suggestion, in a carbonate-evaporite environment, is curve shifting to match known salt and anhydrite densities.

FRACTURE DETECTION

Fractures in the formation effect density measurements. Because the density tool measures effective porosity, if the fractures are connected they show up as a lower formation bulk density. Significant density correction in a relatively smooth borehole (Figure 8-28), can indicate a possible fracture. However, other supporting evidence definitely is needed to confirm this conclusion.

The photoelectron lithology measurement, discussed in the next section, can detect fractures (Figure 8-29) in high-concentration barite muds probably because the barite, with a very high P_e index, concentrates in the fractures. Again, supporting evidence is needed to confirm the fracture conclusion.

DENSITY-PHOTOELECTRON LITHOLOGY TOOL

The third-generation density tool uses the photoelectron effect which previous tools were designed to suppress. This density tool indirectly measures porosity, by measuring bulk density and applying it to a Wyllie time-average equation

$$\phi_D = (\rho_{ma} - \rho_a)(\rho_{ma} - \rho_f)^{-1} \qquad (3\text{-}1)$$

where ρ_a is measured log density, ρ_f is the density of the fluid in the pore spaces, and ρ_{ma} is the density of the matrix rock, graphically presented in Figure 8-30, with the common matrix and fluid densities listed in Table 3-1. The uncertainty in the density-to-porosity transformation is in knowing what matrix to apply to the equation. The matrix can be assumed, as it commonly is for most density logs which present a porosity curve. The historical method has been to crossplot bulk density with a complimentary neutron porosity, either single-detector epithermal or two-detector compensated, to obtain lithology and porosity (Figure 5-21).

The density-photoelectron lithology tool (Figure 8-31), in addition to bulk density, measures the photoelectric cross-section, P_e, which is a function of the electrons per atom or the atomic number, Z, of the material

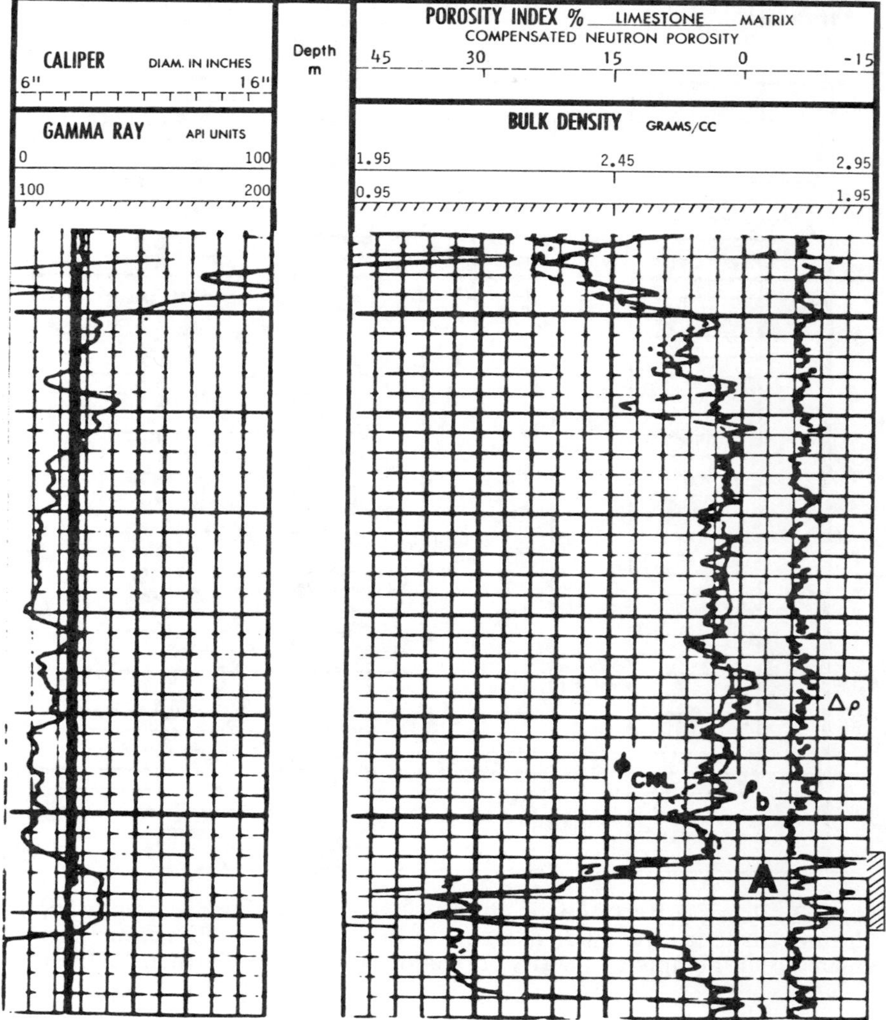

FIG. 8-28. Density log fracture detection showing density deflection (A) in a smooth borehole. In this example there are fracture indications also on the other logs. (Suau and Gartner, 1980).

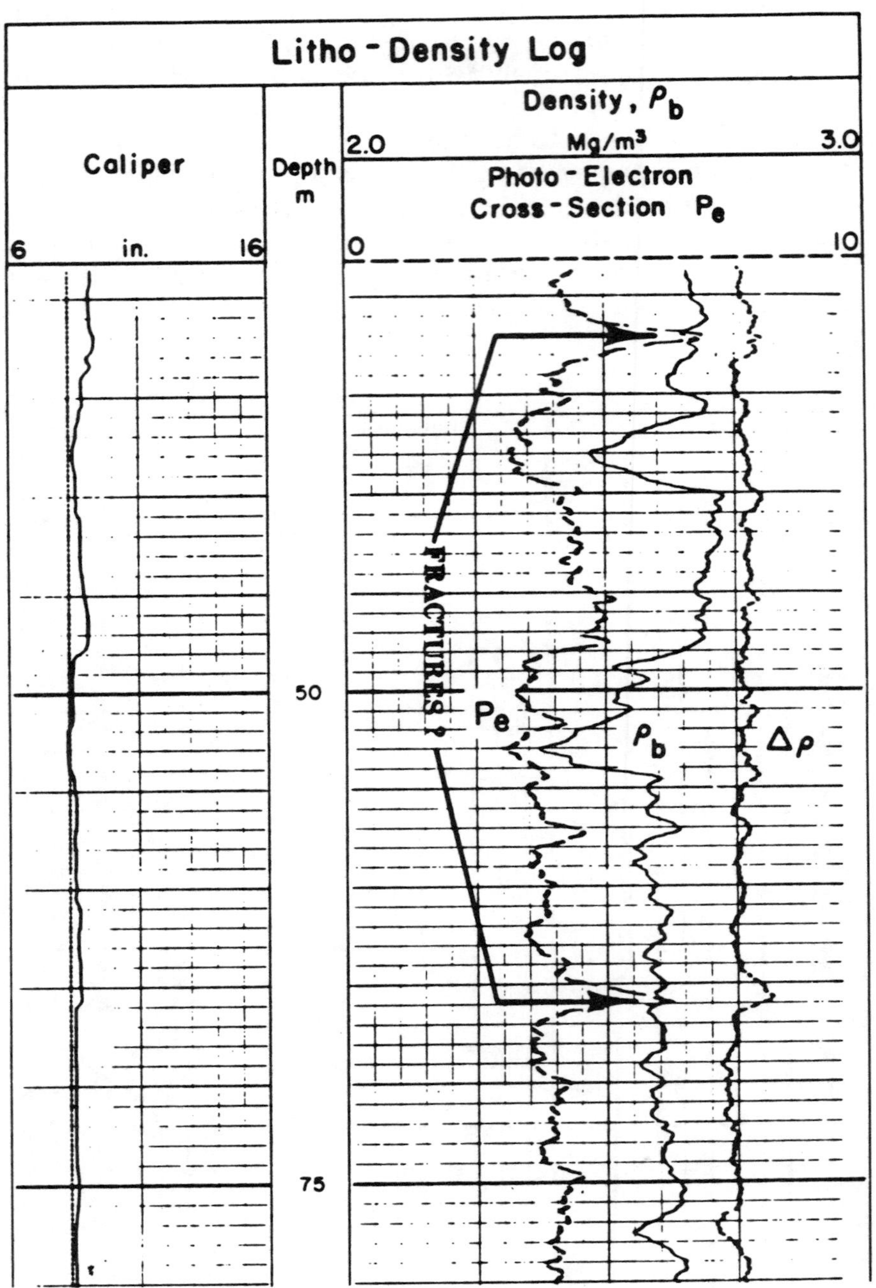

FIG. 8-29. Possible fracture indication on the density-photoelectron lithology measurement in heavy, barite-loaded mud: a dramatic increase in the P_e measurement and an associated increase in the applied density correction. (Suau and Gartner, 1980).

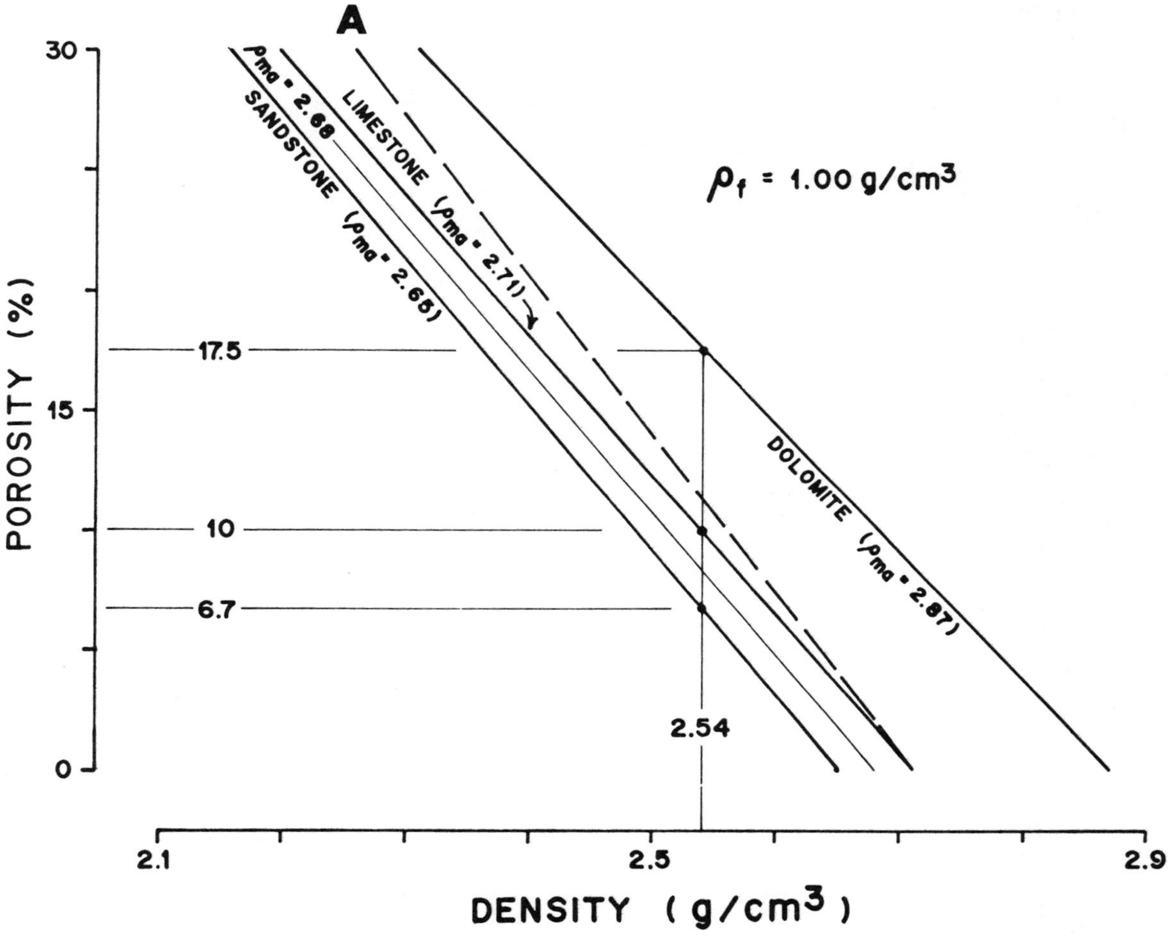

FIG. 8-30. Obtaining porosity from the bulk density measurement by assuming a matrix density and a fluid density of 1.00 g/cm³. Dashed line (A) indicates a limestone with a fluid density of 1.20 g/cm³.

measured. Thus, the photoelectric cross-section measurement is a function of the chemical composition or lithology of the formation (Figure 8-32). There is a small porosity and pore fluid effect on the P_e measurement, but lithology is the prime variable.

One major anomaly is the common mud-weighting material, barite. Because of barite's very high masking P_e of 267, formation P_e cannot be accurately measured in high-concentration barite mud.

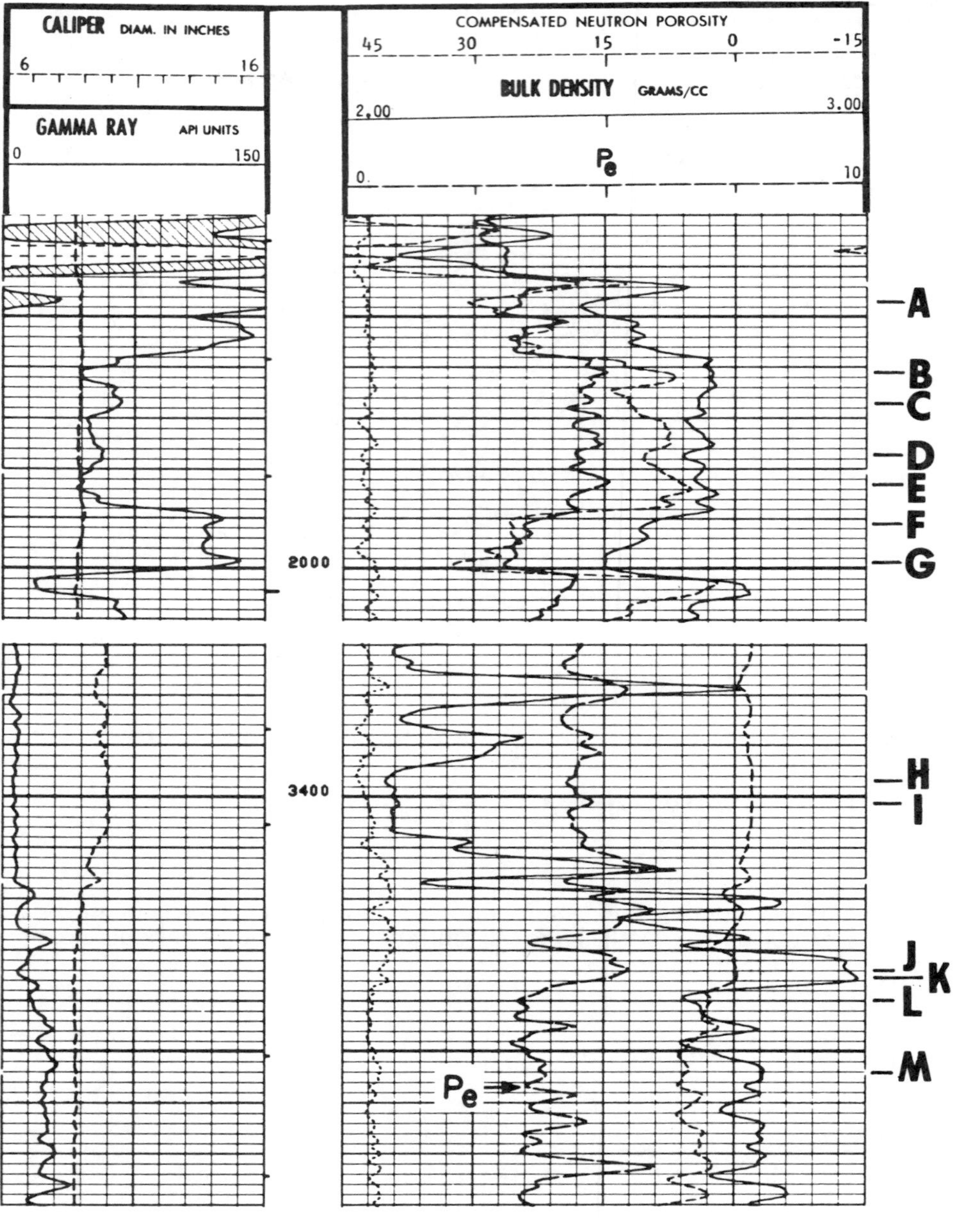

FIG. 8-31. Photoelectron lithology measurement in a number of lithologies. They are crossplotted for matrix determination in Figure 8-33.

Bulk density is crossplotted against P_e (Figure 8-33) to obtain a nonunique solution for lithology and porosity. The constant porosity lines form a triangle in Figure 8-33, as compared to a nearly straight line for the density-neutron (Figure 5-23). There is only a unique solution at the extremes of sandstone or limestone. Combined with the compensated neutron log, this ambiguity is eliminated and it is possible to solve for porosity in three mineral lithologies and to distinguish gas effects.

The photoelectron lithology measurement is especially diagnostic for differentiating changes in lithology from changes caused by gas. In Figure 8-34 the 3 598-3 715 ft interval has a different density-neutron separation than the 3 715-3 800 ft interval. The difference could be caused by either the neutron measurement gas effect or a different lithology. The density-neutron crossplot (Figure 8-35) shows that the upper interval,

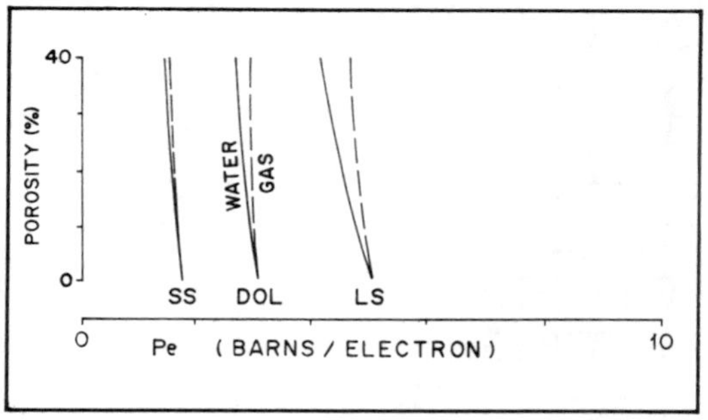

FIG. 8-32. The effects of gas and water versus porosity on the photoelectron lithology measurement (P_e). The predominate variable is lithology. (Adapted from Gardner, 1980.)

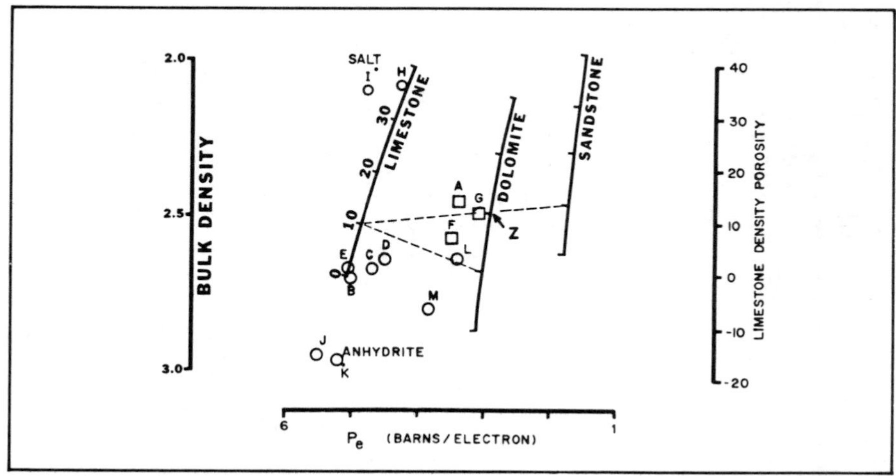

FIG. 8-33. Crossplot of bulk density and the photoelectron measurement (P_e) to obtain a nonunique solution for porosity and lithology. This is illustrated by the point Z being either a 20 percent porosity dolomite or a 10 percent porosity limy sandstone. The other points refer to the log measurements of Figure 8-31. Note that shale, points F and G, fall within the lithology grid. In comparison the density-neutron crossplot (Figure 5-23, point B) normally has shale points outside the lithology grid.

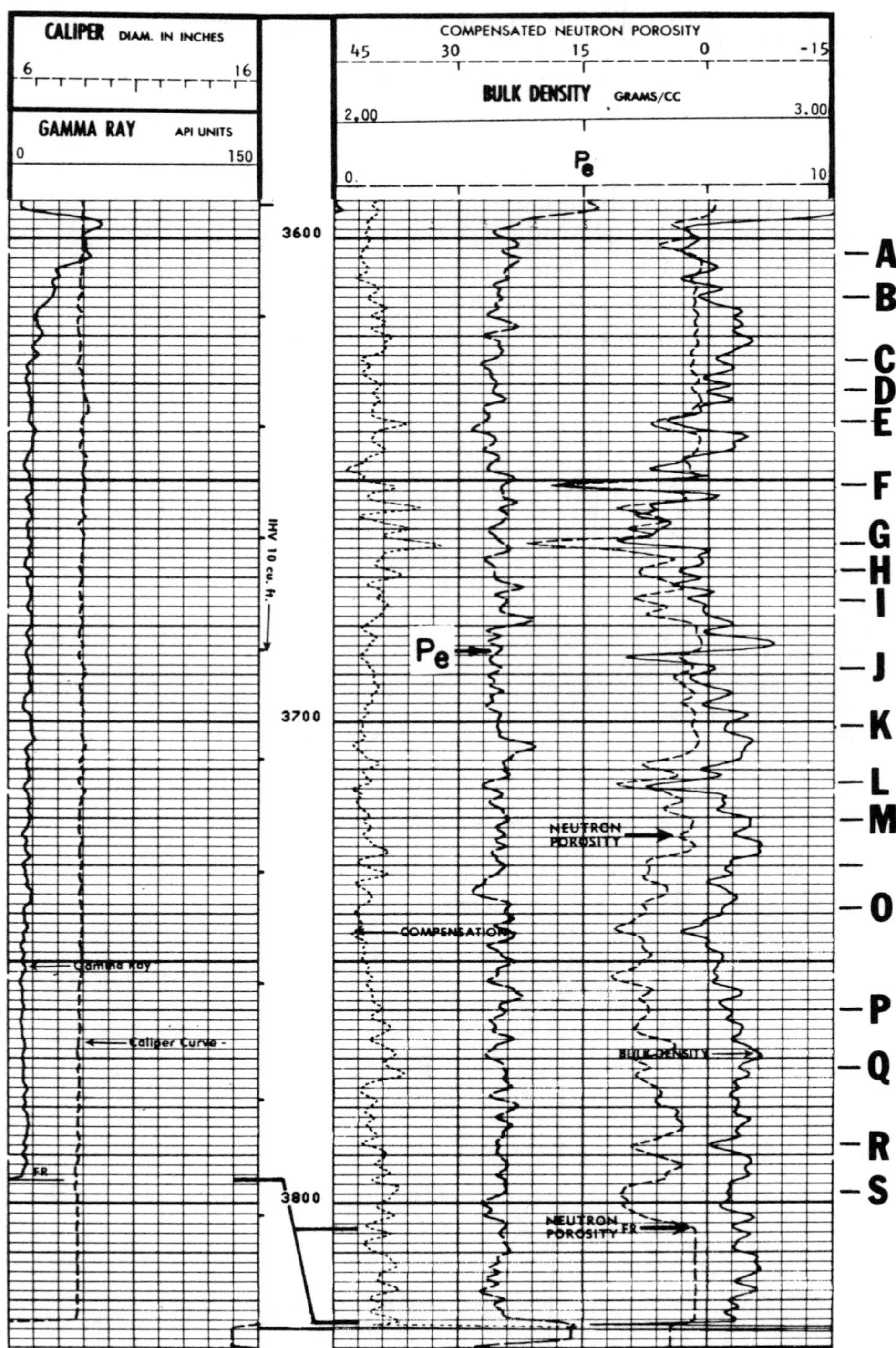

FIG. 8-34. A density-neutron-photo- electron lithology combination log through a Michigan Niagaran reef section.

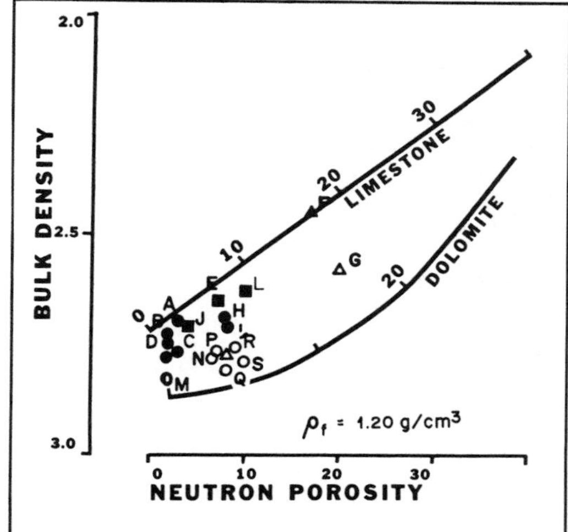

FIG. 8-35. A density-neutron crossplot of the measurements of Figure 8-34. The solid points have an apparent grain density less than 50 percent dolomite, or they could be a gas effected dolomite interval.

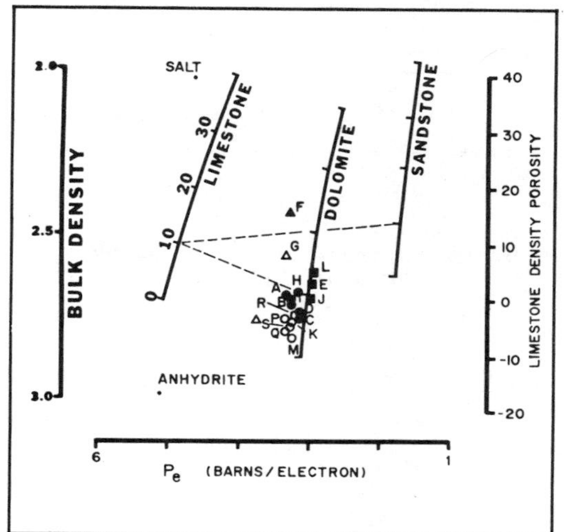

FIG. 8-36. A density-photoelectron lithology measurement crossplot of Figure 8-34 measurements. Because all of the points fall between 70 and 100 percent dolomite, the solid points of Figure 8-35 exhibit a gas effect. A 4 ft interval about point L was perforated and produced 900 Mcfgpd.

points A-L, indicates a less than 50 percent dolomite lithology, or it could be a lithology like points M-S but be gas effected. The density-photoelectron crossplot (Figure 8-36) shows both intervals are highly dolomitic. Thus, the upper interval is probably gas-bearing. The well was perforated from 3 710-3 714 ft, an interval about point L, and produced Mcfgpd.

Figure 8-37 shows Schlumberger's digital calibrations of their photoelectron lithology tool. The after calibration is just a check of the background count rates and does not involve a calibration jig.

P_e can also be obtained from spectral natural gamma-ray measurements. Welex's compensated tool obtains a lithology ratio, R_{LIT}, by optimizing low (photoelectric) and high (Compton) energy window count rates. By correcting the lithology ratio for borehole size and mud weight, P_e is determined (Gadeken et al, 1984, Figure 2).

REFERENCES

Connolly, E. T., 1974, Digital log analysis: Recognition and treatment of field recording errors: 15th Ann. Soc. Prof. Well Log Analysts Logging Symposium, McAllen, paper S.

Dresser Atlas, 1986, Dresser Atlas Computerized Logging Service Calibration Guide.

Fertl, W. H., and Wichmann, P. A., 1977, Open hole porosity logs can be used in cased holes: Oil and Gas J., **75**, 84-86.

CALIBRATION RECORD

AFTER SURVEY TOOL CHECK SUMMARY

PERFORMED: 83/06/27
PROGRAM FILE: LEP18 (VERSION 24.2 00/00/35)

LDTC TOOL CHECK

LDT TOOL SERIAL NUMBER : 1967

	BACKGROUND MEASURED BEFORE	AFTER	UNITS
LL	19.32	19.17	CPS
LU	72.00	71.75	CPS
LS	55.34	55.06	CPS
LITH	5.597	5.69	CPS
SS1	15.94	15.82	CPS
SS2	10.98	11.03	CPS

BEFORE SURVEY CALIBRATION SUMMARY

PERFORMED: 83/06/27
PROGRAM FILE: LEP18 (VERSION 24.2 00/00/35)

LDTC DETECTOR CALIBRATION SUMMARY

LDT TOOL SERIAL NUMBER : 1

	M E A S U R E D			
	BKGD	AL+FE	AL	UNITS
LL	19.3	83.1	93.7	CPS
LU	72.0	127.1	142.7	CPS
LS	55.3	146.7	165.1	CPS
LITH	5.5	36.5	55.8	CPS
SS1	15.9	170.0	192.0	CPS
SS2	10.9	240.7	266.5	CPS

SHOP SUMMARY

PERFORMED: 83/06/14
PROGRAM FILE: CCS18 (VERSION 24.2 83/03/29)
<CR> WHEN DONE
MONITORING: CALI
TURN AC AUX TO 800 MA

LDTC DETECTOR CALIBRATION SUMMARY

LDT TOOL SERIAL NUMBER : 1967

	MASTER CALIBRATED			
	BKGD	AL+FE	AL	UNITS
LL	19.3	83.2	93.8	CPS
LU	72.0	127.2	142.8	CPS
LS	55.3	146.8	165.2	CPS
LITH	5.7	36.5	55.9	CPS
SS1	16.0	170.1	192.0	CPS
SS2	11.0	240.8	266.7	CPS

FIG. 8-37. Photoelectron lithology digital calibrations: the P_e measurement (LITH) uses the FDC aluminum block alone and the block with an iron insert for calibration points.

Gadeken, L. L., Arnold, D. M., and Smith, H. D., Jr., 1984, Applications of the compensated spectral natural gamma tool: Trans. 25th Ann. Soc. Prof. Well Log Analysts Logging Symposium, New Orleans, paper JJJ.

Gardner, J. S., and Dumanoir, J. L., 1980, Litho-density™ log interpretation: 21st Ann. Soc. Prof. Well Log Analysts Logging Symposium, Lafayette, paper N.

Schlumberger, 1984, Log Interpretation charts: Schlumberger Well Services.

Suau, J. and Gartner, J., 1980, Fracture detection from well logs: The log analyst, **21**, no. 2, 3-13.

Tittman, J. and Wahl, J. S., 1965, The physical foundations of formation density logging: Geophysics, **30**, 284.

Wahl, J. S., Tittman, J., Johnstone, C. W., and Alger, R. P., 1964, The dual spacing formation density log: J. Petr. Tech., **16**, 1411-1416.

Waller, W. C., Cram, M. E., and Hall, J. E., 1975, Mechanics of Log Calibration: — Trans. 16th Ann. Soc. Prof. Well Log Analysts Logging Symposium, New Orleans, paper GG.

REFERENCES FOR GENERAL READING

Alger, R. P., Raymer, L. L., Hoyle, W. R., and Tixier, M. P., 1963, Formation density log applications in liquid-filled holes: J. Petr. Tech., **15**, 3, 321-332.

Baker, P. E., 1957, Density logging with gamma rays: Petr. Trans. Am. Inst. Min., Metall., Petr. Eng., **210**, 289-294.

Beck, J., Schultz, A., and Fitzgerald, D., 1977, Reservoir evaluation of fractured cretaceous carbonates in South Texas, 18th Ann. Soc. Prof. Well Log Analysts Logging Symposium, Houston.

Bertozzi, W., Ellis, D. V., and Wahl, J. S., 1981, The physical foundations of formation lithology logging with gamma rays: Geophysics, **46**, 1439-1455.

Campbell, J. L. P., and Wilson, J. C., 1957, Density logging in the Gulf Coast Area: 32nd Ann. Soc. Petr. Eng. Meeting, Dallas, SPE-868-G.

Dupal, L., Gartner, J., and Vivet, B., 1977, Seismic application of well logs: 5th European Logging Symposium (SAID), Paris.

Ellis, D., Flaum, C., Roulet, C., Marienbach, E., and Seeman, B., 1983, The litho-density tool calibration: 57th Ann. Soc. Petr. Eng. Conference, San Francisco, SPE-12048.

Head, M. P., and Barnett, M. E., 1980, Digital log calibration: the compensated density log: 55th Ann. Soc. Petr. Eng. Conference, Dallas, SPE-9343.

Lawson, B. L., 1978, SPWLA Reprint Volume — Gamma ray, neutron, and density logging: Houston, Soc. Prof. Well Log Analysts.

McCall, D. C., Gardner, J. S., and Schulze, R., 1982, Applications of litho-density logs in the Michigan and Illinois basins: Trans. 22nd Ann. Soc. Prof. Well Log Analysts Logging Symposium, Corpus Christi.

Pickell, J. J., and Heacock, J. G., 1960, Density logging: Geophysics, **24**, 891.

Schlumberger Well Services, 1966, Schlumberger Log Calibrations.

Schlumberger Well Services, 1974, Calibration and Quality Standards.

Schlumberger Well Services, C.S.U. Calibration Guide.

Chapter 9

BOREHOLE GRAVIMETER PRINCIPLES

The borehole gravimeter (BHGM) measures, at discrete points along the borehole, the vertical component of the Earth's gravitational field. The gravimeter measurements are used to calculate formation density unaffected by the presence of the borehole or casing with a greater depth of investigation than the continuous gamma-gamma density tool. The tool is used for both formation evaluation and remote sensing applications.

GRAVITY — FORMATION DENSITY RELATIONSHIPS AND MEASUREMENT

For a homogenous, infinite horizontal slab which is cut by a vertical borehole (Figure 9-1A), the gravitational attraction of the slab measured at its upper boundary is

$$g_u = 2 \pi \gamma \rho \Delta Z \qquad (9\text{-}1)$$

where γ is the universal gravity constant, ρ is the slab density, and ΔZ is the slab thickness. At the lower slab boundary

$$g_1 = -2\pi \gamma \rho \Delta Z. \qquad (9\text{-}2)$$

The difference in gravity measurements (Δg) above and below the slab, with a free air correction is

$$\Delta g = F \Delta Z - (g_u - g_1) \qquad (9\text{-}3)$$

where F is a linear with depth free air correction, 94.06 μgals/ft at 45 degree latitude. Equation (9-3) becomes

$$\Delta g = (F - 4\pi\gamma\rho) \Delta Z. \qquad (9\text{-}4)$$

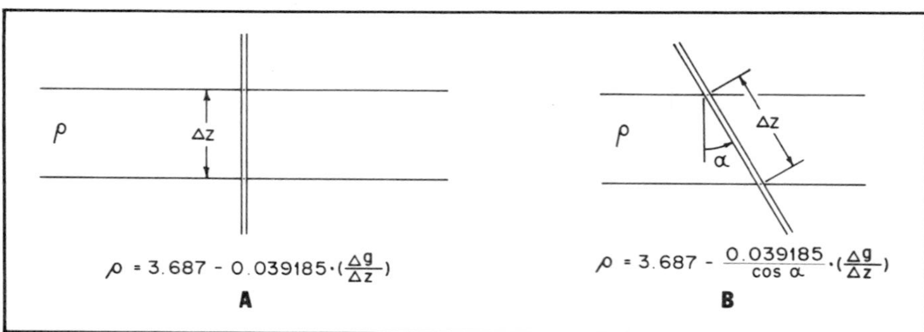

FIG. 9-1. Basic borehole gravimeter (BHGM) density equations for: (A) horizontal slab and vertical borehole, and (B) horizontal slab and deviated borehole.

Equation (9-4) shows that the density of a slab is measurable if the difference in gravity measurements and the difference in station measurement depths (station interval) can be measured:

$$\rho_{BHGM} = [F - (\Delta g/\Delta Z)]\,(4\pi\gamma)^{-1} \qquad (9\text{-}5)$$

which becomes:

$$\rho_{BHGM} = 3.687 - 0.039185\,(\Delta g/\Delta Z) \qquad (9\text{-}6)$$

where Δg is in microGals and ΔZ is in feet. Note that the BHGM measures density directly, and has no Z/A effect as the gamma-gamma density tool does.

The United States Geological Survey uses slightly different constants for equation (9-5) and thus

$$\rho_{BHGM} = 3.680 - 0.039127\,(\Delta g/\Delta Z). \qquad (9\text{-}6a)$$

For consistency, the constants of equation (9-6) are used here.

The basic errors of equation (9-6) hinge primarily on the precision of the gravity measurements and the precision with which the distance between stations is measured. To better understand the precision needed in BHGM density measurement techniques, equation (9-6) is transformed into an error equation:

$$\text{ERROR } \rho = -(0.39185 \text{ ERROR } \Delta g)\,(\Delta Z)^{-1} + (0.039185\,\Delta g \text{ ERROR } \Delta Z)\,(\Delta Z)^{-2}. \qquad (9\text{-}7)$$

If the depth is known absolutely (ERROR $\Delta Z = 0$, and the second term drops out) equation (9-7) becomes:

$$\text{ERROR } \rho = -(0.039185 \text{ ERROR } \Delta g)\,(\Delta Z)^{-1}, \qquad (9\text{-}8)$$

which is plotted in Figure 9-2. With a 6.1 m (20 ft) station interval, and gravity measured with a precision of 5 μGals, density can be measured with a precision of slightly better than 0.01 g/cm^3. Note that the density precision improves for the same gravity meter precision as station interval increases.

If the gravity is known absolutely (ERROR $\Delta g = 0$, then the first term

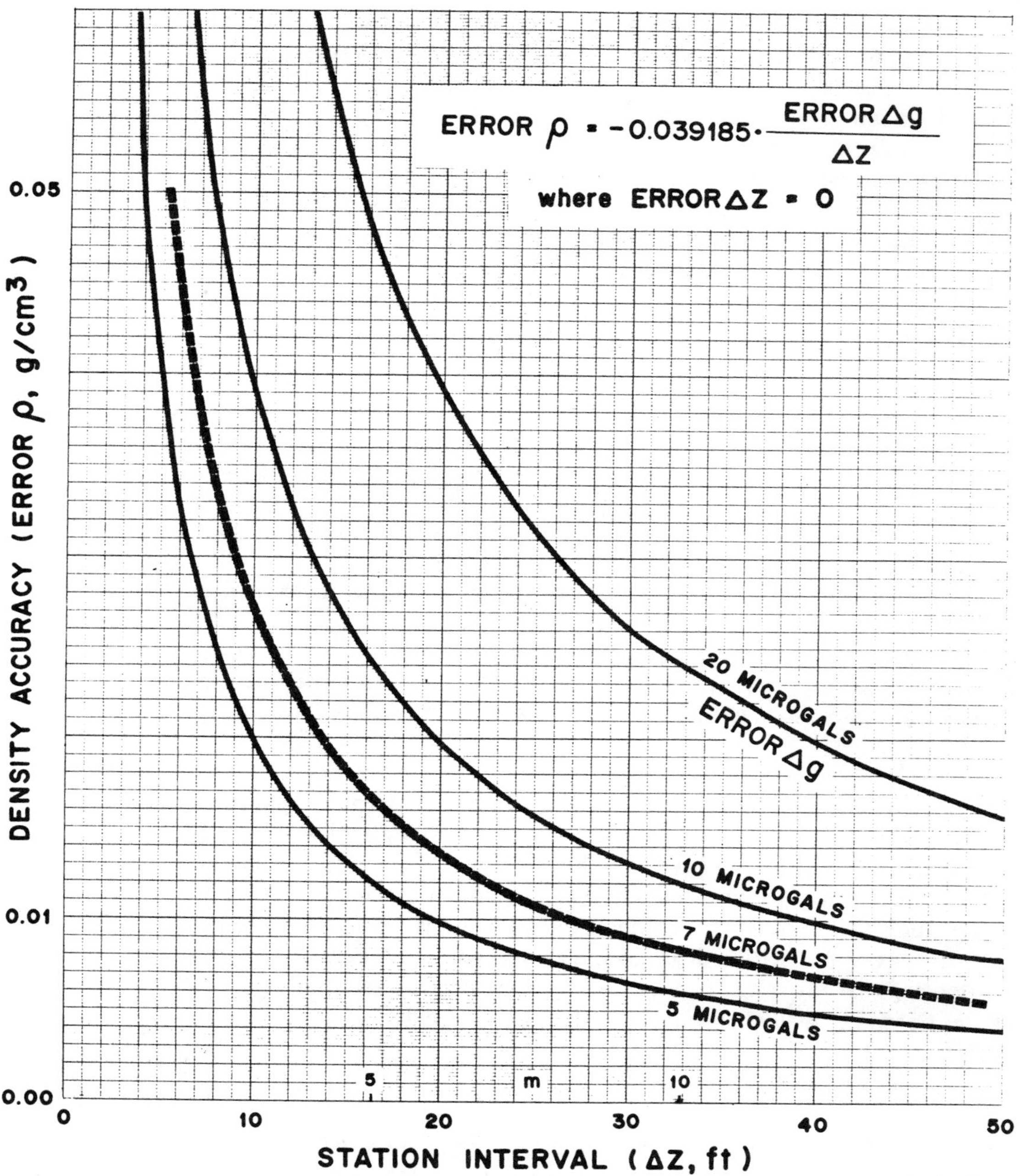

FIG. 9-2. The ability of the borehole gravimeter to measure density to a needed accuracy is a function of station interval and gravity measurement precision.

drops out) equation (9-7) becomes:

$$\text{ERROR } \rho = -(0.039185\ \Delta g \text{ ERROR } \Delta Z)\ (\Delta Z)^{-2} \qquad (9\text{-}9)$$

which is plotted in Figure 9-3. Density precision is dependent on both formation density, a function of Δg, and station interval (ΔZ) precision. For a formation with a 2.50 g/cm³ density, equation (9-6) gives a Δg of

605.84 μGals, with a 20 ft station interval. To measure density within 0.01 g/cm^3, ΔZ must be measured within 2.02 inches (Figure 9-3, point A).

The ΔZ measurement is based on the assumption that the downhole tool moves the same distance the cable moves at the surface. A spot on the cable is optically marked at the surface for accuracy to assure returning to the same surface location. Conventional logging measure wheels may lack the reading accuracy and repeatability required for borehole gravity surveys. The EDCON system allows more accurate cable length measurement to 0.01 ft or 0.12 inches in depth.

EDCON BOREHOLE GRAVIMETER TOOL

EDCON, headquartered in Denver, Colorado, is the prime service company for BHGM surveys. Their tools are LaCoste and Romberg (LC&R) gravity meters packaged for downhole operations (Figure 9-4). The tool is capable of accuracies of 5 to 10 μGals under borehole conditions but is limited somewhat by the present instrumentation and operation techniques.

The meter must be level for proper operation. With the limits of the present internal leveling system, the LC&R borehole gravity meters cannot be used in boreholes deviating more than 14 degrees from vertical. The BHGM does not provide a borehole deviation measurement. If the borehole deviates from vertical, a separate survey, such as the dipmeter, must be run.

The LC&R meter uses a "zero length" spring as the basis of the instrument. Although the spring is made with a low thermal expansion material, the system still needs to be maintained at a very precise constant temperature. Operation of the tool above 400 K, the internal regulated temperature, will cause unacceptable measurement inaccuracies.

The tool is suspended from the surface by the wireline. Because of excessive wave-induced vertical motion the present tools cannot be used on floating offshore rigs. When a tool is developed that can be clamped to the borehole wall, this limitation can be removed.

DEPTH OF INVESTIGATION

The BHGM depth of investigation is large when compared with other logging methods. For a 20 ft station interval, a 90 percent contribution occurs within 100 ft of the borehole, or within approximately five times the station interval. This is considerably greater than the approximate 4 inch depth for a 90 percent response zone for the gamma-gamma density tool.

Some figures used to display response information seem to indicate that by increasing the station interval the lateral depth of investigation can be increased. This is not true because the vertical sampling is increased along with the volume of formation investigated. Hearst (1977) rigorously discussed this topic.

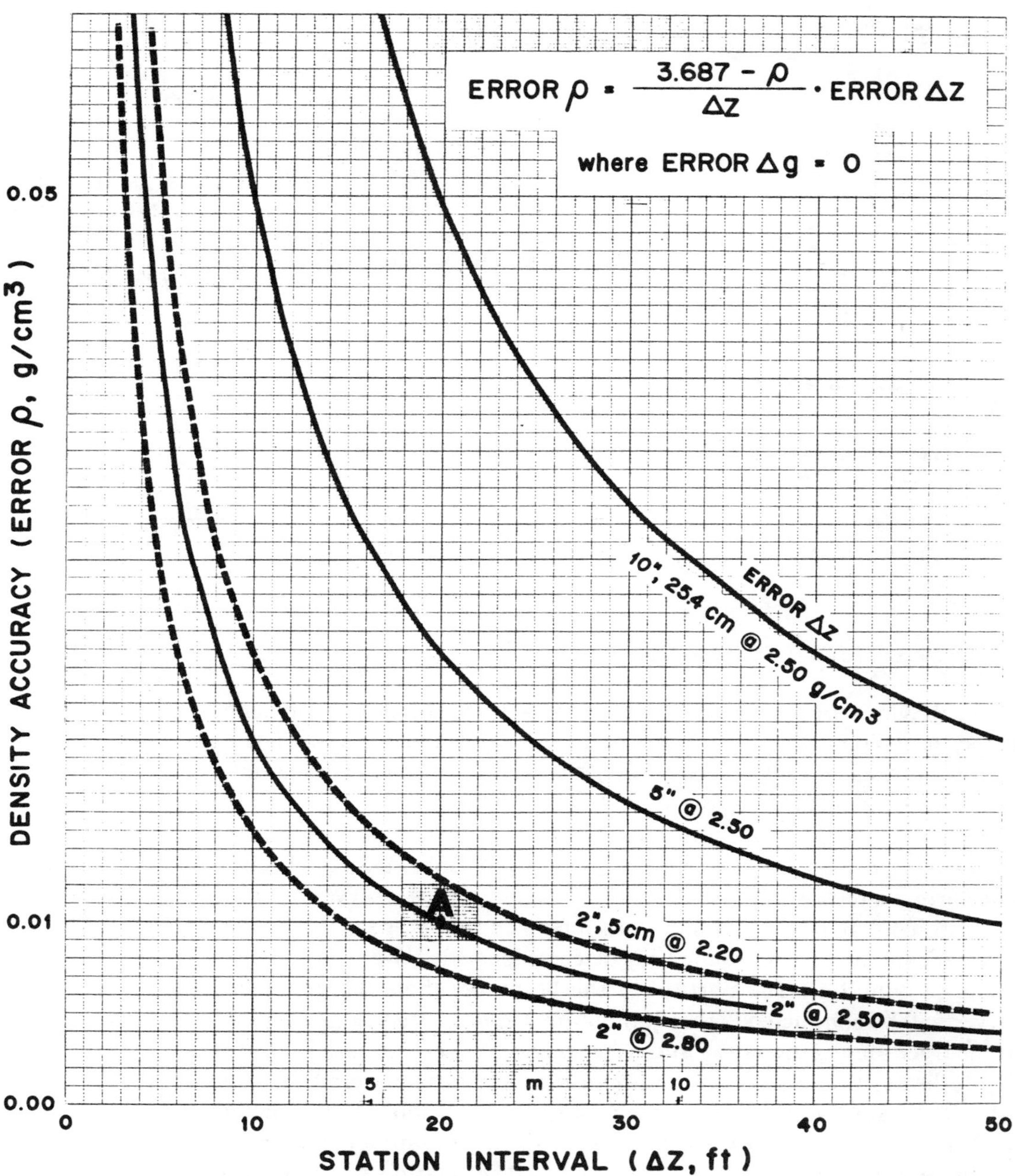

FIG. 9-3. The ability of the borehole gravimeter to measure density to a needed accuracy is a function of station interval, formation density, and depth measurement precision.

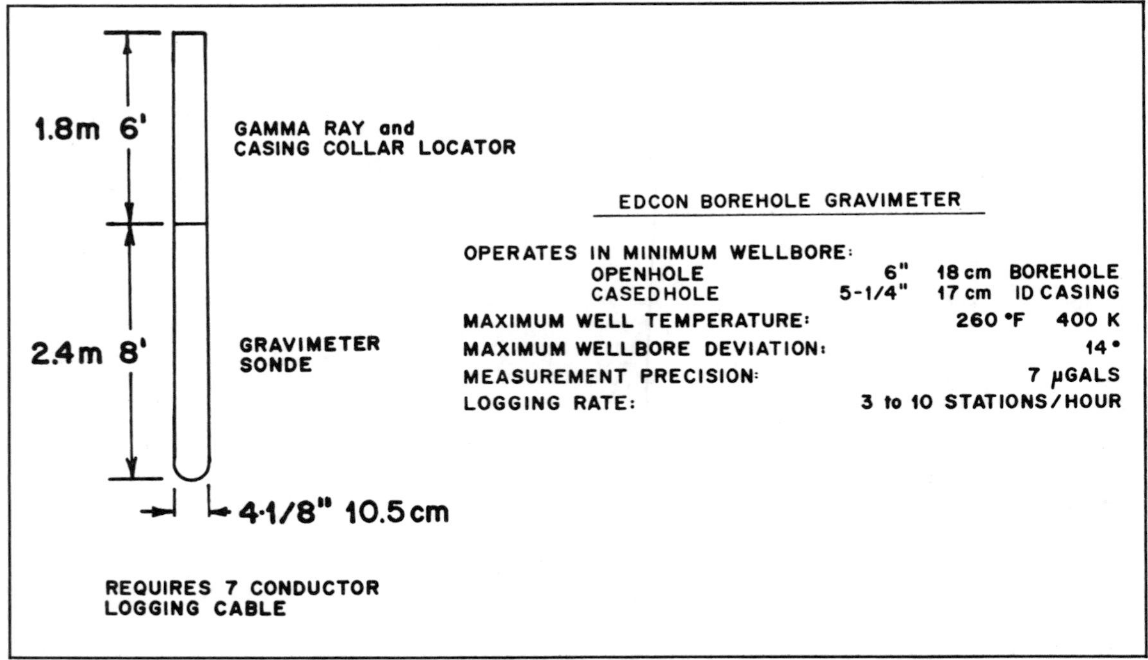

FIG. 9-4. The EDCON LaCoste and Romberg Borehole Gravimeter.

MEASUREMENT TECHNIQUE

The BHGM tool can be used in either open- or cased-hole environments. The tool, which is delicate compared to other logging tools, cannot be spudded or treated roughly. Therefore, it is safer to operate in cased-holes where hole problems are considerably less.

Initially the tool is run a distance into the hole and, after a settling and calming period, the spring is unclamped and tool operation checked. The spring is reclamped and the sonde lowered into the survey interval.

Depth control of the BHGM, with reference to the open-hole logs, is achieved with an add-on gamma-ray detector located above the gravimeter sonde (Figure 9-4). The portion of the borehole where the gravimeter survey will be conducted is first gamma-ray logged. This log is compared to the open-hole gamma ray and other measurements (Figure 9-5). Any depth discrepancies are corrected by using the odometer crank (Figure 3-1). If the survey is run in a cased hole, a casing collar measurement (CCL) can also be run for depth reference during the survey. The collars are often recorded off-depth (unmemorized) in relation to the gamma ray and hand depth corrected (Figure 9-5).

After the necessary depth adjustments are made, the tool is run to below the deepest station and allowed to stabilize to borehole conditions, typically a 30 to 40 min wait. The tool is raised gently to the first station, which is also the drift check station. As in all logging, each station is approached from below to remove all slack from the depth measurement system. As

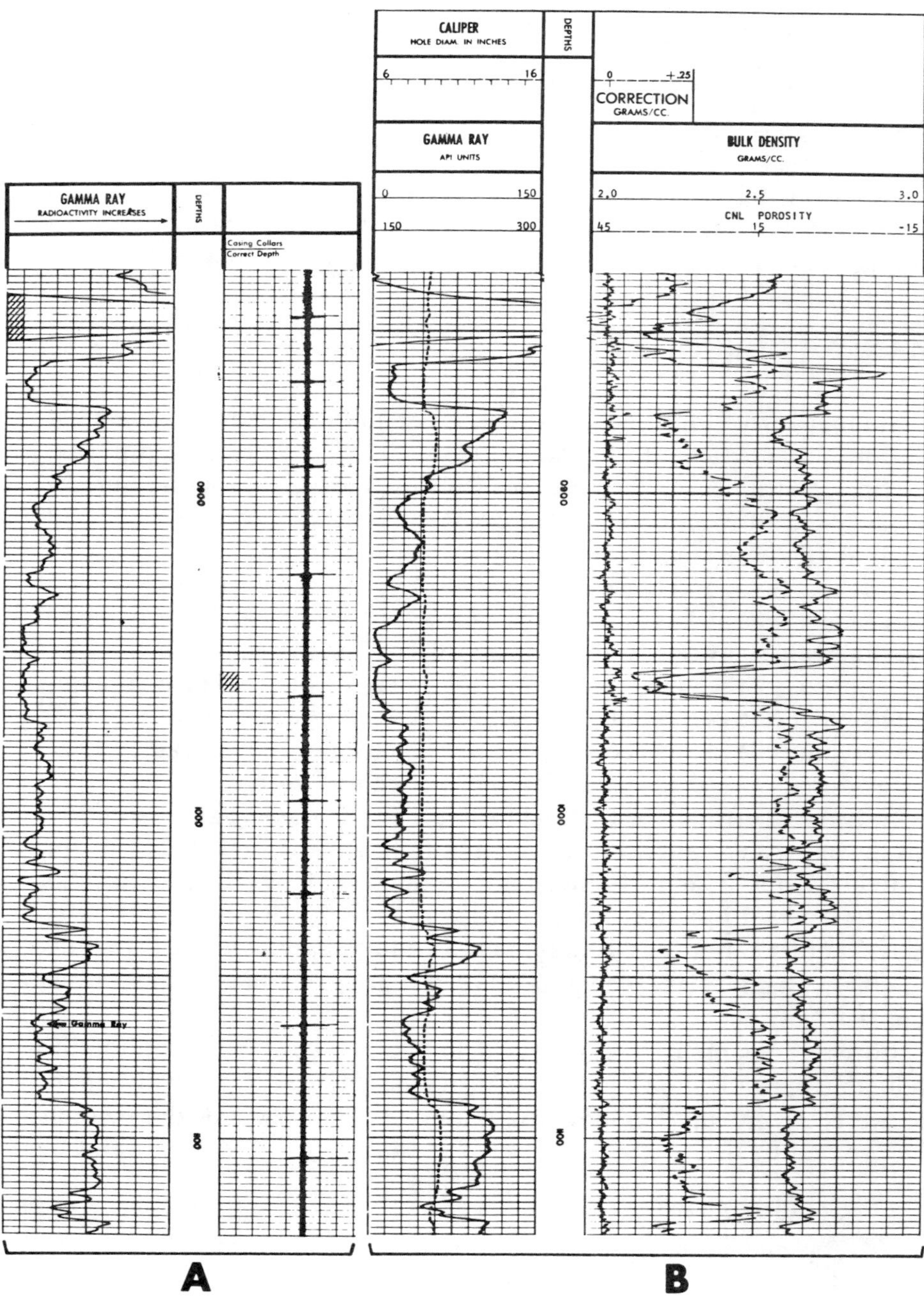

FIG. 9-5. Tying the cased-hole BHGM gamma ray and casing collar log (A) to the open-hole measurements (B).

each station is approached the cable velocity is reduced gradually. A maximum velocity of 100 ft/min for 1 000 ft station changes and grading down gently to a maximum of 3 ft/min for the last few feet below a station is recommended. On the first traverse of the stations the cable is also optically marked at the surface. On subsequent traverses an attempt is made to gradually approach and reoccupy each station. However, once the cable is stopped, no readjustment of the sonde depth should be attempted. The difference between the desired, initial-depth station and the actual-depth station is measured and used in data reduction to the desired initial station location.

Approximately three to six stations can be measured per hour. Repeated traverses are made, with periodic return to the initial station for drift checks, until the necessary degree of repeatability is obtained.

MANUAL CORRECTIONS

Manual corrections, usually small, are required to reduce the gravity station measurements to formation density. The major corrections are for tides, instrument drift, borehole geometry, borehole deviation, surface terrain, structure, and hole deviation. Rasmussen (1973) gives a detailed analysis of these corrections.

Tides. — Solar and lunar tides greatly effect the measurements. Because these changes can constitute a large percentage of the differences in measurements between stations this time dependent correction must always be applied to the measured data. To generate the needed computer-determined tide corrections table, the latitude and longitude of the well (within one mile) are needed. During measurement operations an accurate time source is used to determine the proper tide correction.

Instrument drift. — Instrument drift is highly variable and is dependent on the age, recent physical treatment, and recent thermal history of the individual tool. Rates of 1 μGal/min are not uncommon. Drift corrections are determined by repeatedly reoccupying one depth station and after making other corrections, computing a drift rate per unit time. Corrections to gravity readings are then applied on a time basis to all readings since the last drift check.

Borehole geometry. — Variation in the gravitational effect of the borehole depends on irregularities of the borehole and is usually small. However, significant borehole conditions and the distances they should be avoided by are:

Condition	Feet
Top and bottom of the borehole	20
Air fluid contact (top of mud)	1
Change in casing string (shoe, liner, etc.)	2
Borehole enlargements greater than 10 inches	2-3.

Borehole deviation. — Because the borehole was assumed vertical, if there is any deviation, a correction must be applied. The basic equation (9-6) then becomes:

$$\rho_{BHGM} = 3.687 - 0.039185\ (\text{Cos}\alpha)^{-1}\ (\Delta g/\Delta Z), \qquad (9\text{-}10)$$

where α is deviation of the borehole from vertical (Figure 9-1B).

Surface terrain. — Surface terrain effects are normally insignificant for measurements at borehole depths greater than 3 000 ft. The surface terrain effects on measurements made less than 3 000 ft will depend on the degree of surface relief. The surface geometry and density need to be modeled to determine a surface terrain correction with depth. Beyer (1979) and Hearst, et al (1980) give more complete discussions.

Dipping layers. — The original equation (9-6) assumes a horizontal, flat infinite slab. If the rock layers are dipping by more than 7 degrees, equation (9-6) is no longer correct. Brown and Lautzenhiser (1982) offer a practical solution to the dipping bed problem.

Some texts show an equation corrected for dipping layers

$$\rho_{BHGM} = 3.687 (\cos \phi)^{-2} - 0.039185 (\cos \phi)^{-2} (\Delta g/\Delta Z), \quad (9\text{-}11)$$

where ϕ is the dip of the bed from horizontal. Practically, this is not correct because it is invalid where other layers are adjacent and where the layer is not infinitely-extended. Thus, equation (9-11) should not be used for dipping bed correction.

FORMATION EVALUATION

The BHGM's primary advantages, greater depth of investigation, minimum effect of near borehole conditions, and ability to make measurements in cased wellbores, can be used in formation evaluation.

Improved density determination. — Because the BHGM tool samples a larger volume of formation, it can also be used when the gamma-gamma tool measurements are invalid due to near-borehole formation damage. And because it samples a much greater volume that is uneffected by the drilling of the borehole, its measurement gives densities closer to undisturbed formation conditions.

Because the BHGM has a larger statistical sample it can better measure the density of nonuniformly distributed porosity formations, such as fractured formations. With a better density value a better porosity value can be determined using the density-porosity transform equation (3-1).

The BHGM's greater depth of investigation also allows area measurements that correspond to those of the seismic signal. A 20 to 30 ft station interval corresponds approximately to 1 ms of one-way time in a carbonate/evaporite environment. Thus, the BHGM formation density measurements are more accurate input for synthetic seismogram generating programs.

Cased hole gas detection. — The BHGM's ability to measure through casing can be used to detect by-passed gas in older cased wells. Most older wells have few logs and were drilled when gas was cheap, and therefore gas was left untested behind casing.

When cased-hole density and cased-hole neutron measurements are crossplotted, the points fall above the formation lithology line, because the neutron measurement is significantly reduced in gas-bearing formations. If the crossplot points fall near the sandstone lithology line, it would

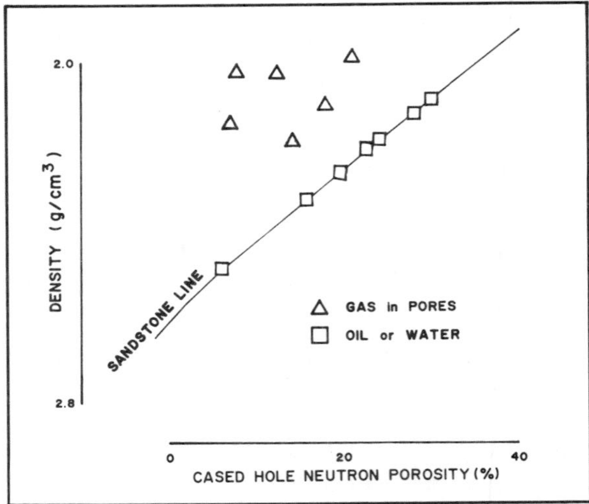

FIG. 9-6. Gas detection through casing using a BHGM density and a neutron crossplot. In a sand sequence, points that fall on the sand lithology line (□) are wet or oil bearing, and points above (Δ) are gas bearing.

be ambiguous in that it can be interpreted as either a wet- or oil-bearing sandstone or a gas-bearing limestone (Figure 9-6, □ points). However, if the points fall significantly above the sandstone lithology line (Figure 9-6, Δ points) probably the interval is a gas-bearing sandstone. The easiest lithology in which to determine gas is sandstone. Cased-hole neutron logs alone can often be used to detect gas-bearing sands if the sands contain high salinity connate water. But, if the connate water is fresh, the interpretation falls in a gray area. BHGM density measurements make possible a more definitive solution. Therefore, the BHGM is most useful for gas detection in fresh water sandstones. Several techniques can solve this problem.

The first technique is to run a cased-hole neutron log (Figure 9-7, item A) over shallower intervals of interest shown on the original electric log (item B). Low readings could be either low-porosity or "suppressed-reading" gas-bearing sands. The suspected interval is then logged with the BHGM to eliminate the ambiguity.

The second technique, when more information is available, is the determination of the formation pore fluid density. This technique requires independent determination of the lithology and porosity of an interval. If the two items are known, charts for the lithology can be determined that crossplot the porosity with a density measurement, to determine the pore fluid density (Figure 9-8). Obviously, it is difficult to differentiate oil from water in low porosity formations, but gas easily can be detected from oil or water for most porosity values.

A third technique is use of off-setting well log studies. Some more recently drilled wells have more complete formation information. This data can be used to determine the most likely cases for a specific

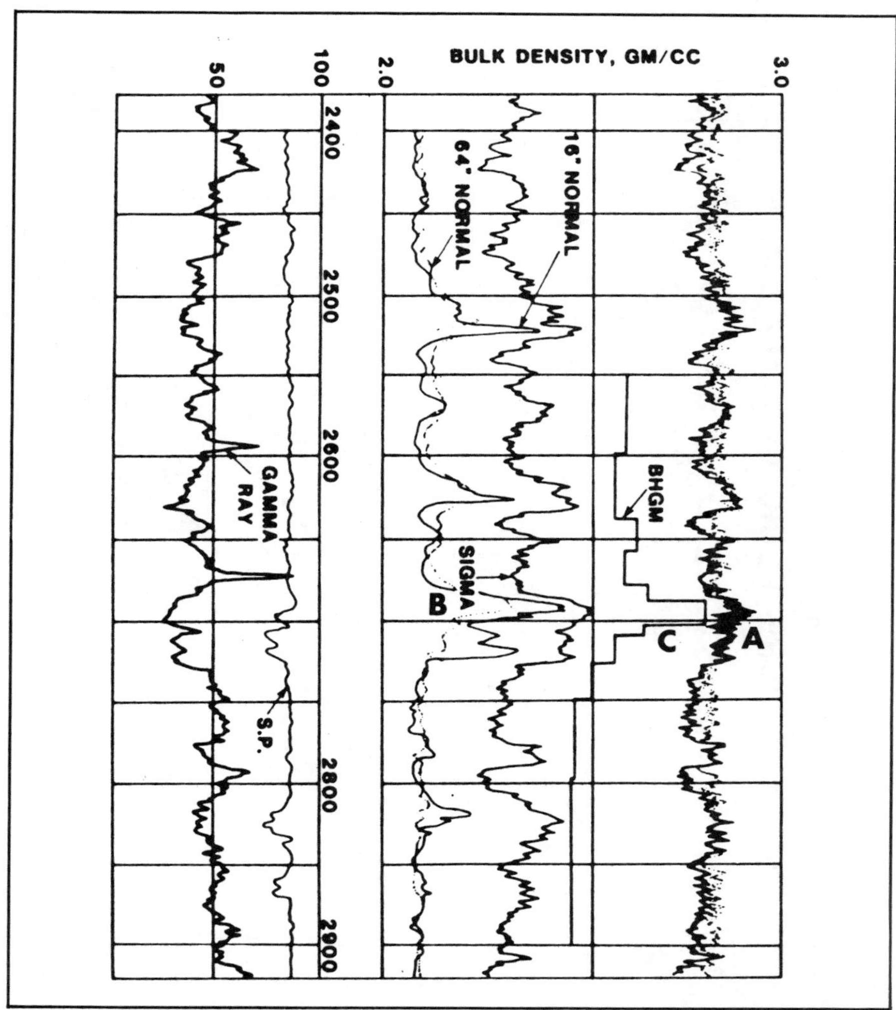

FIG. 9-7. A cased-hole pulsed neutron log indicated a potential bypassed gas interval (A). Examining the electric log run when the well was drilled (B) reinforces the speculation. However, the cased-hole BHGM (C) indicates the interval is tight, not gas bearing. (Gournay, 1982).

formation. For example, a formation can be a porous gas-bearing sand, a porous wet sand, or a low-porosity wet sand. From these models the expected density can be determined using

$$\rho = \rho_{ma} (1-\phi] + \phi \, \rho_f \, S_w + \phi \, \rho_{hc} (1-S_w) \qquad (9\text{-}12)$$

where,

ρ_{ma} is density of formation matrix rock,
ρ_f is density of connate water in pores,
ρ_{hc} is density of hydrocarbons in pores,
ϕ is total porosity,
and S_w is water saturation

Intervals can be picked specifically from previously run old *e*-logs and

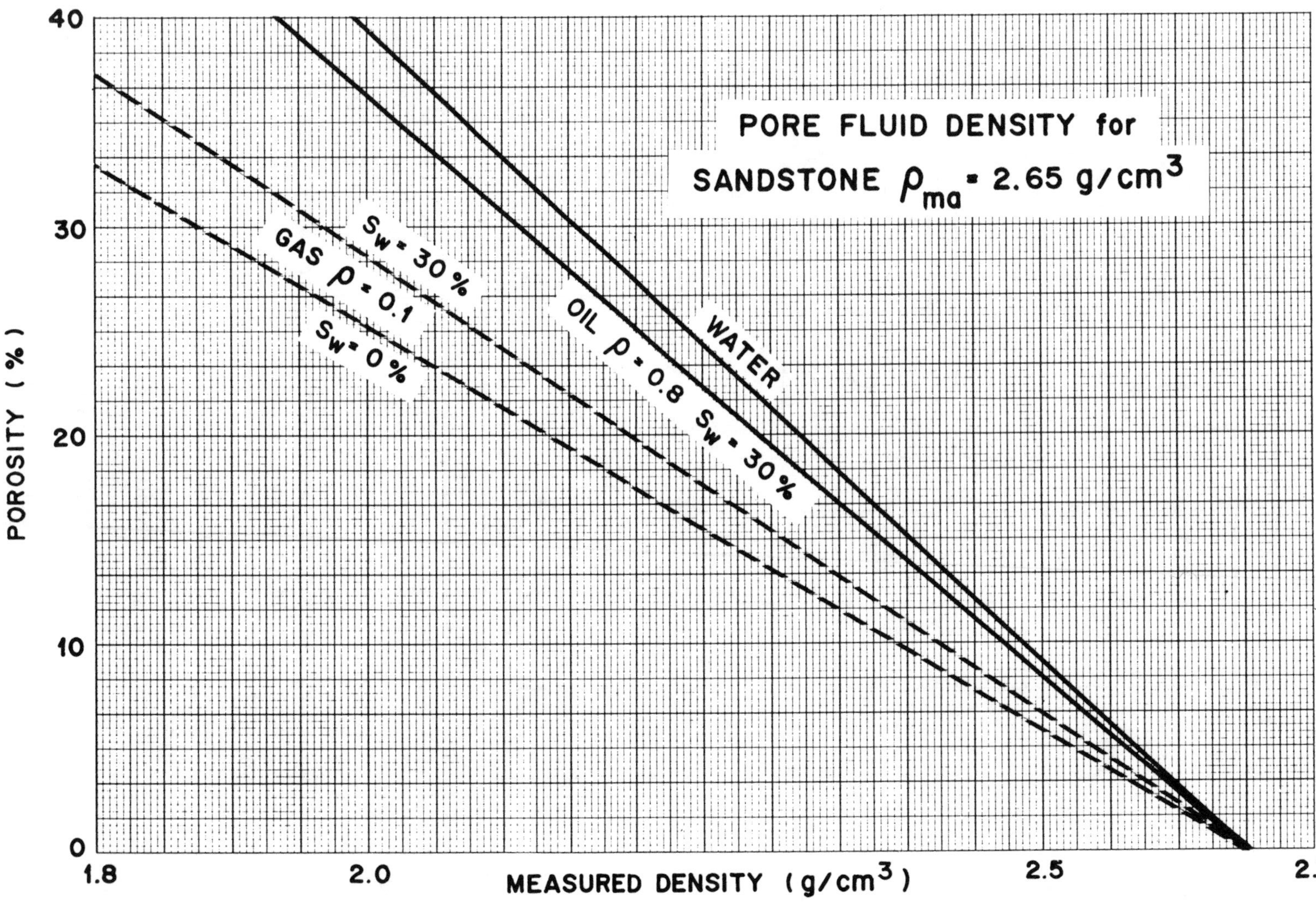

FIG. 9-8. Pore fluid determination if lithology, porosity, and density are independently determined.

then logged with the BHGM. These measurements are compared with the expected models to determine the probable formation contents.

REMOTE SENSING

With its greater depth of investigation, the BHGM responds to changes in lateral density contrasts, especially where rocks with large density contrasts such as carbonate or shale and salt are present. By subtracting the near borehole responding integrated gamma-gamma density from the BHGM density, the lateral density changes are emphasized. Two basic applications are suggested; (1) using the density difference *curve shape* to detect lateral stratigraphic differences, and (2) using the density difference *values* to detect lateral formation (mostly porosity) differences. In both applications the effect can be measured but *not* the direction to its cause.

Stratigraphic differences. — The presence of the less dense salt beds surrounding the carbonate Niagaran reefs in the Michigan Basin influences the deep measuring BHGM. The closer to the edge of the reef, and thus closer to the salt beds, the stronger their influence.

Modeling of the difference between BHGM and near borehole density measurements (Figure 9-9) in a typical reef shows that the curve shape can suggest where the borehole lies on the reef. Thus, a flank-edge well can be differentiated from an off-reef well, by the curve shape. However, curve shape does not indicate the direction to drill to get higher onto the reef crest. The dipmeter technique to determine this direction and distance is discussed in Chapter 12.

A similar modeling comparison could be used in salt dome and salt overhang areas. The technique may be used to detect the truncation of beds by faulting not cut by the borehole if there is a density contrast between beds on each side of the fault, and could also determine if the borehole has encountered basement granite or a finite-thickness sill or dike.

Formation differences. — The BHGM and gamma-gamma density difference can also be used to detect porosity remote from the borehole. These lateral differences in small intervals can be significant porosity development or, unfortunately, salt-plugged porosity development. Figure 9-10 shows BHGM and gamma-gamma density measurements in a northern Michigan Niagaran reef well. There is reasonable porosity development shown by the gamma-gamma density measurements above 6 560 ft. However, notice that the interval from 6 665-6 685 ft shows more porosity in the BHGM measurements, indicated by the density difference increase. This interval was perforated and potential was tested at 2 000 Mfgpd and 79 BCPD on a 7/64 inch choke with 2 750 lbs pressure. Obviously, the completion broke into a porosity interval that did not cut the borehole.

An alternate approach to use in remote sensing applications of the BHGM is the use of the Bouguer anomaly instead of the density difference. This approach was proposed in Snyder (1976). The borehole

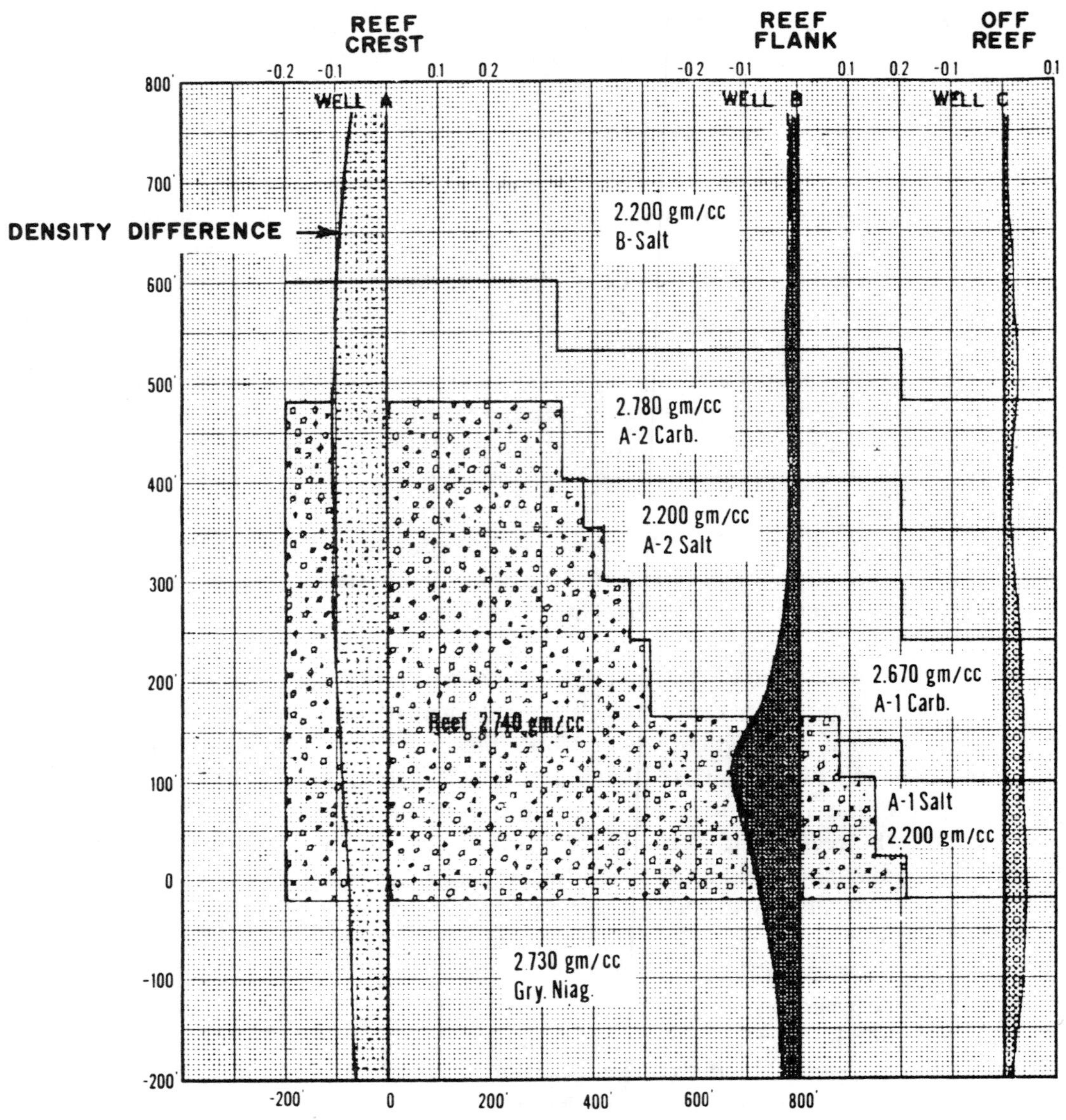

FIG. 9-9. Calculated density difference profiles expedited : (A) on crest of a Michigan reef, (B) on the flank of the reef, and (C) off the reef. The profiles are primarily responding to the salt. (Adapted from Bradley, 1975.)

Bouguer anomaly is determined by:

$$g_{Bouguer} = g_{obs} - (Z\,F) - B, \tag{9-13}$$

where,

g_{obs} is observed gravity at depth Z,
Z is depth of the station,
F is a linear with depth free air correction, and
B is Bouguer effect.

The last term, B, must be determined from an independent source of

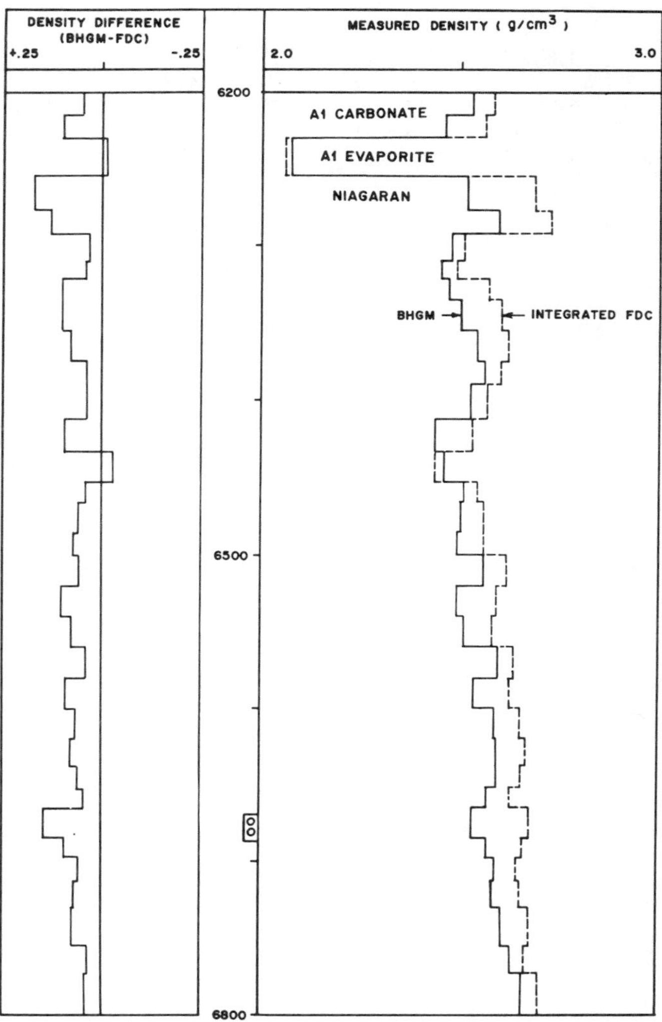

FIG. 9-10. Michigan reef remote porosity detection using the density difference method. (Adapted from Bradley, 1975.)

density information, such as the gamma-gamma density tool. It is a compensation for the densities above and below the measurement station. This approach offers an alternative to the density difference method. Figure 9-11 is a comparison of the two approaches.

REFERENCES

Beyer, L. A., 1979, Terrain corrections for borehole gravity measurements: Geophysics, **44**, 1584-1587.

Bradley, J. W., 1975, The application of the borehole gravimeter to the evaluation and exploration of oil and gas reserves: 45th Ann. Soc. Explor. Geophys. Meeting, Denver.

Brown, A. R., and Lautzenhiser, T. V., 1982, The effects of dipping beds on a borehole gravimeter survey: Geophysics, **47**, 25-30.

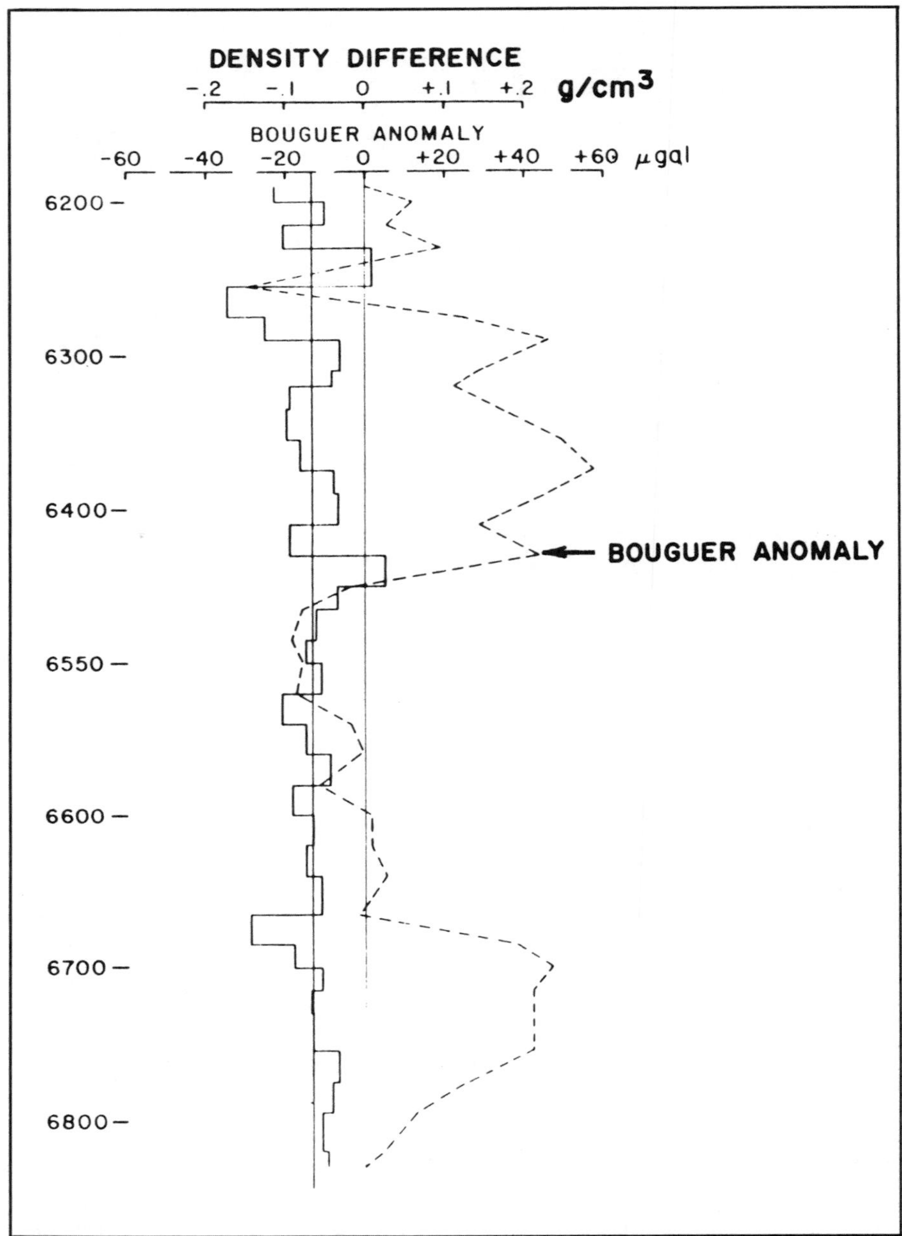

FIG. 9-11. Michigan reef remote porosity detection using the Bouguer gravity method. (Adapted from Snyder, 1976.)

Gournay, L. S., and Maute, R. E., 1982, Detection of bypassed gas using borehole gravimeter and pulsed neutron capture logs: The log analyst, **23**, 3, 27-32.

Hearst, J. R., 1977, On the range of investigation of a borehole gravimeter: Trans. 18th Ann. Soc. Prof. Well Log Analysts Logging Symposium, Houston, paper E.

Hearst, J. R., Schmoker, J. W., and Carlson, R. G., 1980, Effects of terrain on borehole gravity data: Geophysics, **45**, 234-243.

Rasmussen, N. F., 1973, Borehole gravity survey planning and operations; Trans. 14th Ann. Soc. Prof. Well Log Analysts Logging Symposium, Lafayette.

Snyder, D. D., 1976, The borehole Bouger gravity anomaly — Application to interpreting borehole gravity surveys: Trans. 17th Ann. Soc. Prof. Well Log Analysts Logging Symposium, Denver, paper AA.

REFERENCES FOR GENERAL READING

Brown, A. R., Rasmussen, N. F., Garner, C. O., and Clement, W. G., 1975, Borehole gravimeter logging fundamentals: 45th Ann. Soc. Explor. Geophys. Meeting, Denver.

EDCON, 1977, Borehole gravity meter operations and interpretation manual.

Jones, B. R., 1972, The use of downhole gravity data in formation evaluation: Trans. 13th Ann. Soc. Prof. Well Log Analysts Logging Symposium, Tulsa.

Rasmussen, N. F., 1975, The successful use of the borehole gravimeter in Northern Michigan, Trans. 5th CWLS Formation Evaluation Symposium, Calgary, paper X.

Chapter 10

ACOUSTIC LOG PRINCIPLES

An acoustic log was experimented with in 1935 by the Schlumberger brothers using an automobile horn as a source and a two-receiver system that attempted to detect the phase shift between the receivers (Allaud and Martin, 1977). However the first commercial acoustic log service was made available in 1954 when Seismograph Service Corporation began marketing a continuous velocity log (CVL). This log was a single-receiver system under patent license use (Figure 10-1) from Magnolia Petroleum Company, now Mobil Oil Corporation. Schlumberger acquired the patent rights to the Humble Oil two-receiver system in 1955 and began marketing the service in 1958.

The log primarily was used in interpreting seismic data. It was not used in log analysis until the correlation between recorded traveltime and the formation factor, and thus porosity, became apparent. With this formation evaluation use wireline service companies showed more interest in tool development.

The acoustic log measures vertical transit time, the reciprocal of vertical velocity. To eliminate borehole transit time a single-transmitter, two-receiver system was initially developed. To further reduce problems of changing borehole size on the desired formation measurement, a compensated acoustic log system measures downgoing and upgoing times.

SOUND PULSE — FORMATION INTERACTION

When a source emits a compressional sound pulse into the borehole there are four modes of propagation to a distant receiver in the borehole (Figure 10-2):

1. Compressional (P)-wave — The pulse is refracted at the borehole wall

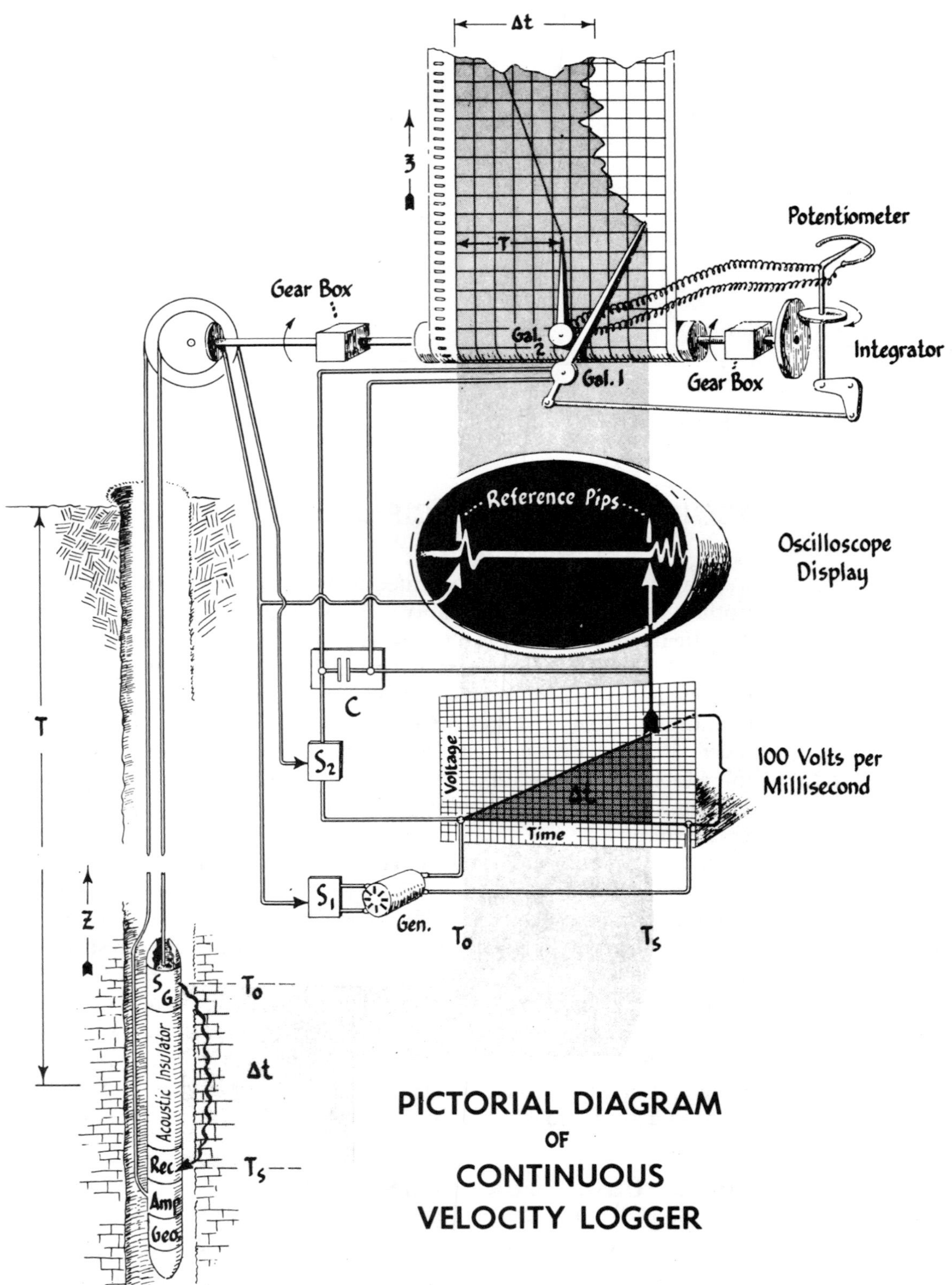

FIG. 10-1. The early Seismograph Service Corporation Continuous Velocity Log (CVL) system under the Magnolia patent. Courtesy Seismograph Service Corporation.

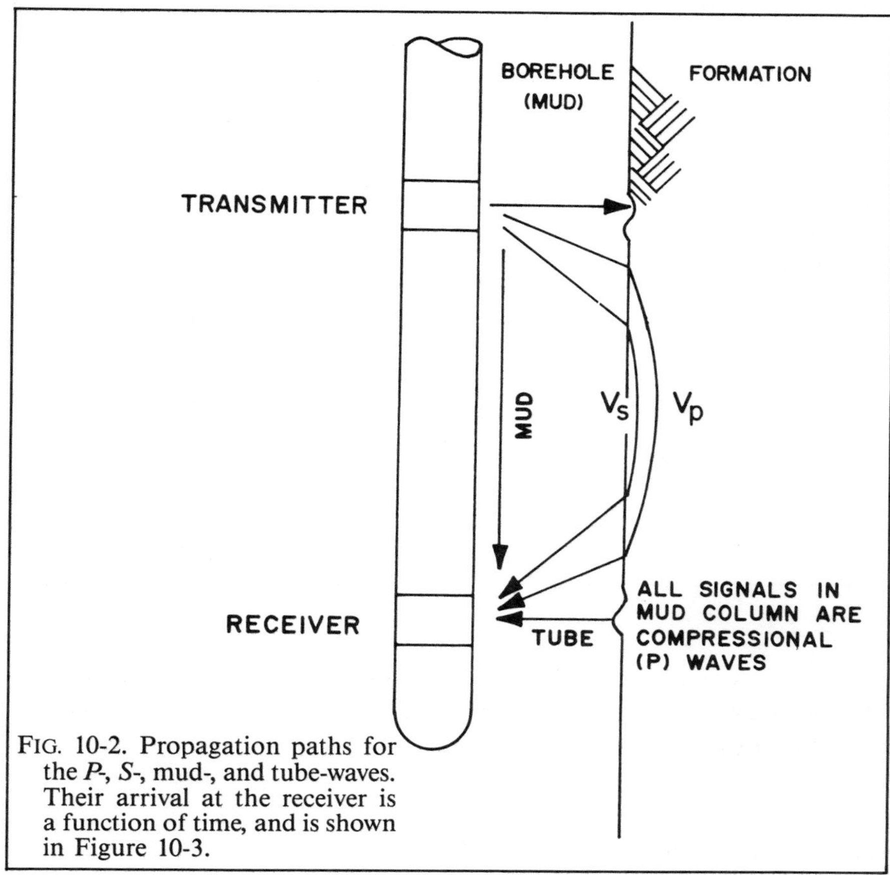

FIG. 10-2. Propagation paths for the *P*-, *S*-, mud-, and tube-waves. Their arrival at the receiver is a function of time, and is shown in Figure 10-3.

into the formation at a critical angle determined by Snell's law and travels through the formation at the formational *P*-wave velocity (V_p). It is then refracted back as a *P*-wave to the receiver where it is changed to an electric signal for transmission to the surface. The *P*-wave is normally the first arrival at the distant receiver.

2. Pseudo-Rayleigh, shear (*S*) wave — The compressional pulse is refracted into the formation at a different critical angle and converts to a *S*-wave where it travels down the formation at near the formation *S*-wave velocity (V_s). It is refracted back into the borehole and reconverts to a *P*-wave to reach the receiver. The *S*-wave velocity is less than the *P*-wave velocity.

3. Mud or direct wave — The pulse travels down the borehole fluid directly to the receiver. This wave is of relatively high frequency.

4. Boundary, Stonely, or tube wave — The pulse strikes the formation at normal incidence and sets up a standing wave that propagates down the borehole/formation interface where it is transmitted back normally to the receiver. It is a high-amplitude and low-frequency wave with a velocity dependent on borehole size and fluid and formation properties.

With sufficient transmitter-receiver spacing the four signals are

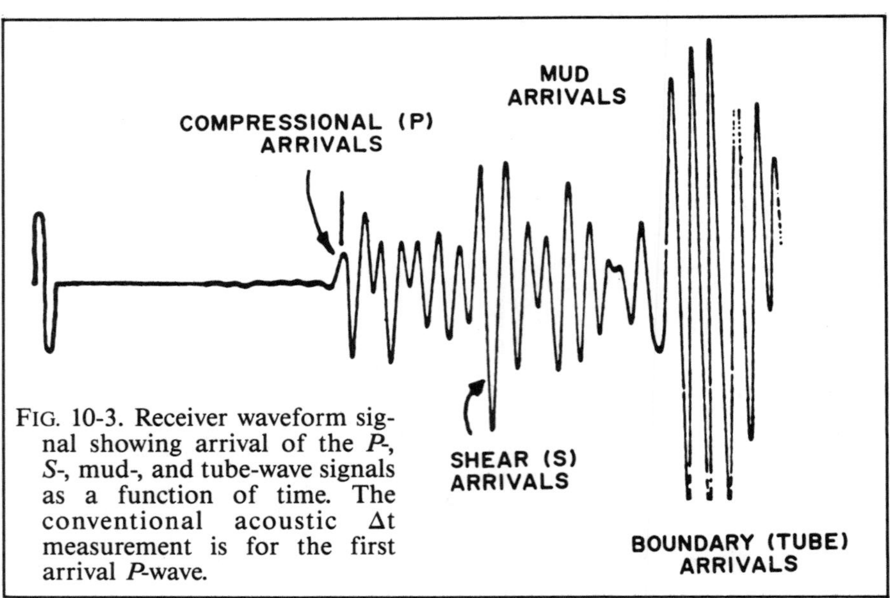

FIG. 10-3. Receiver waveform signal showing arrival of the *P*-, *S*-, mud-, and tube-wave signals as a function of time. The conventional acoustic Δt measurement is for the first arrival *P*-wave.

distinctively separated in arrival time (Figure 10-3). A common acoustic sonde has spacings of 3 and 5 ft. The former is borderline with the latter providing adequate time separation.

ACOUSTIC (SONIC) LOG TOOL

The acoustic or sonic logging tool measures and presents the *first arrival* or *P*-wave measurements as interval (vertical) traveltime, the reciprocal of vertical velocity (Figure 10-4). A secondary feature allows a total elapsed vertical traveltime [integrated travel time (TTI)] to assist in the tool's original purpose of aiding interpretation of seismic data.

The interval traveltime can be presented in any number of scales but the more common scales are 80-60-40, 100-70-40, 140-90-40, and 150-100-50 μs/ft. Common Canadian metric scales are 300-200-100 and 500-300-100 μs/m. Because the scales selection is variable each log should be carefully checked for both scales and scale changes.

Gamma-ray (GR), SP, caliper, and acoustic derived porosity (ϕ_{sv}) measurements also can be presented with the acoustic measurements. Porosity is derived from the Δt measurement using Wyllie's time average transformation

$$\phi_{sv} = (\Delta t_a - \Delta t_{ma}) (\Delta t_f - \Delta t_{ma})^{-1}, \tag{3-2}$$

which is normally only valid for sandstones with primary, intergranular porosity (Figure 5-20). A more general equation is:

$$\Delta t = A + B \phi_{sv}, \tag{10-1}$$

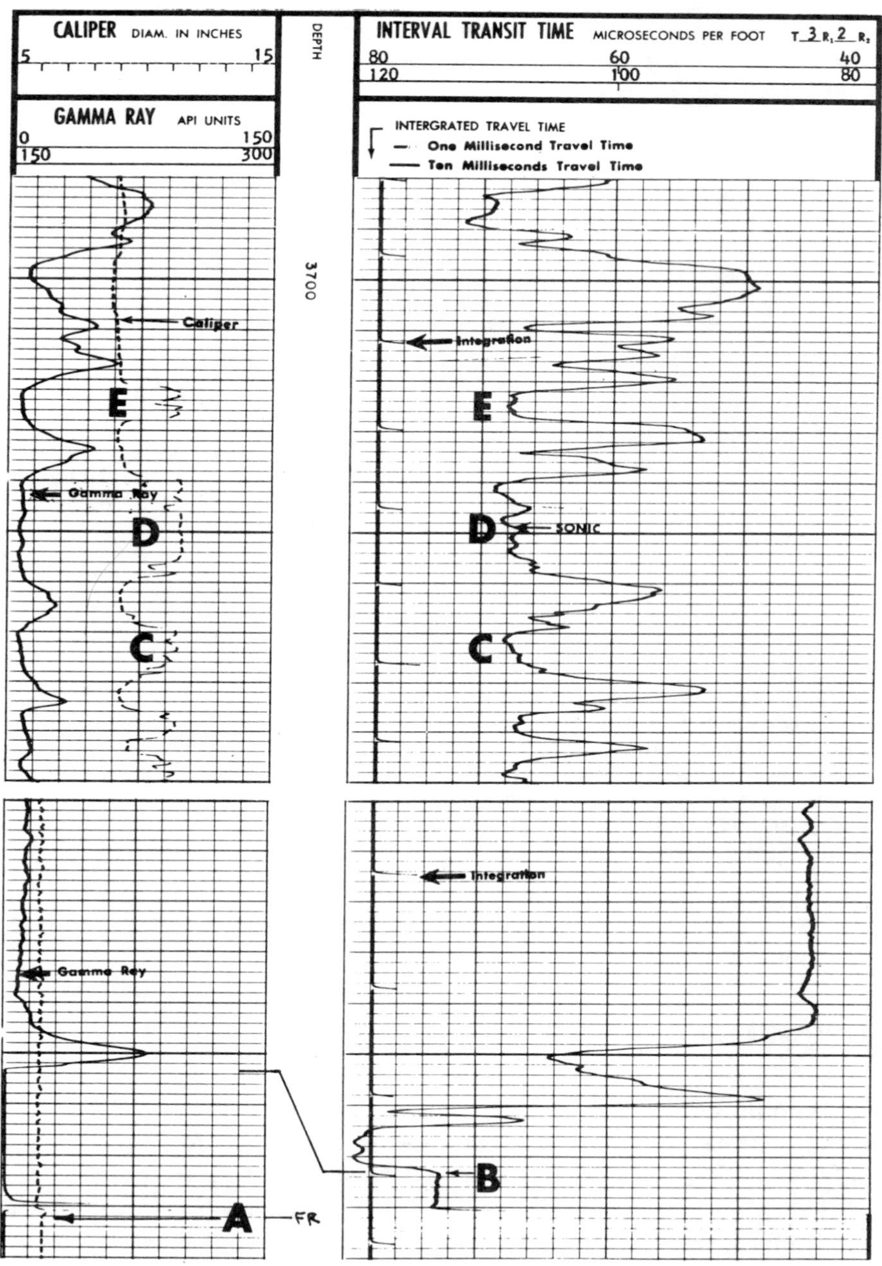

FIG. 10-4. Schlumberger's acoustic log presentation with the gamma ray and Δt memorized to the caliper measurement. This is indicated by the difference in caliper (A) and Δt (B) first measurements, plus their being in step through the salt interval (C-E). The TTI is recorded 7.5 ft too deep in this presentation. With Δt memorized, acoustic spikes are more rounded and may be less obvious.

which becomes,

$$\phi_{sv} = (\Delta t - A)\, B^{-1} \tag{10-2}$$

where A and B are empirically derived constants for a formation.

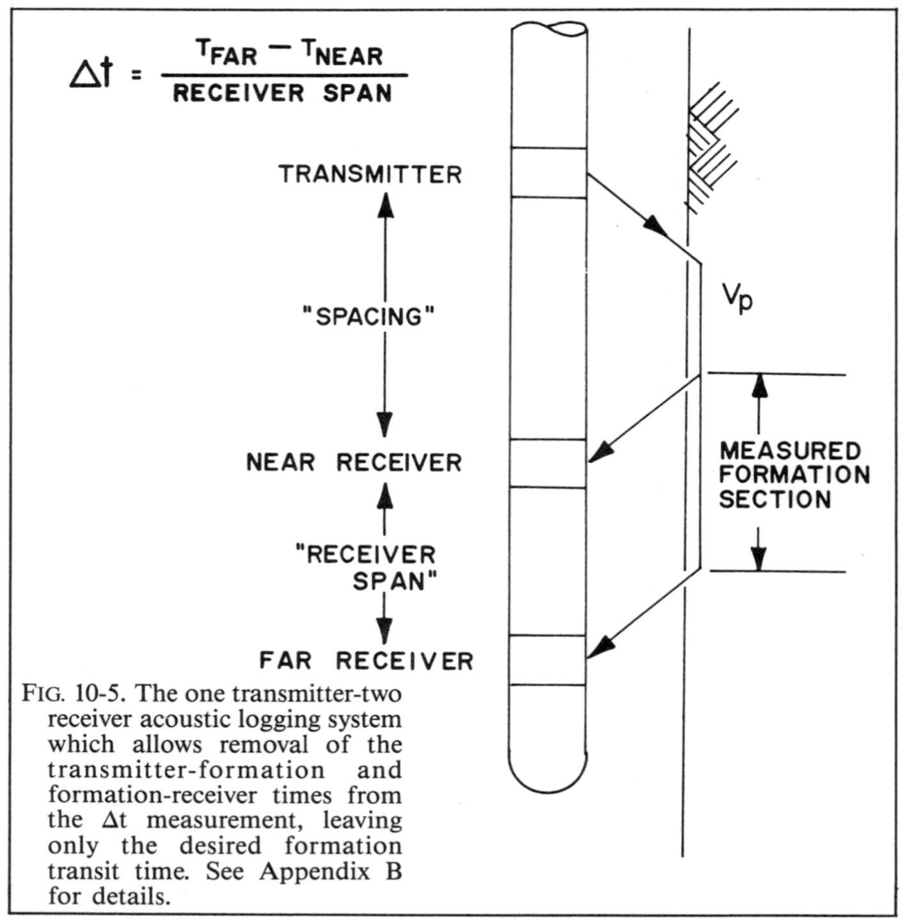

FIG. 10-5. The one transmitter-two receiver acoustic logging system which allows removal of the transmitter-formation and formation-receiver times from the Δt measurement, leaving only the desired formation transit time. See Appendix B for details.

BOREHOLE COMPENSATION

To minimize the effect of tool formation-tool traveltime on the desired formation time measurement, a two-receiver system was developed (Figure 10-5). The time for the first arrival *P*-wavefront to pass between the receivers is measured. This system eliminates the traveltime through the mud from the measurement and details are given in Appendix B. However, it does introduce a problem at the interface of enlarging or shrinking borehole diameter (Figure 10-6).

To eliminate the changing diameter spike effect a time measurement is taken downgoing and upgoing through the same interval and the measurements are averaged. This is normally accomplished with a two-transmitter, two-receiver system (Figure 10-7), but can also be done with the single-transmitter, two-receiver system (see Appendix B). This compensation is not quite absolute however. Schlumberger goes one step further with their two-transmitter, four-receiver overlapping configuration (Figure 10-8).

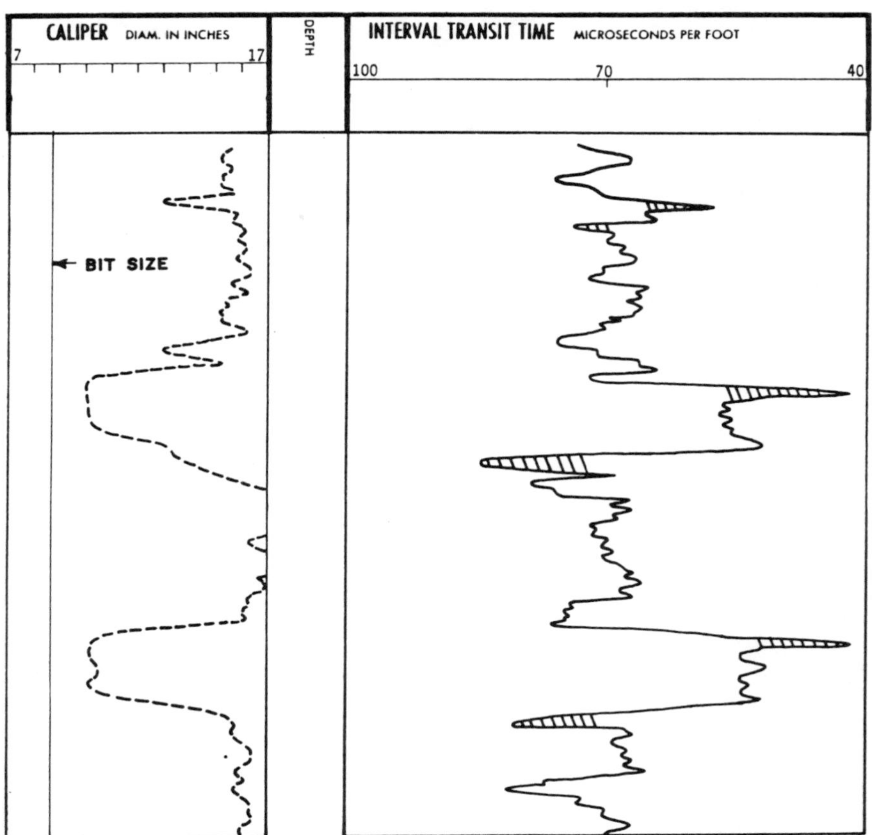

FIG. 10-6. Δt spikes at the boundaries of borehole diameter changes when using the one transmitter-two receiver system. See Appendix B for details. (Adapted from Doh and Alger and Tixier et al., 1959, © 1959 SPE AIME.)

To determine which system is being used a careful check of the log heading for sonde model information and the heading scales area for spacing and receiver span information is necessary (Figure 10-9). Appendix C covers the more common acoustic tools.

TIMING THE FIRST *P*-WAVE

The receiver wave trains are transmitted to the surface panel to be timed using a crystal oscillator time base and an amplitude threshold detection system (Figure 10-10). The measured times are a function of both the arriving signal amplitude and the timing trigger level, the later controlled by the logging engineer. The timing trigger level is completely variable, if it is set at A in Figure 10-11, the measured arrival times are t_a and t'_a, and if the trigger level is raised to B the measured times become t_b and t'_b. Note that because the same arrival cycles of both receivers is timed, the difference time, which partly determines Δt, is the same for the

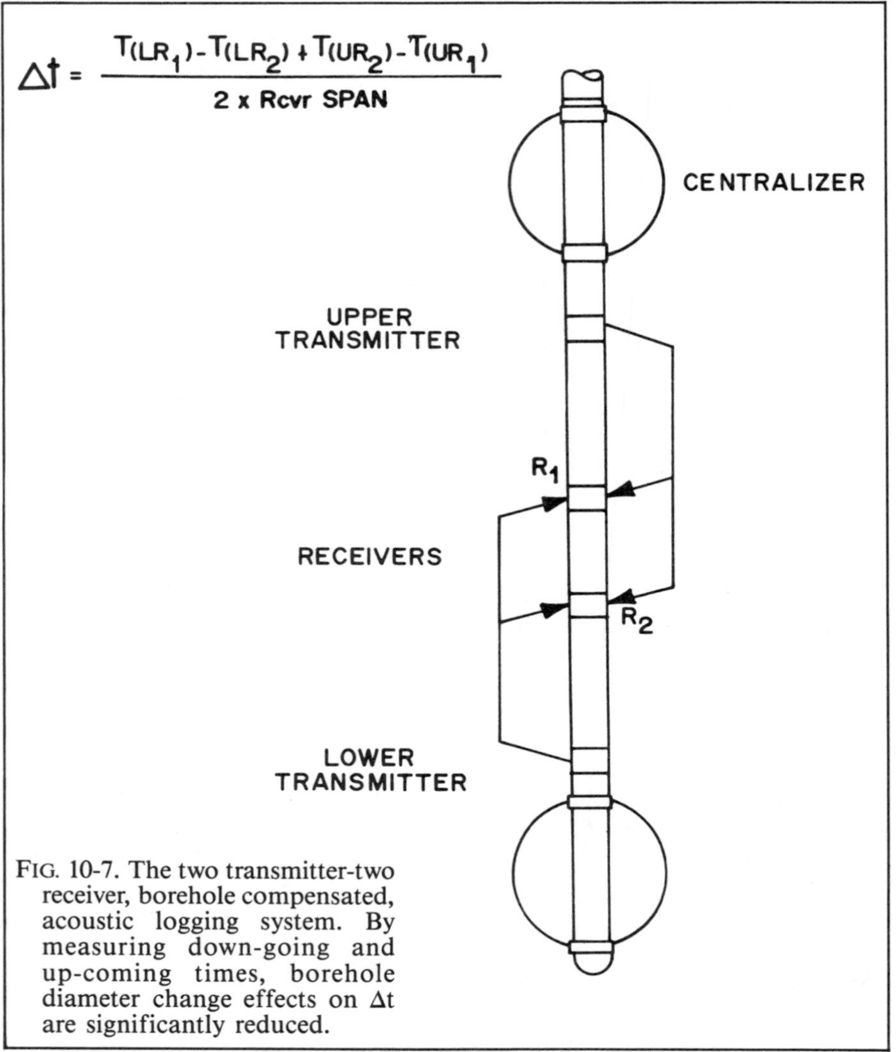

FIG. 10-7. The two transmitter-two receiver, borehole compensated, acoustic logging system. By measuring down-going and up-coming times, borehole diameter change effects on Δt are significantly reduced.

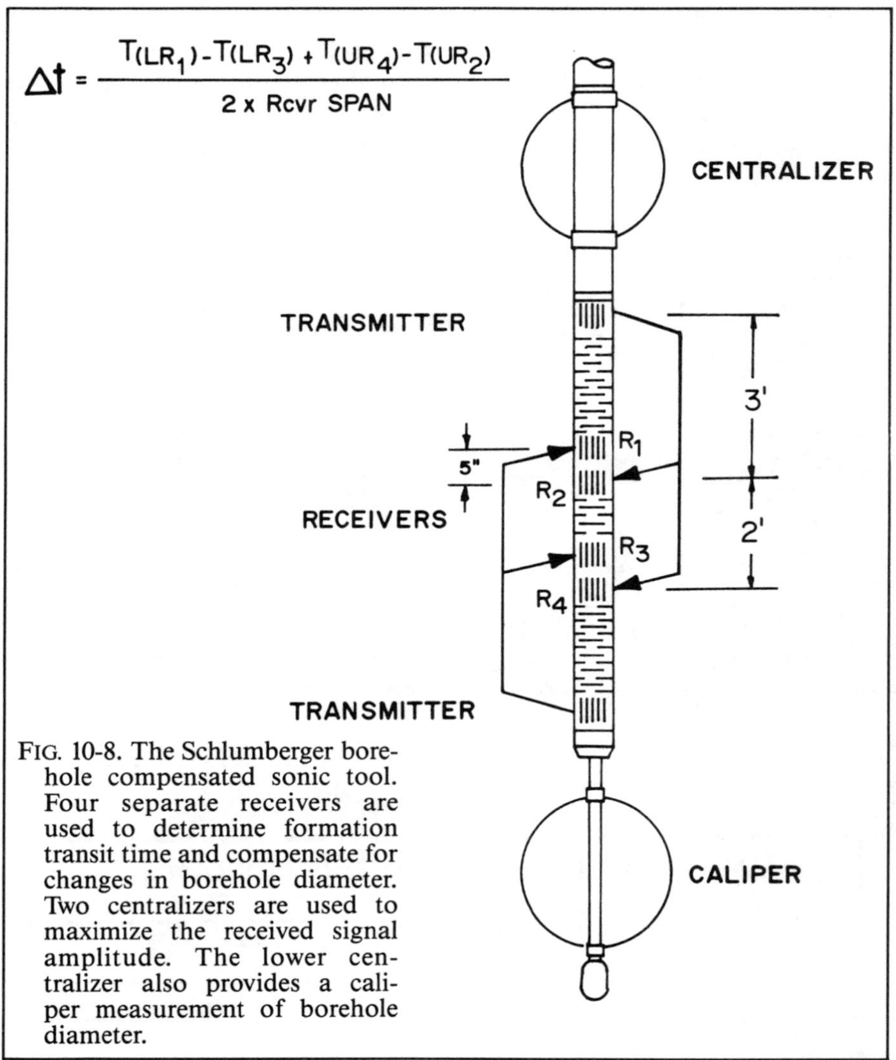

FIG. 10-8. The Schlumberger borehole compensated sonic tool. Four separate receivers are used to determine formation transit time and compensate for changes in borehole diameter. Two centralizers are used to maximize the received signal amplitude. The lower centralizer also provides a caliper measurement of borehole diameter.

different trigger levels. If a level between A and B is chosen, then the far receiver will be timing a later arrival and the partical Δt is in error. An accurate Δt measurement can be influenced by noise spikes before the formation signal (Figure 10-12) or the difference in amplitude of the signal at each receiver. As the signal amplitude decreases there are two potential problems, Δt stretch and cycle skip. The measured time of a signal near the detection level differs from that of a signal with an amplitude well above the detection level (Figure 10-13). This difference, easily up to 10 μs/ft (Willis and Toksoz, 1983), is very difficult to detect on the finished log. If the far receiver signal amplitude drops below the detection level (Figure 10-14), the time measurement is erroneous because it is of two different arrival cycles, and therefore is called cycle skip. Noise

(Text continued on page 210)

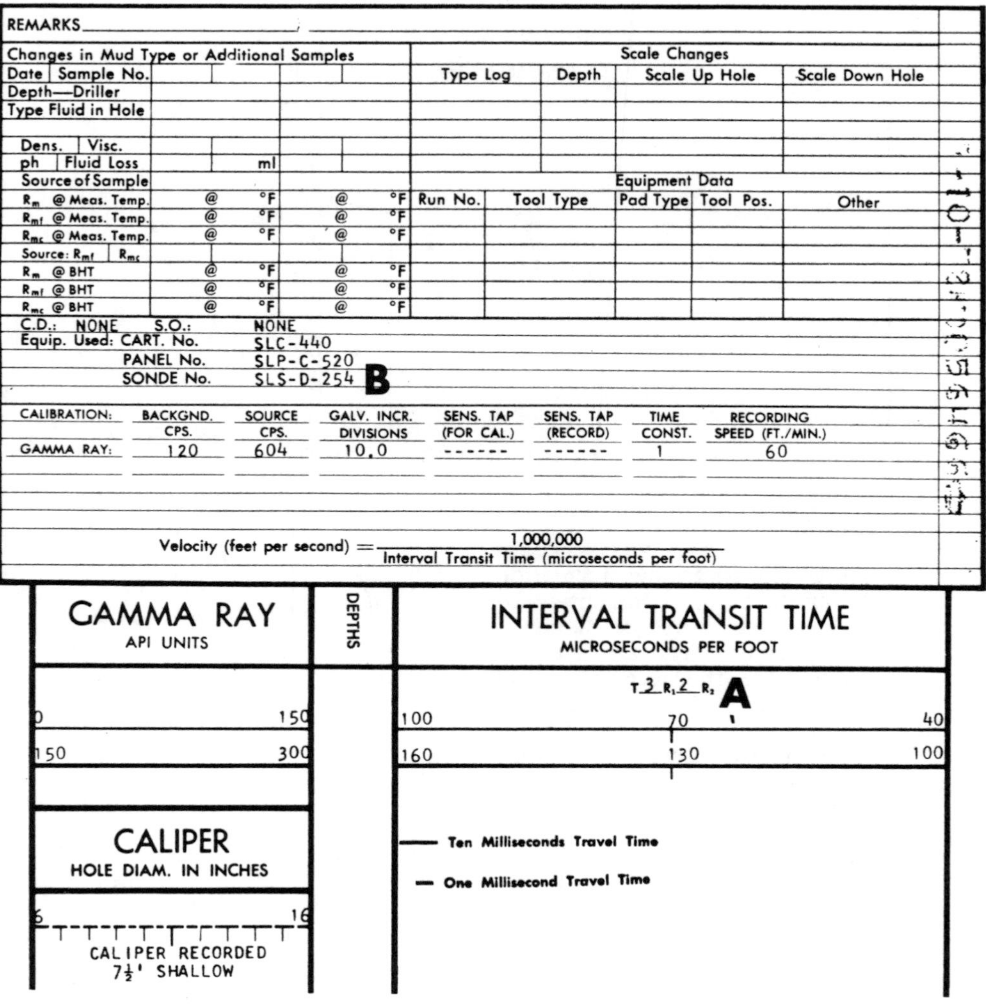

FIG. 10-9. Locating (A) acoustic spacing and receiver span dimensions, and (B) sonde model information from the scale insert and heading, respectively. Also see Figure 3-11, items G and H.

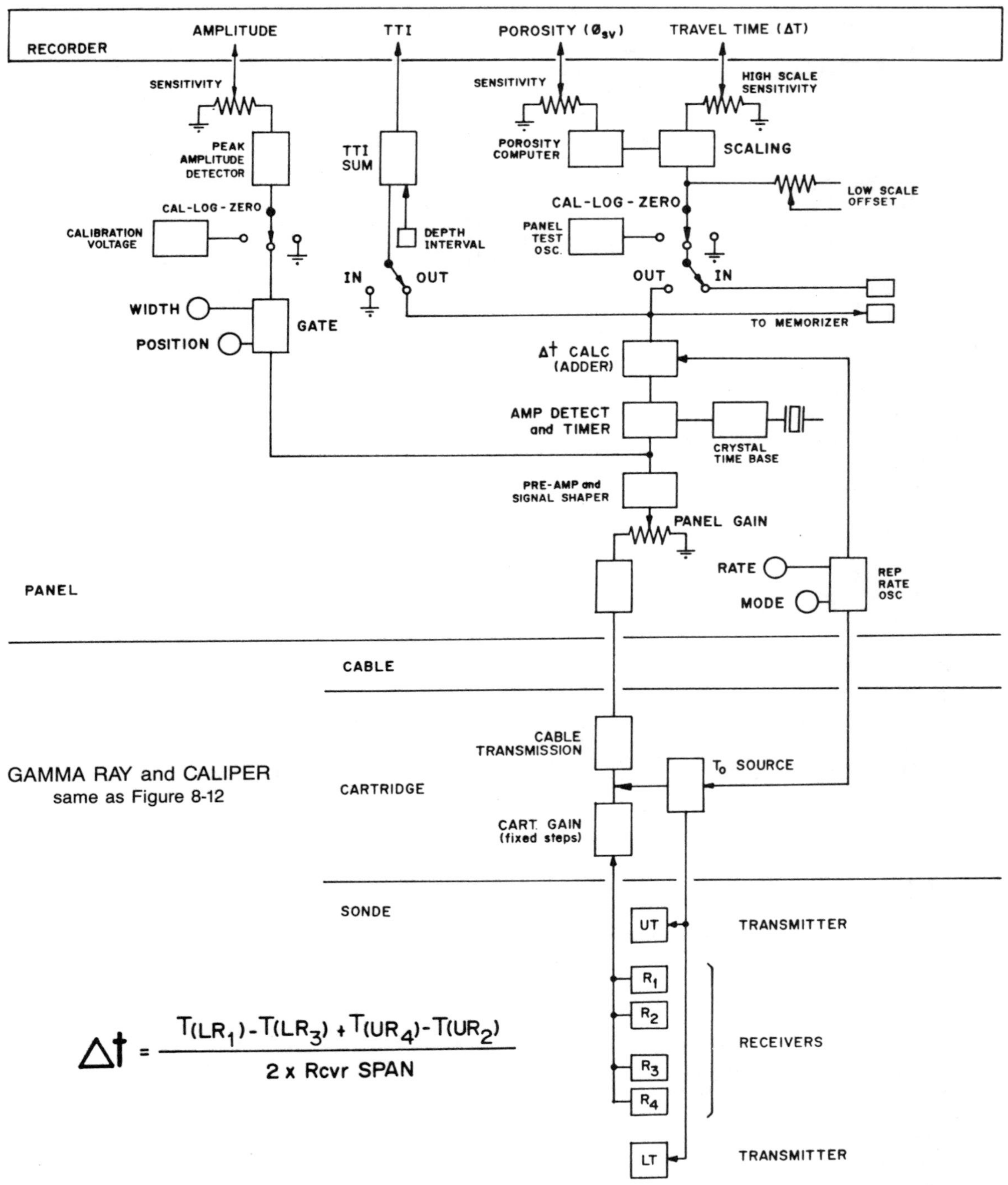

$$\Delta t = \frac{T(LR_1) - T(LR_3) + T(UR_4) - T(UR_2)}{2 \times \text{Rcvr SPAN}}$$

FIG. 10-10. The borehole compensated sonic tool block diagram. Each receiver signal is sent uphole and timed in the surface panel where Δt is computed.

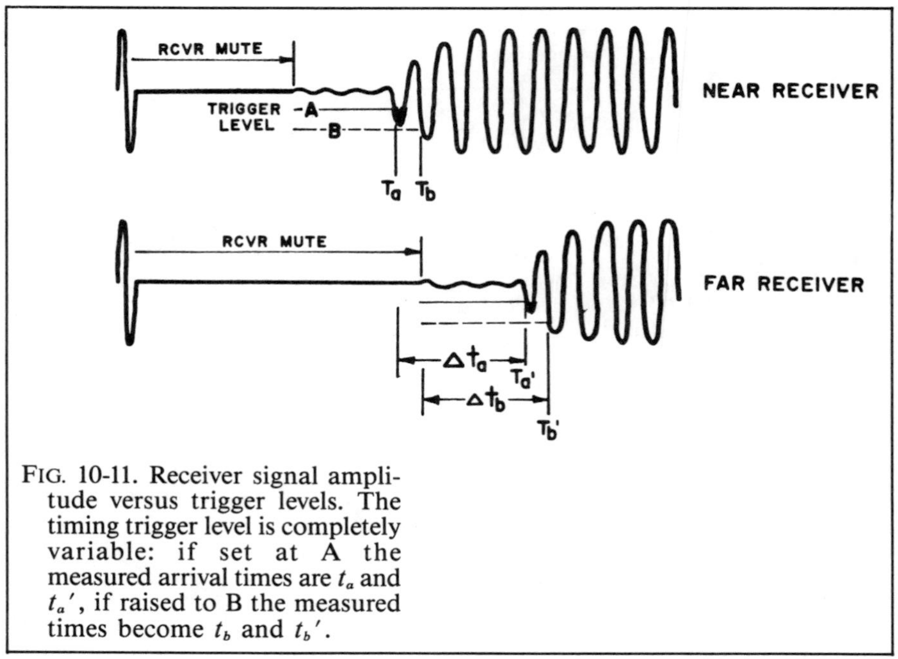

FIG. 10-11. Receiver signal amplitude versus trigger levels. The timing trigger level is completely variable: if set at A the measured arrival times are t_a and t_a', if raised to B the measured times become t_b and t_b'.

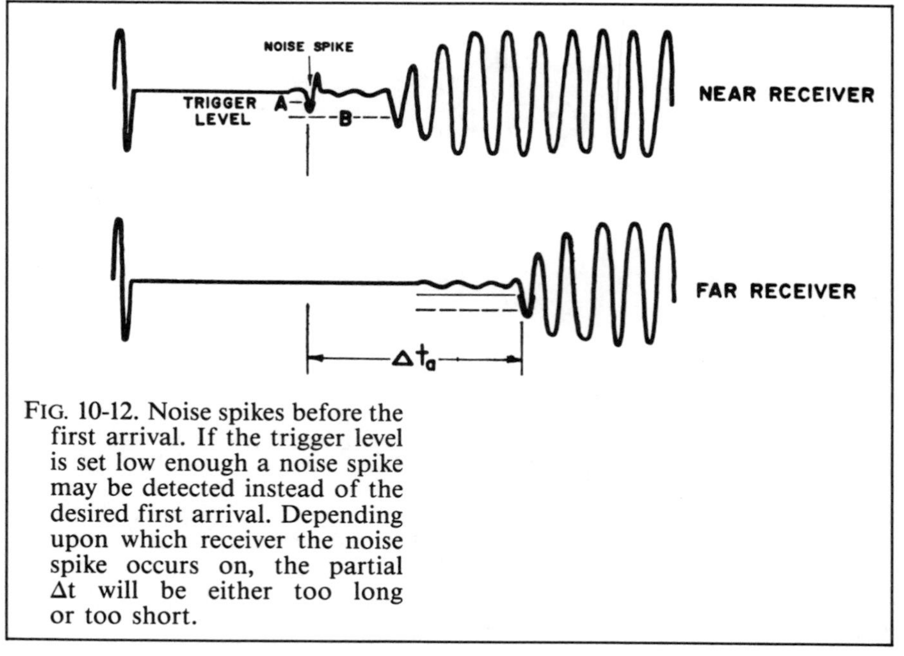

FIG. 10-12. Noise spikes before the first arrival. If the trigger level is set low enough a noise spike may be detected instead of the desired first arrival. Depending upon which receiver the noise spike occurs on, the partial Δt will be either too long or too short.

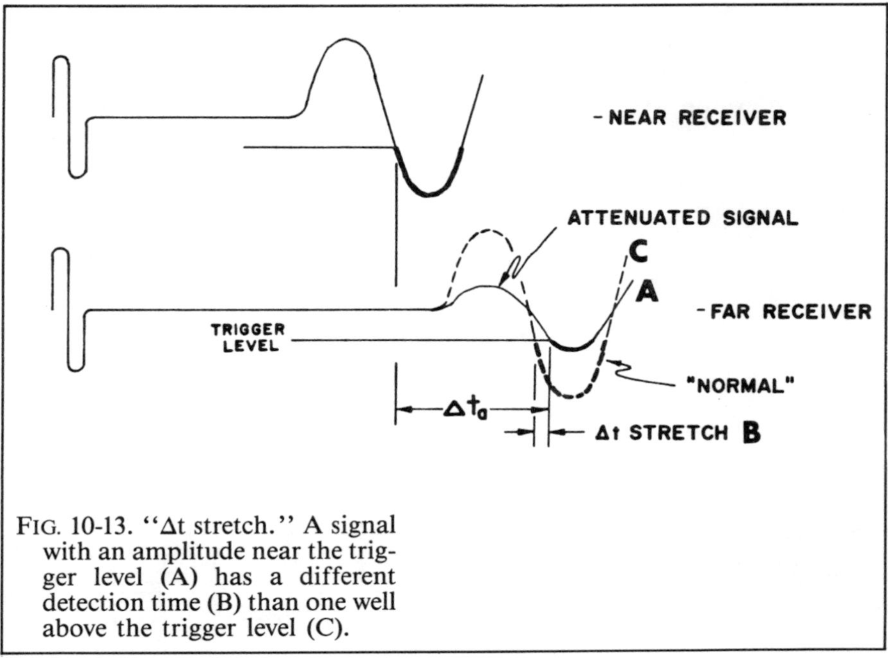

FIG. 10-13. "Δt stretch." A signal with an amplitude near the trigger level (A) has a different detection time (B) than one well above the trigger level (C).

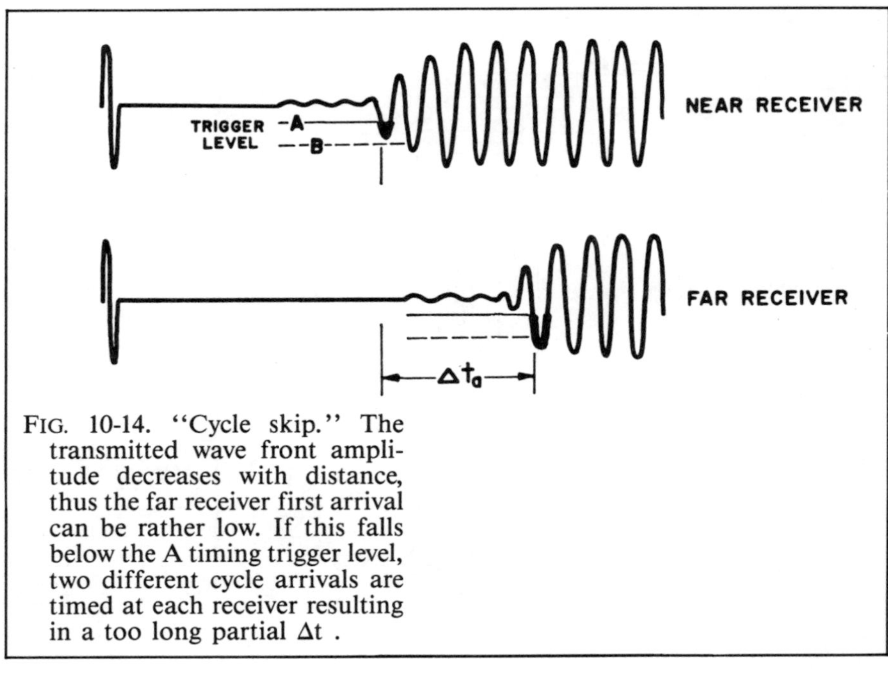

FIG. 10-14. "Cycle skip." The transmitted wave front amplitude decreases with distance, thus the far receiver first arrival can be rather low. If this falls below the A timing trigger level, two different cycle arrivals are timed at each receiver resulting in a too long partial Δt .

spikes and cycle skip problems are shown as a spike on the analog recording (Figure 10-15) and as a much broader spike on Schlumberger's digital recording (Figure 10-16), or as an abrupt parallel shift in readings if the conditions persist (Figure 10-17). The spikes of noise or cycle skip on analog recordings are very sharp unless the Δt measurement has been memorized.

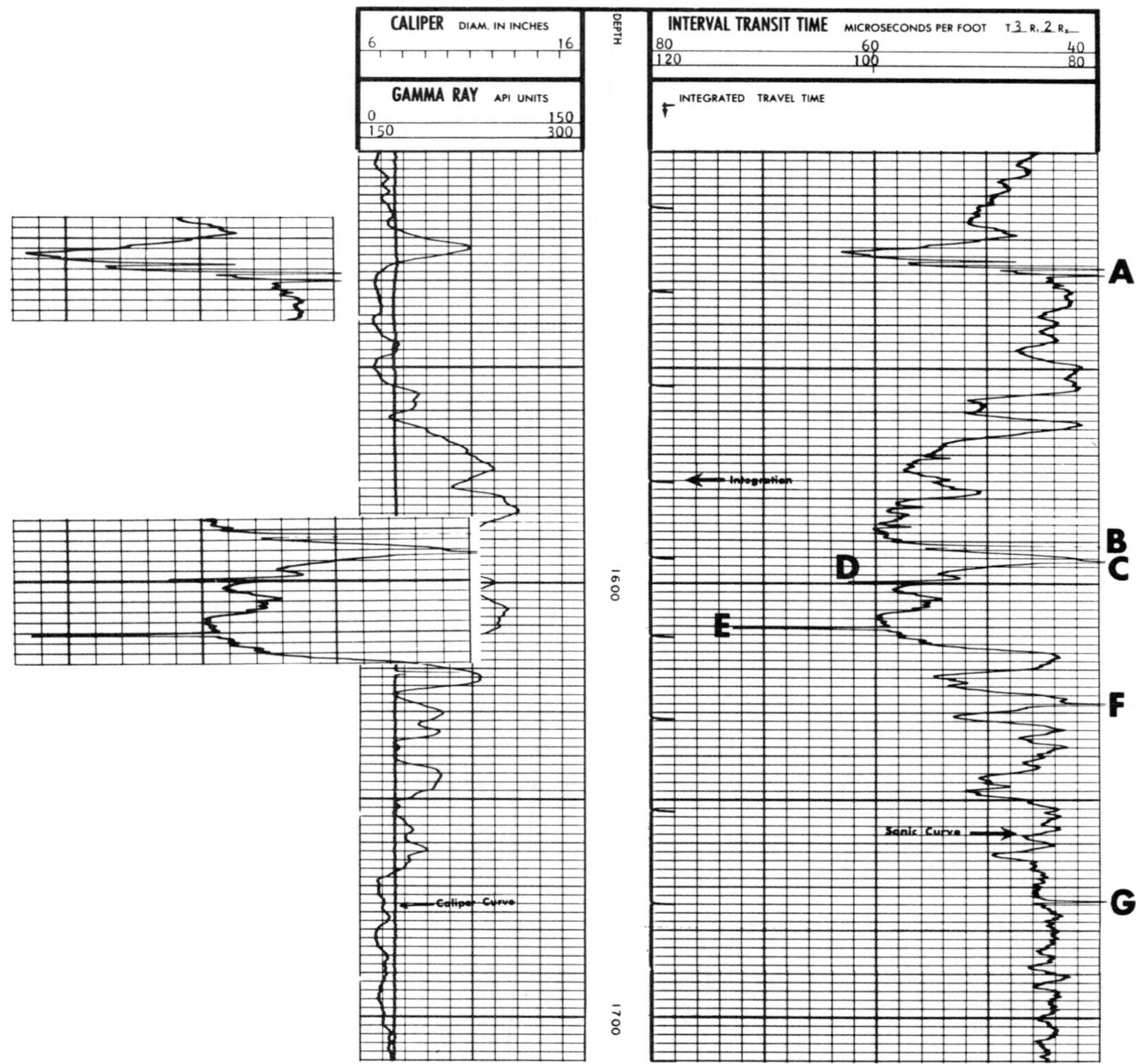

FIG. 10-15. Spikes on the acoustic log caused by cycle skips and noise (A through G). Note the characteristics of these unmemorized analog recorded spikes are fast breaks and slightly slower decay on the shallow side. The traveltime integrator (TTI) integrates the times under the spikes also, contributing to total TTI error.

MEMORIZATION

Schlumberger has the caliper beneath the acoustic sonde (Figure 10-8) leading to two memorization and presentation modes. First the GR and Δt curves can be memorized to the caliper (Figure 10-4) and only the

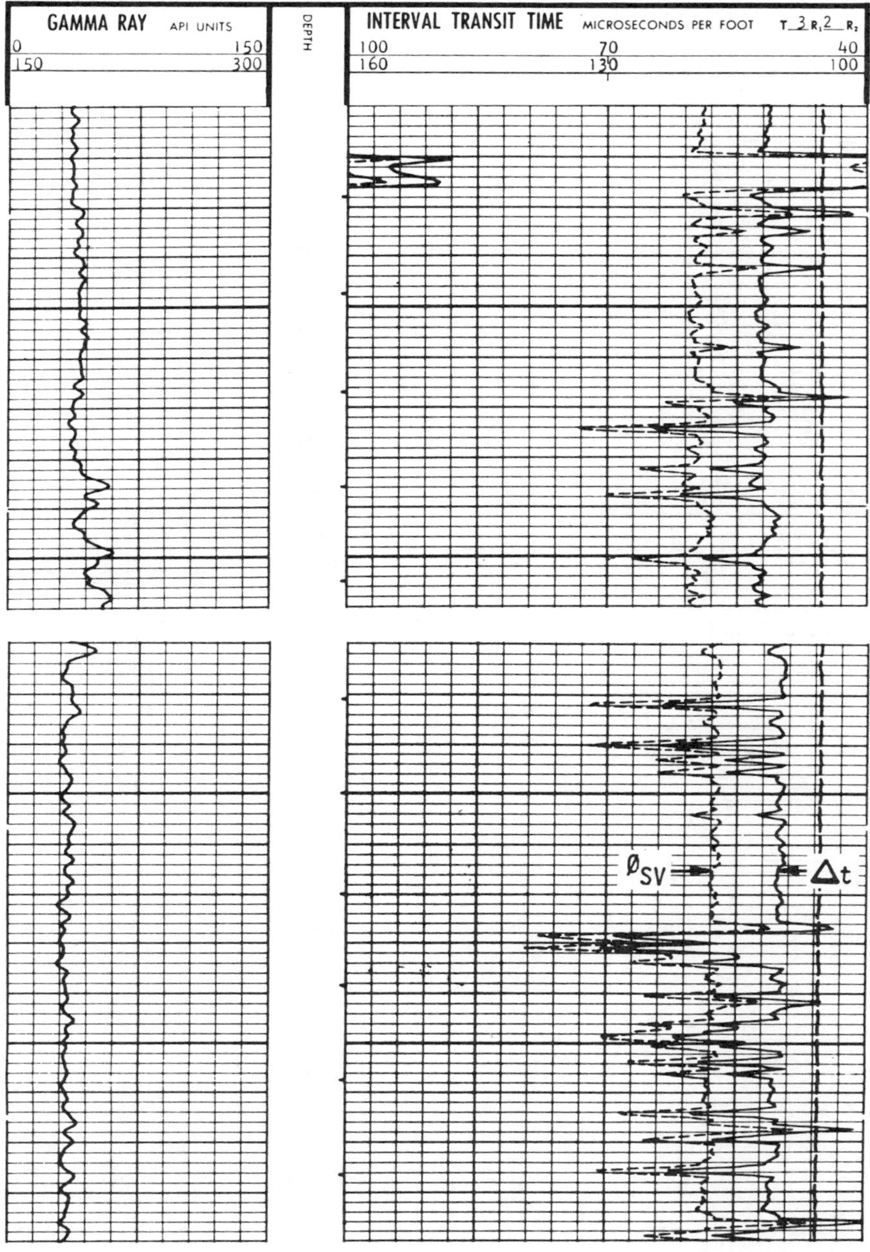

FIG. 10-16. Acoustic log spikes on the Schlumberger digital processed BHC Sonic log. Note that the spikes are broader and of smaller amplitude than those of Figure 10-15. These are more difficult to distinguish because they require comparison with other logs over the same interval.

TTI curve presented off-depth. Or, the Δt curve is left unmemorized and the GR memorized to the Δt measure point. This presents the caliper off-depth (Figure 10-18). The other service companies have their calipers above the acoustic section (see Appendix C) which eliminates this minor problem.

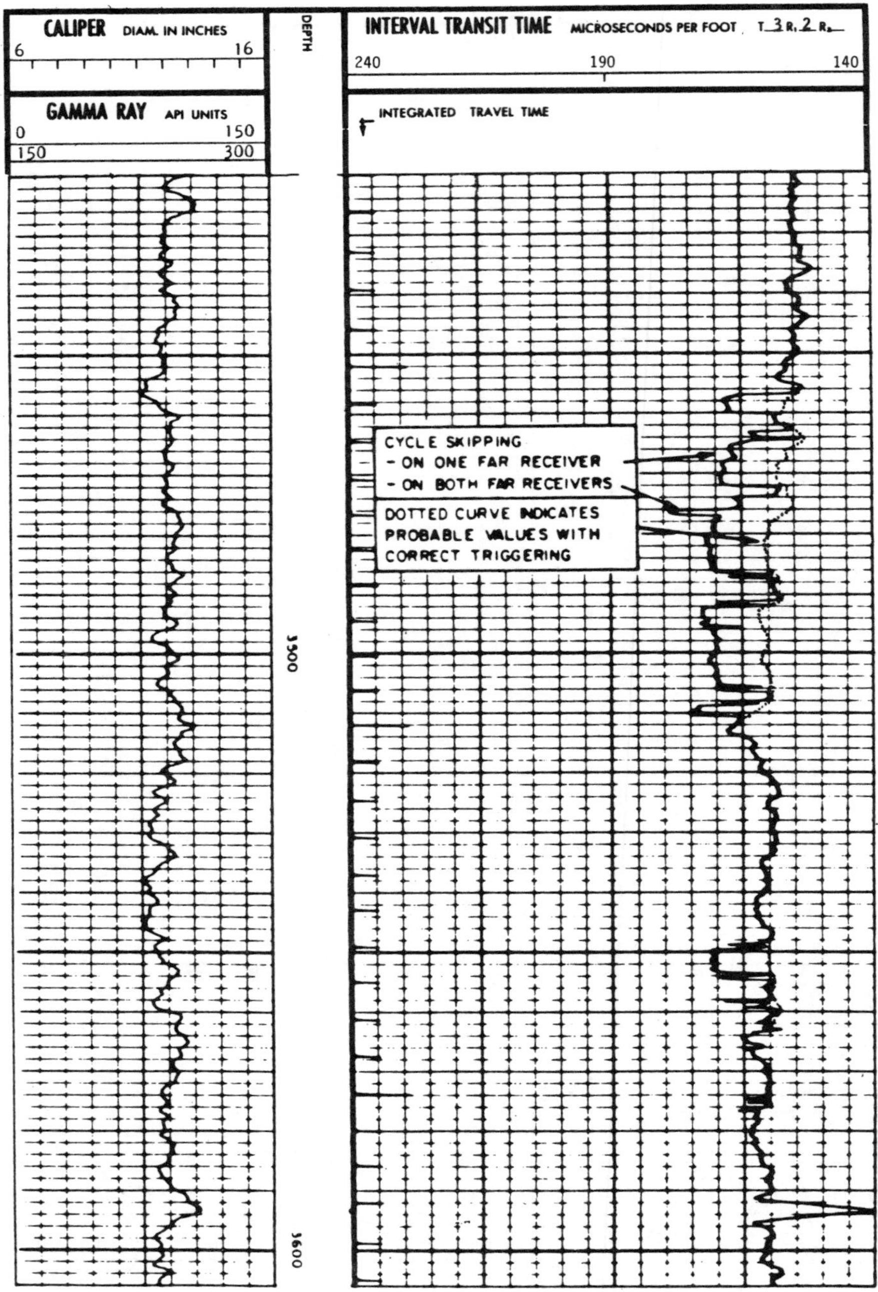

FIG. 10-17. If the conditions of Figures 10-12 or 10-14 persist, the Δt curve will exhibit a shifted, incorrect value. This must be recognized and removed during editing (from Thomas, 1978).

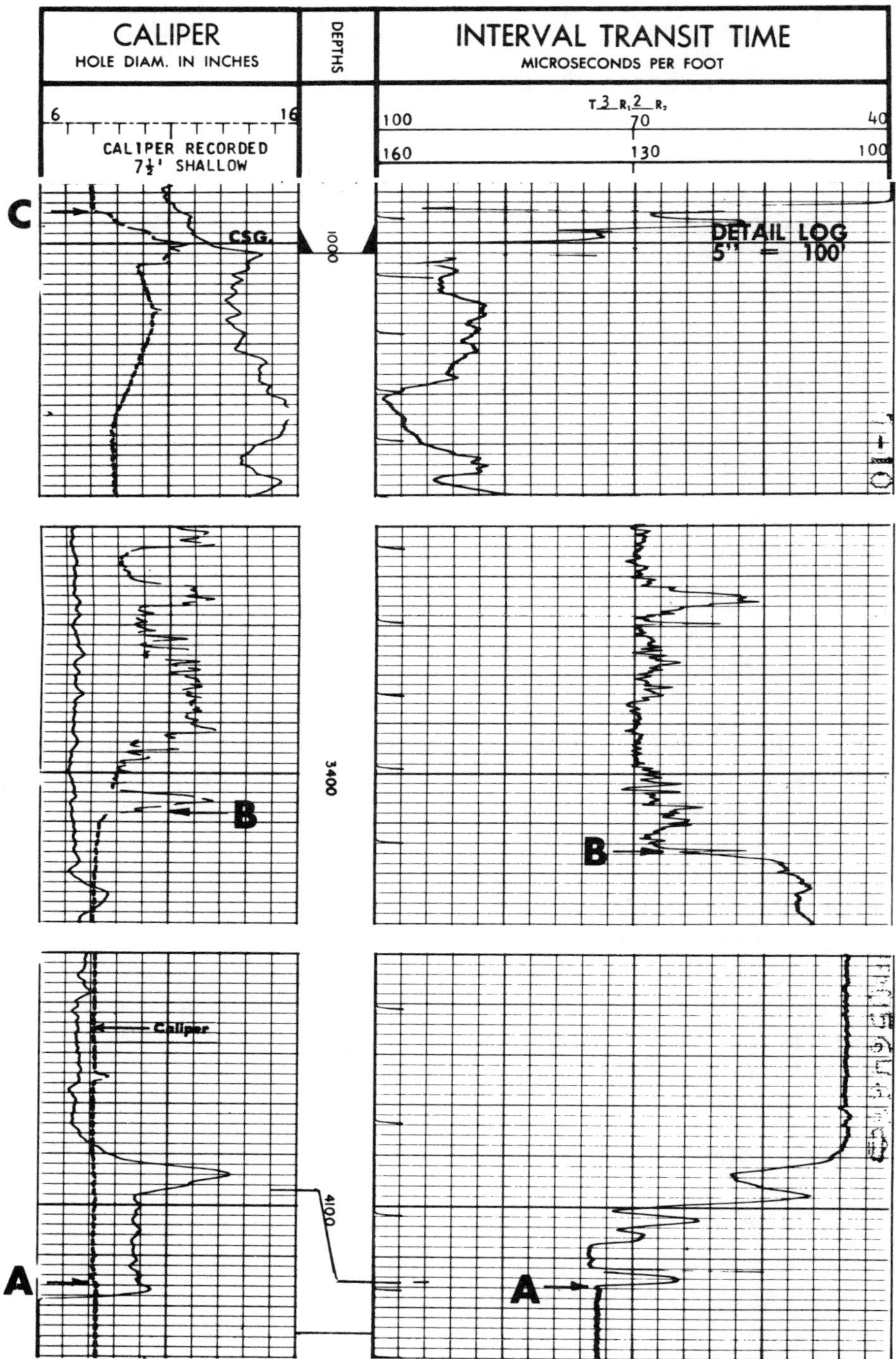

FIG. 10-18. Alternative Schlumberger acoustic log presentation: the gamma ray is memorized to the Δt measurement and the caliper presented 7.5 ft too shallow. This is indicated by (A) the first movement of the caliper and Δt at the same depth, (B) the difference in depth of inter-salt beds, and (C) upon entering casing. The TTI curve is on depth with the Δt curve in this presentation.

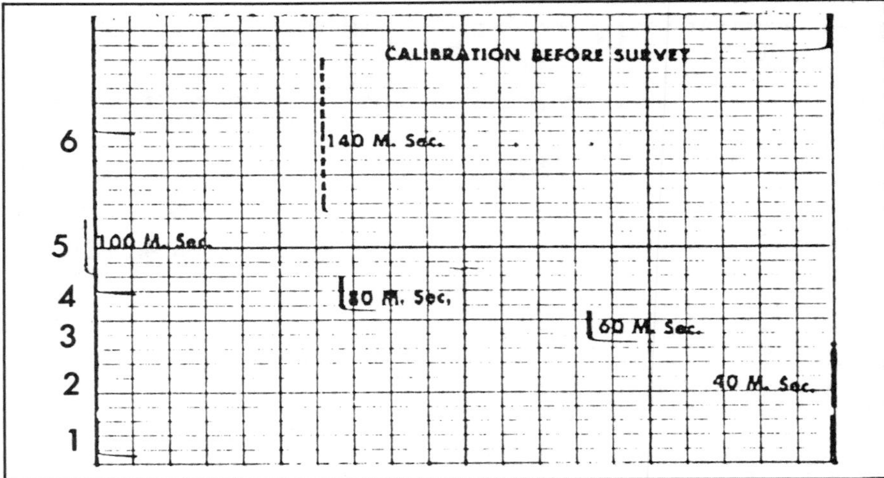

FIG. 10-19. Schlumberger sonic log calibrations are entirely panel generated. The downhole tool does not need to be connected. The test consists of a mechanical zero (1), and five test oscillator inputs to the Δt measuring circuitry (2-6).

CALIBRATION

In practice there is no calibration to the acoustic log. Assuming the time base for measuring Δt is accurate, the times measured should be absolute. The analog acoustic calibrations shown (Figures 10-19 and 10-20) are panel checks, and they can be made without the acoustic downhole tool attached to the cable. The computer logging system recognizes this limitation and has eliminated the Δt calibration presentation entirely.

Thus the calibration sequence does not check the ability of the receivers to detect the formation signals. Birdwell and Welex (Figure 10-20, step 6) do an environmental check to see if the downhole sonde is working under very high signal-to-noise conditions, by placing a casing sleeve about the sonde and checking for a 57 μs/ft reading.

MEASUREMENT QUALITY CHECKS

Because there are no real acoustic log calibrations the log itself must be checked carefully. In carbonate environments the tool should read 50 μs/ft in anhydrites and 67 μs/ft in salts or halites.

One check of tool operation is the signal recorded in casing after the open-hole logged section (Figures 10-21, 22, and 23), if there is little cement behind the casing. Inability of the tool to read the casing signal at 57 μs/ft, especially if the tool locks consistently on a later arrival, leaves the log quality in question. The casing check in Figure 10-22 shows that the sonde had some difficulty staying locked onto the proper first arrival (A). This log should be carefully checked for similar problems in the open-

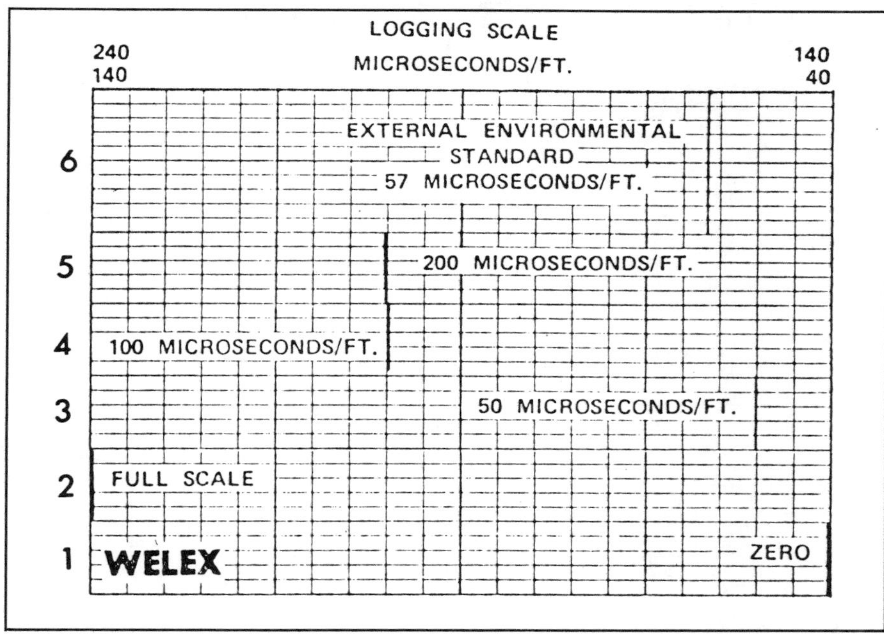

FIG. 10-20. Welex acoustic log calibrations: (1) recorder mechanical zero, (2) setting the recorder sensitivity, (3-5) panel checks at three different Δt rates, and a sonde environmental check. This last is inside a water filled tube that simulates the downhole casing at 57 μs/ft (6).

hole portion. A wandering value about the 57 μs/ft point indicates poor receiver sensitivity, a weak transmitter signal, or improper centralization (Figure 10-24).

For proper quality check of the acoustic log the tool should be run into intermediate casing 200 ft with the Δt and GR recording, to obtain both a Δt signal stability indication and a GR overlap with the previous logging runs.

LIGHT HYDROCARBON AND BOREHOLE EFFECTS

The presence of light hydrocarbons or gas in the drilling mud attenuates the acoustic signals, resulting in cycle skip. Often the signal is attenuated to the point where a Δt measurement cannot be made (Figure 10-25). If the gas becomes dispersed throughout the mud, it becomes very difficult to determine the source interval of the gas.

Like the gamma-gamma density tool, a prime borehole effect is due to loss of contact with the formation (Figure 10-26). The formation path becomes so long around washouts that the through-the-mud signal is the first receiver arrival. If the log of Figure 10-26 were used unedited in a synthetic seismogram, it would produce reflections from those washouts that might or might not be geologically valid.

The next major borehole effect is formation alteration, primarily altered shales. Some shales, primarily those with more clay content (Blakeman,

(Text continued on page 222)

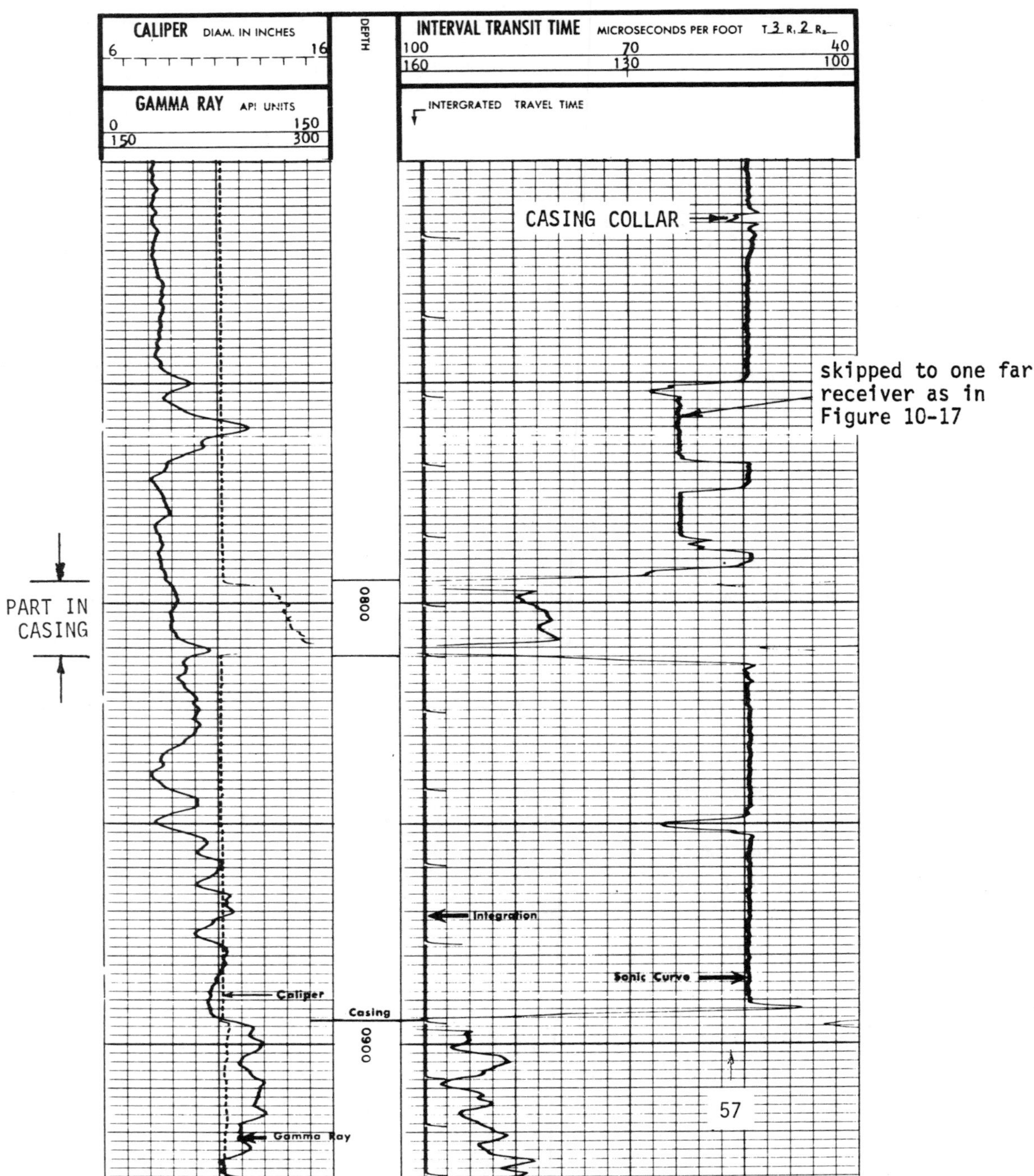

FIG. 10-21. The acoustic log measurement in casing is a good check of sonde transmitter-receiver "reliability." The reading should stabilize to 57 ± 1 μs/ft when above the cemented casing shoe. The acoustic log should be run at least 200 ft into the casing for this check, with the gamma ray still recording.

FIG. 10-22. This casing check example shows (A) the sonde having some difficulty staying locked into the proper first arrival. The log should be checked carefully for similar problems in the open hole portion. To avoid later confusion, the after survey calibrations (B) were run after the casing check and left attached to the top of the log.

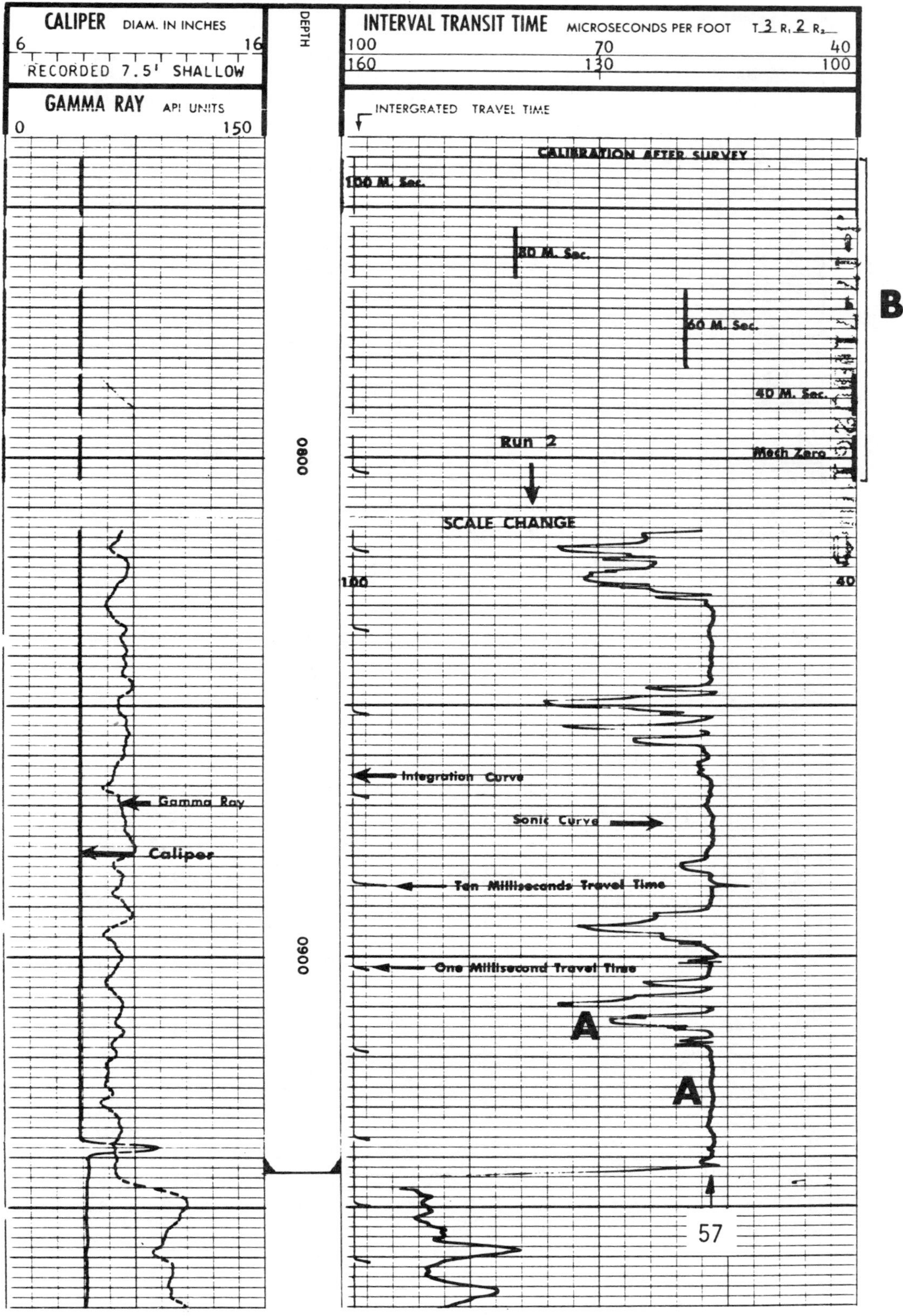
CALIPER DIAM. IN INCHES
6
16
RECORDED 7.5' SHALLOW
GAMMA RAY API UNITS
0
150
DEPTH
INTERVAL TRANSIT TIME MICROSECONDS PER FOOT T 3 R1 2 R2
100
70
40
160
130
100
INTERGRATED TRAVEL TIME
CALIBRATION AFTER SURVEY
100 M. Sec.
80 M. Sec.
60 M. Sec.
40 M. Sec.
Run 2
Mech Zero
0800
SCALE CHANGE
100
40
B
Integration Curve
Gamma Ray
Sonic Curve
Caliper
Ten Milliseconds Travel Time
0900
One Millisecond Travel Time
A
A
57

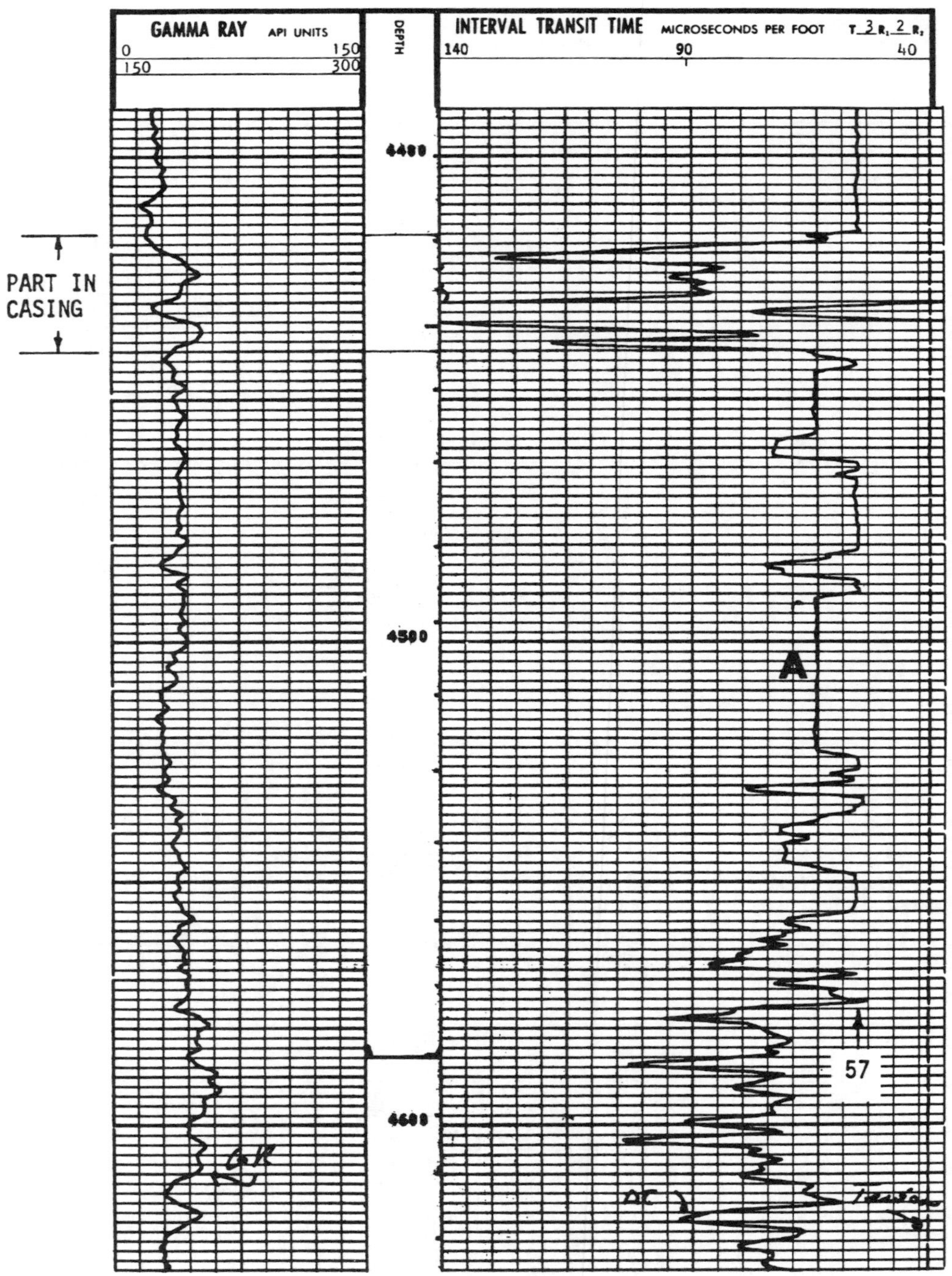

FIG. 10-23. Schlumberger digital processed BHC sonic log entering casing. Note the consistent Δt shift (A), indicating inability to lock onto the correct arrival.

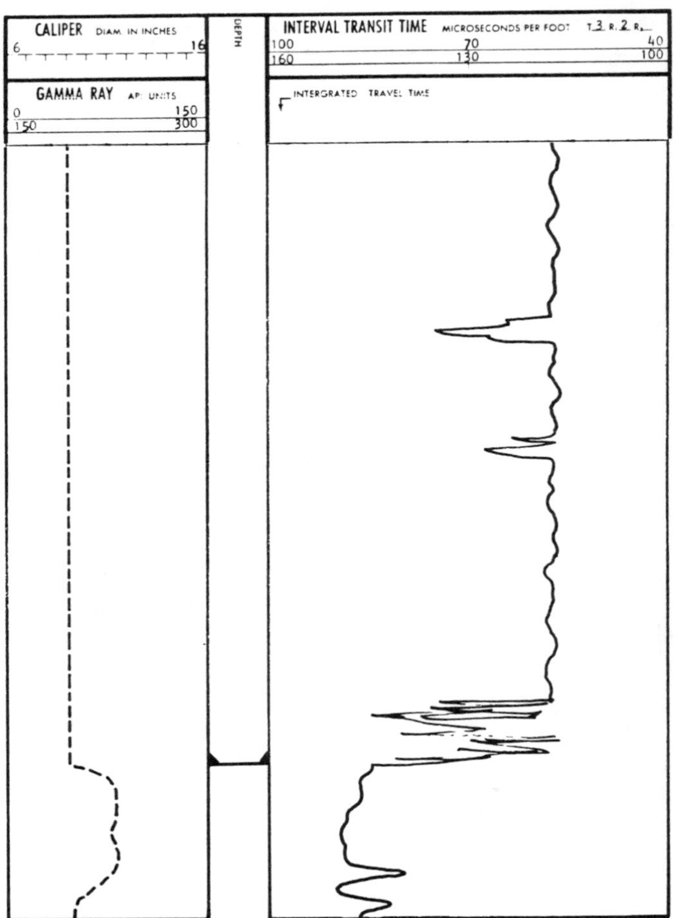

FIG. 10-24. Wandering Δt in casing indicating low-signal amplitudes due to either poor transmitter or receiver performance or both or inadequate centralization of the acoustic sonde.

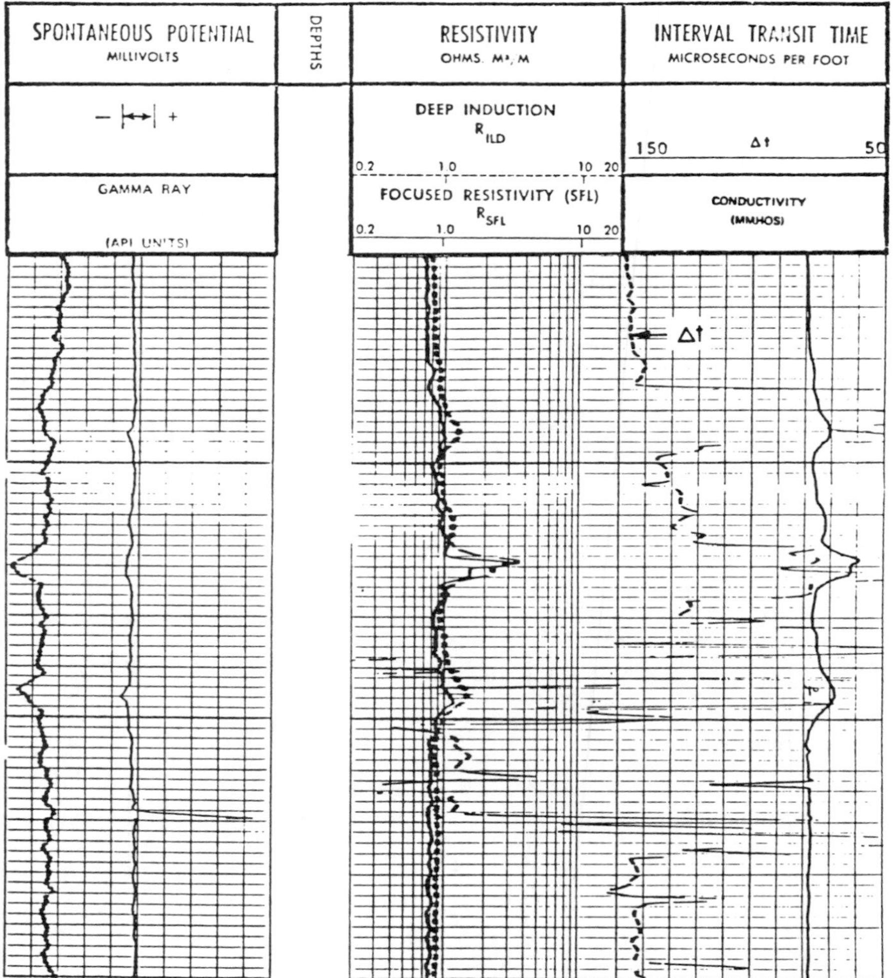

FIG. 10-25. Cycle skipping caused by gas from the formation getting into the drilling mud and attenuating the acoustic signal.

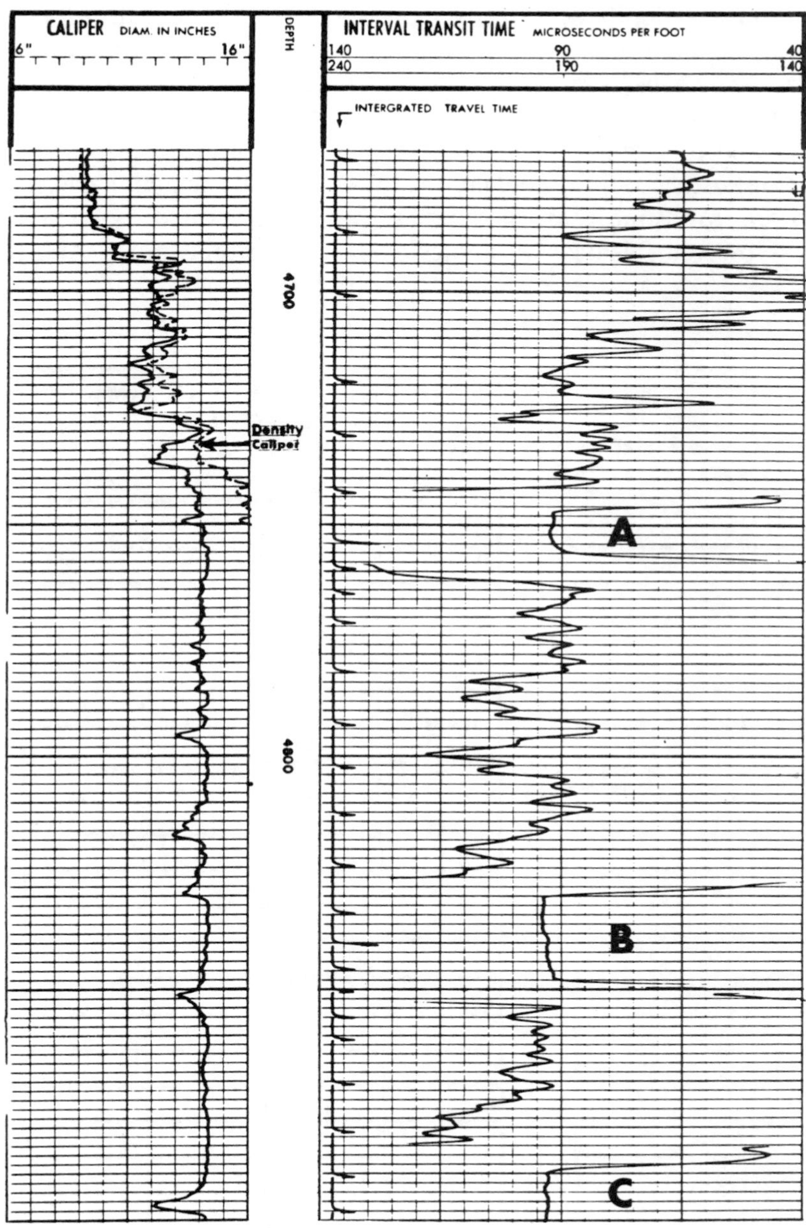

FIG. 10-26. Acoustic log sonde standoff in large diameter boreholes. The tool is no longer measuring formation Δt in intervals A-C, but the mud transit time, about 190 μs/ft. Intervals A-C would generate strong synthetic seismogram reflections which may or may not be valid.

1982), change their properties around the borehole when exposed to mud drilling fluids (Figure 10-27). The amount of Δt change is a function of the shale, mud-fluid properties, and time. The conventional acoustic tool has a shallow depth of investigation and therefore usually does not penetrate to the undisturbed formation under such conditions (Figure 10-28). To achieve a greater penetration and measure the unaltered, undisturbed zone, a longer spaced acoustic tool is needed. This procedure is discussed in more detail in the Long Spaced Sonic Tool section.

The opposite condition, where the invaded zone has a higher velocity, a shorter traveltime than the undisturbed formation does sometimes occur. However, this near-borehole velocity inversion is much rarer than

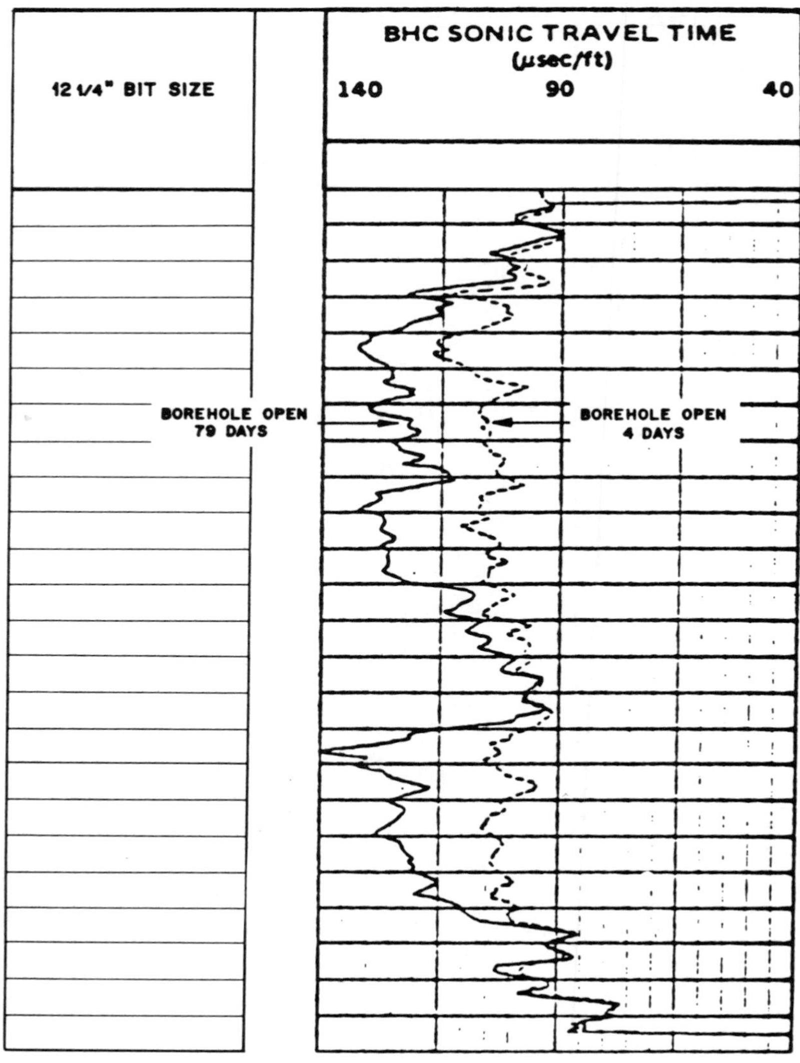

FIG. 10-27. Change of convertional spaced Δt measurement due to formation alteration around the borehole because of time exposure to drilling fluids. (Misk et al., 1977.)

formation alteration. It is usually restricted to shallow formations and possibly very overbalanced mud conditions.

The caliper on the Schlumberger tool, a three-arm bowspring system, easily fills with drilling debris and becomes sluggish which prevents it from opening fully to measure the true borehole diameter. Therefore, its

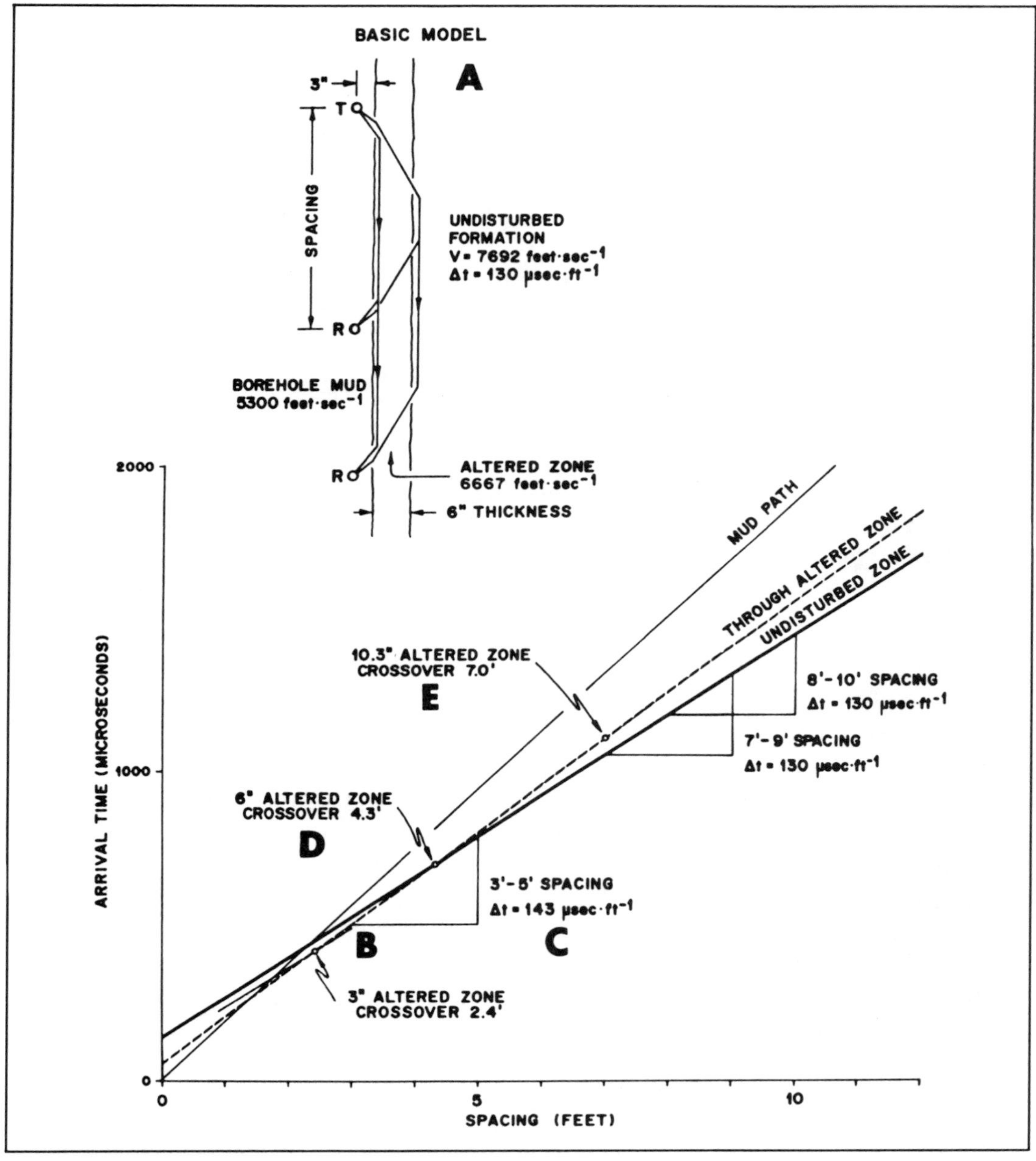

FIG. 10-28. Incorrect Δt due to near-borehole formation alteration: analysis of (A) a two-layer model shows that the 3 ft first arrival is from (B) the altered zone. Thus (C) the computed Δt is too high. To measure the undisturbed formation (D) a near receiver spacing of at best 4.3 ft is needed in this example. For altered zones greater than 10.3 inches (E) the longer spaced tool measurements are effected. [Created from Goetz et al. (1979) equations (6) and (7)].

measurement quality is often questionable. Note in Figure 10-26 that the acoustic caliper (solid curve) reaches a maximum diameter, while the traced density caliper (dashed curve) indicates considerably larger borehole diameter.

LOGGING THROUGH CASING

The acoustic log detects the formation signals through casing under conditions where the formation Δt is less than 57 μs/ft, primarily in fast carbonates (Figure 10-29), if both the casing is adequately bonded to the cement and the cement is adequately bonded to the formation. There is also evidence (Fons, 1968, and Fertl and Wichmann, 1977) that with proper cementing the formation can be logged through casing for longer traveltimes or slower velocities.

Again, to assist in run-to-run depth control, the Δt curve should be left on for 200 ft, especially in carbonate environments. This overlap is important in final editing of multiple run acoustic logs as the measurements for shallower runs do not reach the bottom of the borehole. The acoustic log has one of the longest distances between first reading and total depth of the borehole, leaving a large unrecorded gap at the bottom of each logging run (Figure 4-13).

If the open-hole mud system is very gaseous, a good acoustic log is difficult to obtain. Logging through casing, after the casing has been completely cemented and the mud degassed, makes a good acoustic log possible.

TRAVELTIME INTEGRATOR

The integrated traveltime (TTI) curve was added to the acoustic log to aid in interpretation of seismic data. This was done by adding successive interval traveltime measurements until the sum reaches 1 000 μs, and then a 1 ms pip is indicated on the TTI trace (Figure 10-4). Every tenth pip is made larger for counting convenience. The result is a *one-way* vertical traveltime measurement scaled in milliseconds.

Because integrated times are used to help identify seismic events, the TTI is considered proprietary information. Thus, logs from commercial log services commonly have the TTI taped or whitened out (Figure 10-30, item A). If this information has been eliminated, it can be obtained through digitizing the Δt measurement. However, examine the log carefully before assuming all this information has been lost. Often if the taping were careless, tips of the 10 ms pips are still visible (Figure 10-30, item B) or if the pips were whitened out individually (item C), they can be identified. By careful examination of both small scale and detail scale the gross curve can often be reconstructed. On older logs often a summary of TTI measurements is in the remarks area of the heading (Figure 10-30, item D). This indicates overall integration. If the exact footage is not

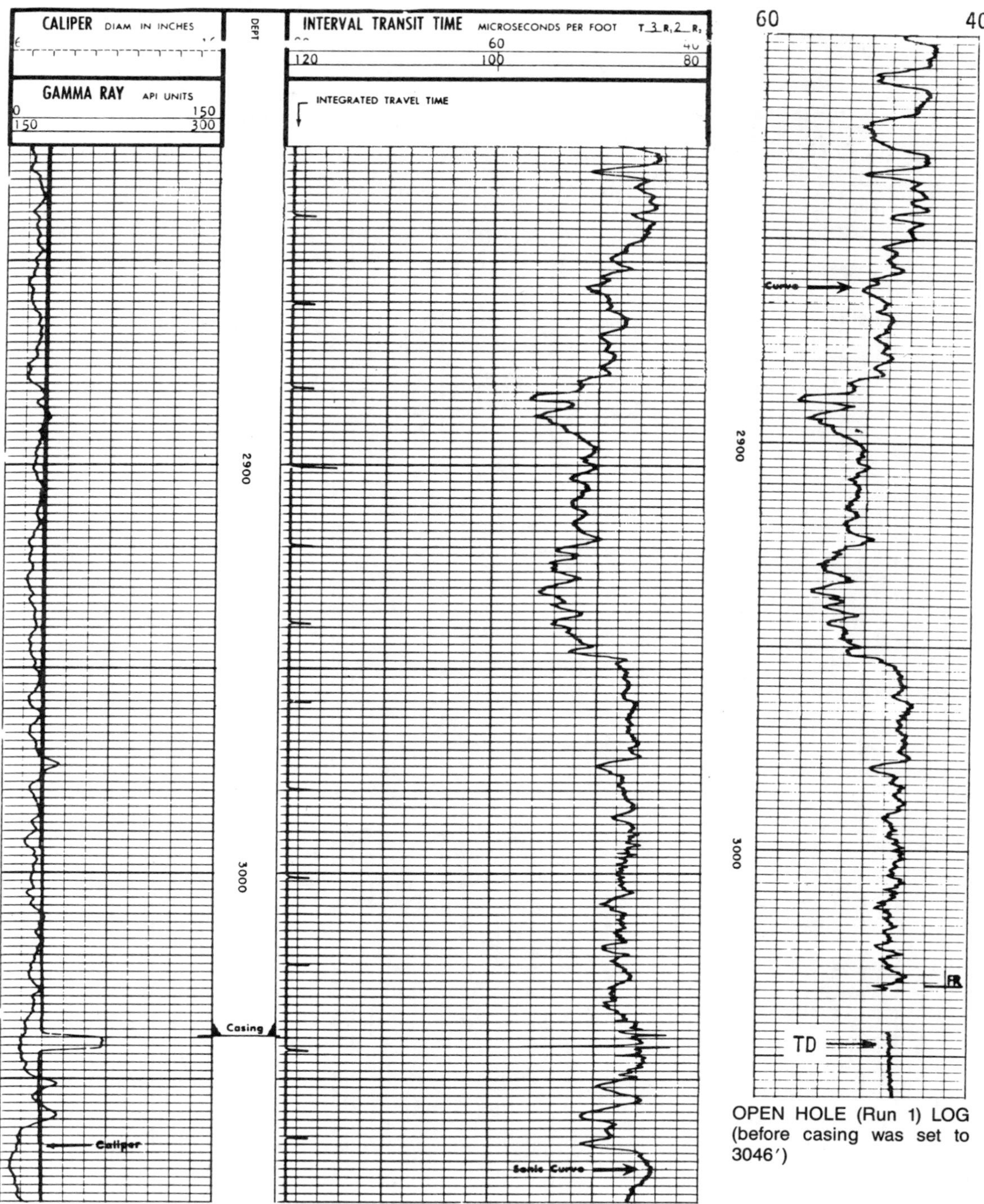

FIG. 10-29. Logging formation behind casing. In well-cemented casing (good bond between casing and cement and also between cement and the formation) the first formation arrival is measurable. Note the 6 ft difference in depth between the two runs. The cased-hole gamma ray does not have enough character to detect this depth difference.

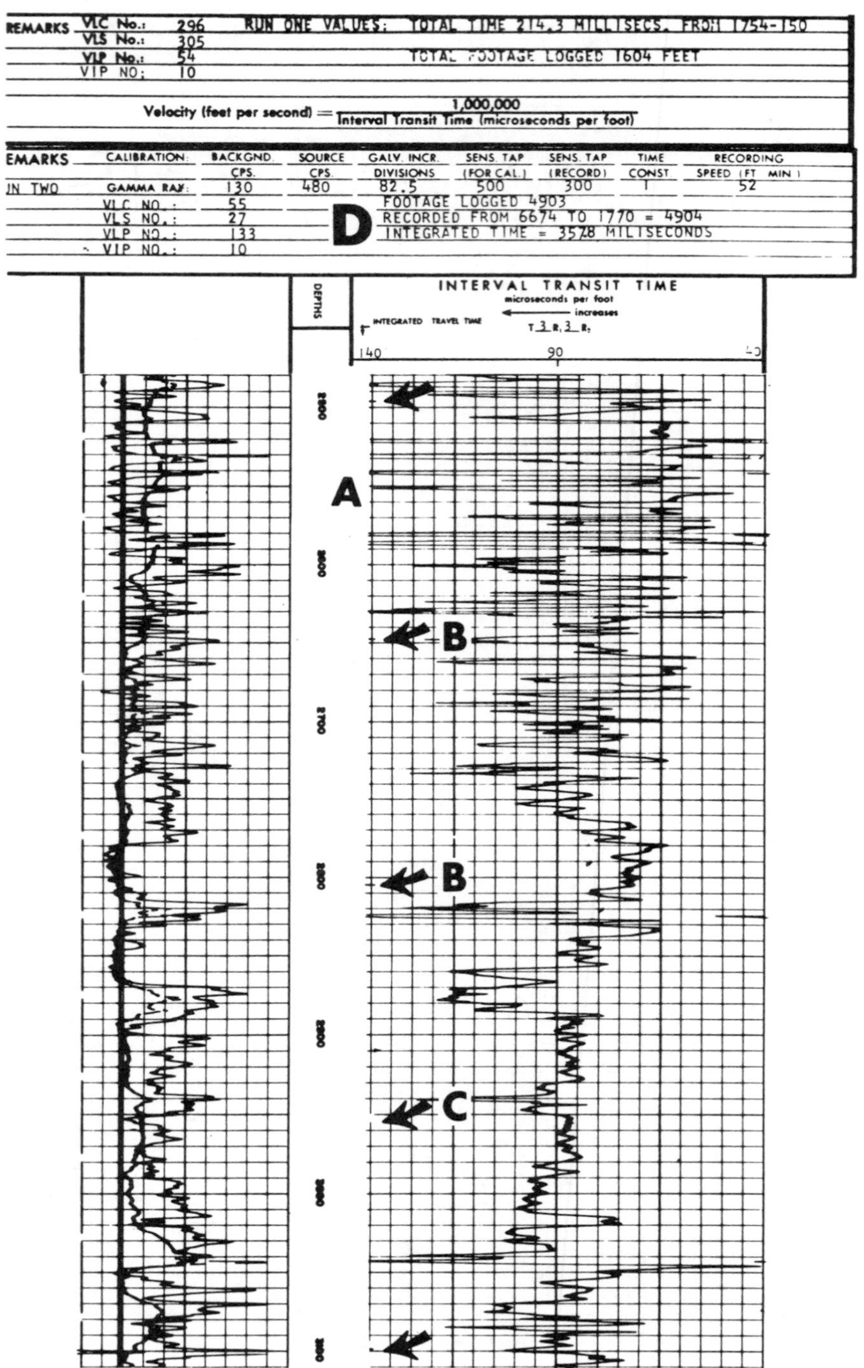

FIG. 10-30. Obtaining missing TTI measurements. Most commercial log services tape out the TTI (A). However, the tips of the 10 ms pips may still be visible (B) or were whitened out individually (C). Often the remarks summary of TTI information has not been deleted (D).

indicated, it is for Δt first reading depth to either casing or last reading depth.

Like all log measurements the TTI is subject to error. First, the method of summing measures the traveltime under all spikes caused by cycle skip or noise. These errors are accumulated over the entire log. Second, there can be error in the summing method. The log should have in the calibration tail a TTI check. Unfortunately this check is usually made at the completion of the acoustic log. The tool is stopped in casing where Δt is approximately 57 μs/ft in free pipe, and the recorder hand cranked until four or five pips are recorded. A Δt is then computed from the pips and compared with the logged value. The example in Figure 10-31 shows five pips were recorded over 88.25 ft.

$$5\ 000\ \mu s\ /\ 88.25\ \text{ft} = 56.65\ \mu s/\text{ft},$$

which is very close to the recorded value. If this check is done after the

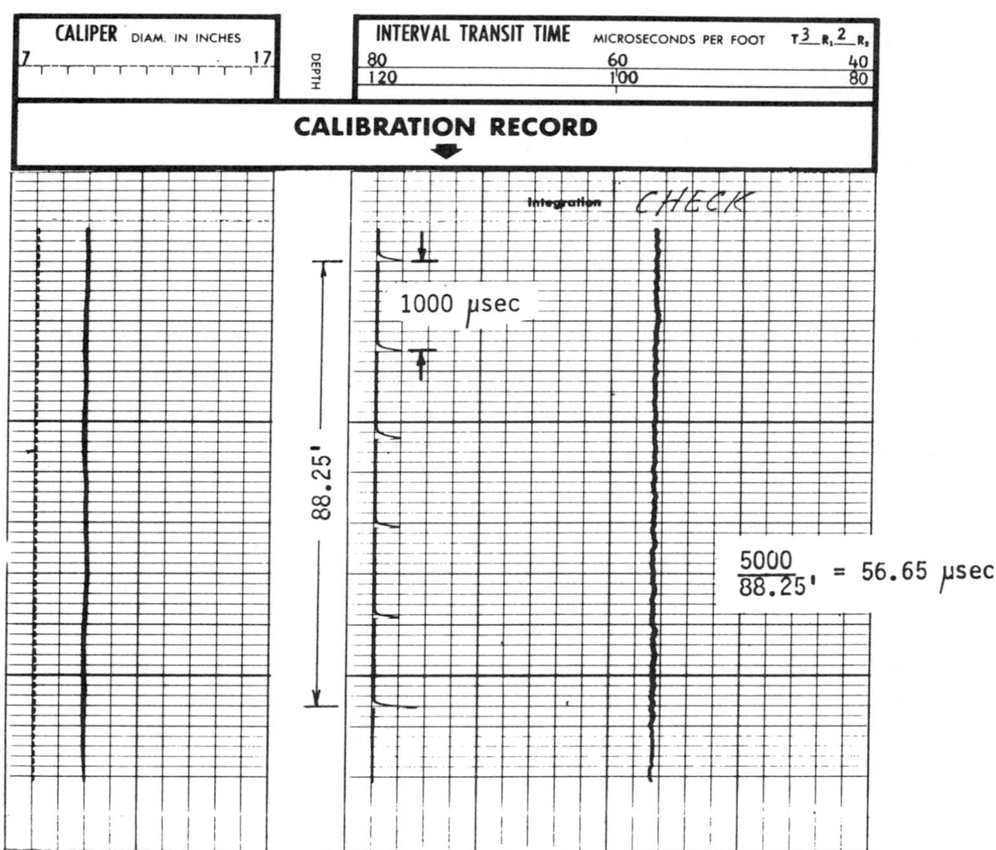

FIG. 10-31. An integration check made during the logging operation. Whenever a traveltime integrator (TTI) is presented with the acoustic log, a TTI check should be inclueded in the calibrations. The tool is normally stopped going into the hole, before the logging pass, in unbonded casing. Time drive or hand cranking introduces at least four TTI pips. Then the pips are used to back-calculate the uncemented casing Δt.

log is run and the TTI is incorrect, its invalidity may only be noted in the log heading remarks space. If the check is done while going into the borehole before emerging from casing, corrective measures can be made before the logging pass.

If a TTI check has not been made, the log itself should be examined in a thick consistent interval. In Figure 10-32 there are four pips from approximately 3 518 to 3 433 ft:

$$4\ 000\ \mu s\ /\ 84.5\ \text{ft} = 47.3\ \mu s/\text{ft},$$

which is a good agreement (line A) with the actual measurements for the interval. However, significant breaks as shown in the salt interval from 3 294 to 3 414, create a difference between interval average (line B) and actual measurements.

Of course, even with an obviously wrong TTI, there is enough variation in Δt log scales that the TTI could be correct and the Δt curve be scaled wrong. In Figure 10-33 the TTI from 1 614 to 1 805 ft indicates:

$$50\ 000\ \mu s\ /\ 191\ \text{ft} = 261.8\ \mu s/\text{ft},$$

which is not near the average of the presented Δt, even if the curve were actually on the 250-150 μs/ft back-up scale. Thus a TTI check done going into the hole and developed with the repeat section film can assure both the TTI and the Δt scales are correct. If the TTI in Figure 10-33 had been checked before the main logging pass, the error would have been obvious and either corrected or the TTI not presented.

Another source of TTI errors, in either the analog or digital acoustic log systems, is film splices. If the recorder ran out of film, the analog tool is lowered to record again, or the digital data tape is partially rewound. For either the TTI integration begins again with a different zero reference. When the film parts are spliced a TTI time discontinuity is introduced.

LONG SPACED SONIC TOOL

The conventional, 3 and 5 ft transmitter-receiver spacing, acoustic log is one of the shallowest reading "porosity" logs, with a depth of investigation of only a few inches into the formation. Thus it is highly susceptible to the effects of drilling damage and to the effect of property changes of shales after they have been drilled (Figure 10-27). To get a better depth of investigation a longer transmitter-to-receiver spacing is needed, the Long Spaced Sonic (LSS). The arrangement can be either a single transmitter and two-receiver system or a borehole-compensated system. Schlumberger uses a close receiver span, the same as their conventional acoustic tool. They have offered 7 and 9 ft, 8 and 10 ft, and 10 and 12 ft spacings.

With longer spacing resulting in a greater depth of investigation, a better transit time is possible in such environments as drilling altered shales (Figure 10-34). With the longer spacing it is especially advantageous to get more information than just the first arrival *P*-wave transit time.

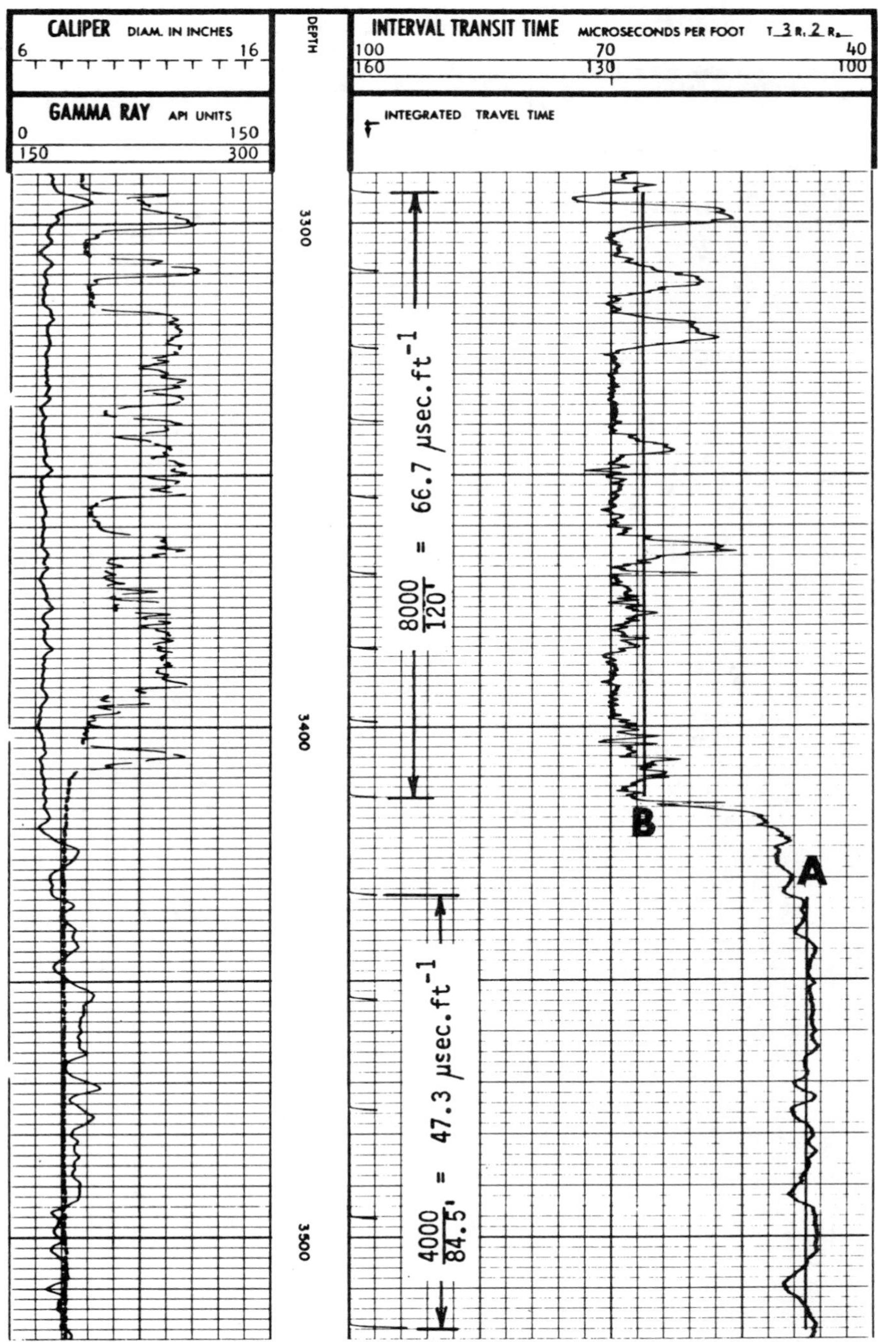

FIG. 10-32. If a TTI check (Figure 10-31) was not made or presented, then the TTI can be checked over a fairly consistent interval. There is a good agreement between average and actual measurements for interval A. Significant breaks, as shown in the salt section from 3 294 to 3 414 ft, can effect the comparison between average for interval B and the actual measurements.

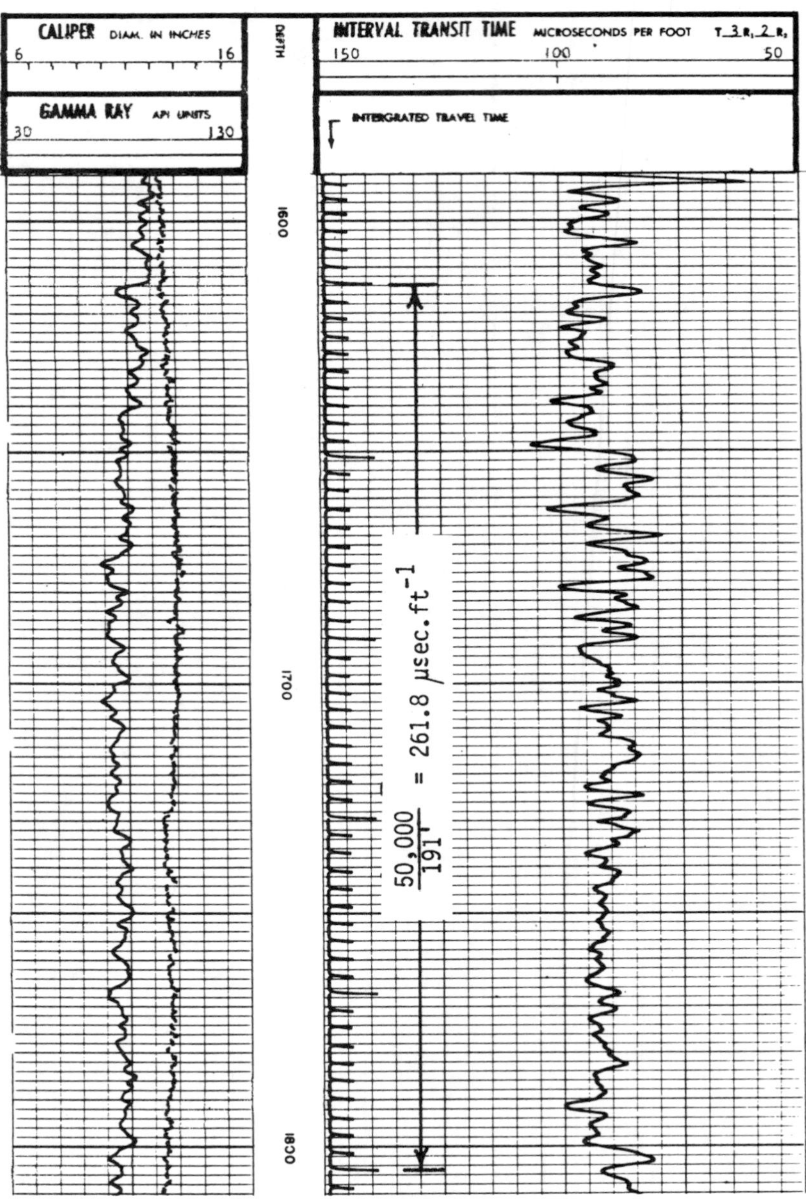

FIG. 10-33. An incorrect TTI curve. The TTI from 1 614 to 1 805 ft indicates an average transit time of 262 μs/ft which is not near the presented Δt. If a check had been run going into the borehole, the incorrect TTI would have been obvious and either corrected or not presented.

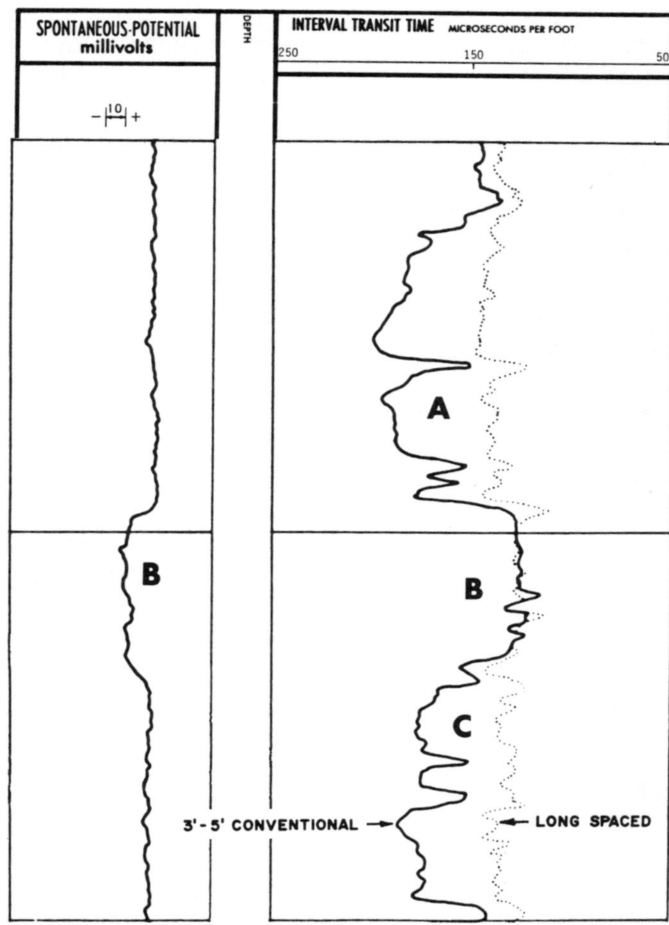

FIG. 10-34. A long spaced sonic versus the conventional acoustic log. The conventional spaced, 3 ft and 5 ft to receivers, acoustic log shows a considerable difference in the shale intervals (A and C). In the non-shale interval (B), as shown by the SP deflection, there is good agreement between the two logs. (From Thomas, 1978.)

FULL WAVEFORM RECORDING

With the recording of the full acoustic log wavetrain *P*-, *S*-, and tube-wave velocities and amplitudes are measurable. This measurement can lead to more complete determination of formation physical properties such as the computed Birdwell's Elastic Properties Log and Schlumberger's Mechanical Properties Log.

To obtain the complete wave train, the receiver signals can be recorded digitally or on video tape. A wiggle trace recording (Figure 10-35) can be made of one of the receiver outputs. However, to physically accommodate the waveform wiggle this can only be presented at periodic depth intervals, often a trace only every one or two feet.

If the absolute amplitude information is not necessary, then a continuous variable density recording can be made by clipping the receiver signal and the resultant signal can be used to intensity modulate a "curve" (Figure 10-36). The variable density presentation is easily comparable to a wiggle trace (Figure 10-37) with the disadvantage of only relative amplitude information. Full wave train recording is available under several trade names as shown in Table 2-1.

Most waveform presentations are for only a single-transmitter, single-receiver configuration. These presentations contain some transmitter-formation and formation-receiver time in their measurement. However, this is less of the path as the spacing used is commonly longer for wave-form recording than for first arrival transit time measurements. The full waveform log needs the caliper as an aid in interpretation to tell when Δt changes are due to borehole or formation changes (Figure 10-38). It should also be helpful to record the single-receiver, first-arrival time as an aid in interpreting amplitude changes in the waveform.

FRACTURE DETECTION

Acoustic wave train recordings are used for the detection of formation fractures by analysis of wave train amplitudes and wave train pattern changes with depth. Early fracture detection techniques used only amplitude measurements of the acoustic receiver signals. Full wave train recording greatly aids in the interpretation of the amplitude measurements and provides significant additional indications of formation fracturing.

Early acoustic fracture detection techniques used only amplitude measurements over specific time intervals [Anderson and Walter (1961) and Morris, et al. (1963)]. Morris' empirical studies show that low-angle, near-horizontal fractures were indicated by the reduction in amplitude of the *S*-wave arrivals, while higher angle fractures caused the *P*-wave amplitude to decrease. The more recent model studies of Dzeban (1970) show that for fractures at angles up to 70 degrees from the horizontal,

FIG. 10-35. XY waveform or wiggle trace presentation of a full wavetrain acoustic log.

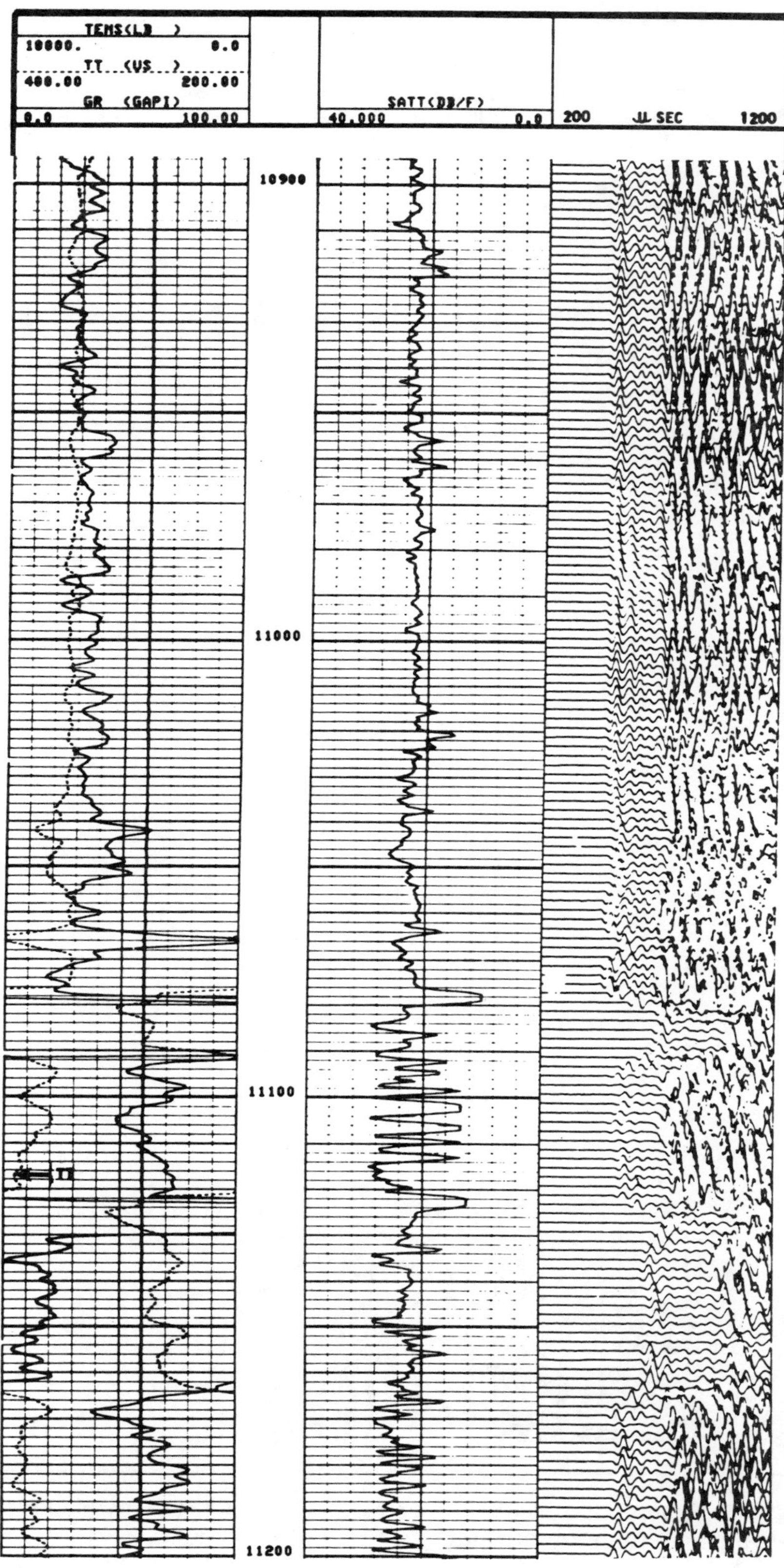
TENS(LB)
10000.
0.0
TT (US)
400.00
200.00
GR (GAPI)
0.0
100.00
SATT(DB/F)
40.000
0.0
200
μ SEC
1200
10900
11000
11100
11200

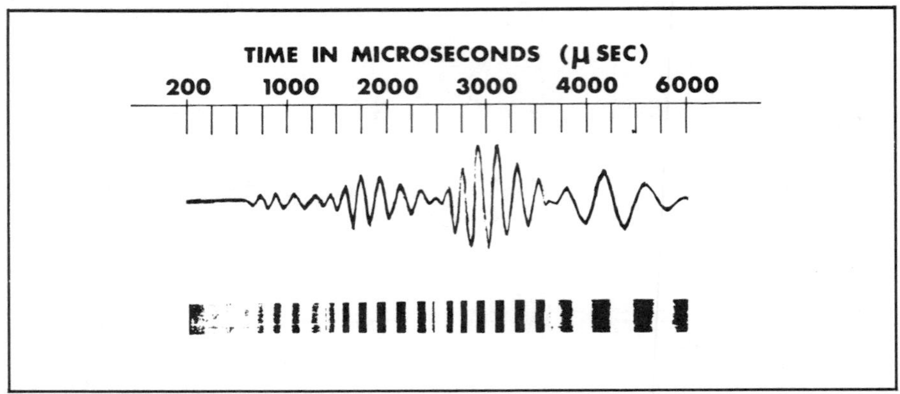

FIG. 10-36. The variable density presentation concept. The waveform is clipped and intensity modulates the display, with white for all negative amplitudes and blacker for greater positive amplitudes. Courtesy of Birdwell.

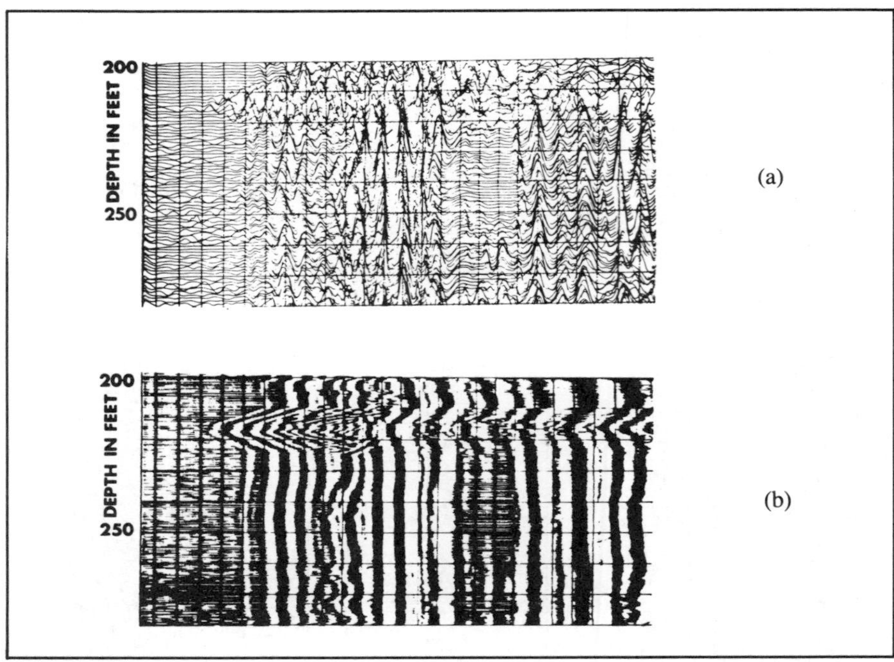

FIG. 10-37. Comparison of (A) X-Y or wiggle trace and (B) variable density presentations. Courtesy of Birdwell.

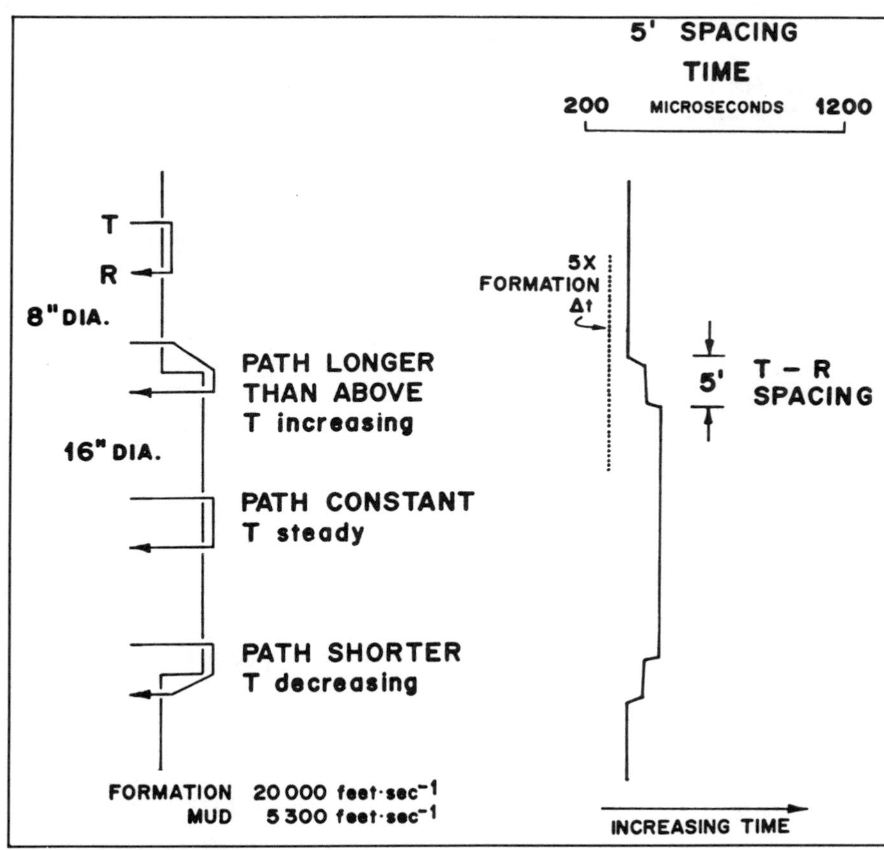

FIG. 10-38. The single transmitter-single receiver system time measurements respond to mud time and borehole diameter changes in addition to changes in formation transit time. This system is commonly used in recording of the full acoustic waveform.

there is a 20 to 30 percent reduction in *P*-wave amplitude, as compared with the amplitude of unfractured models. There is also a 60 to 80 percent reduction in the *S*-wave amplitude.

The measurement of the acoustic wave amplitudes can use a very narrow gate, for example the peak amplitude of a half-cycle either of the first-arrival wave or at a time delay after the first-arrival *P*-wave. Or the gate can be widened to measure amplitudes of *P*- or *S*- waves or both. Welex uses the last, a very wide gate, for their "fracture finder" amplitude measurement (Figure 10-39). Unfortunately, an amplitude decrease is also caused by things besides fractures. Acoustic amplitude measurement depends on the centering of the sonde within the borehole and the roundness of the borehole. A sonde off-centered by 1/4 inch can cause a 50 percent amplitude reduction in the received signal (Morris et al., 1963). This reduction is caused by the stacking of different arrivals from the formation due to the differences in path lengths. A similar destructive interference potential is caused by lack of roundness of the borehole. To better understand acoustic amplitude reductions, the compressional arrival transit time

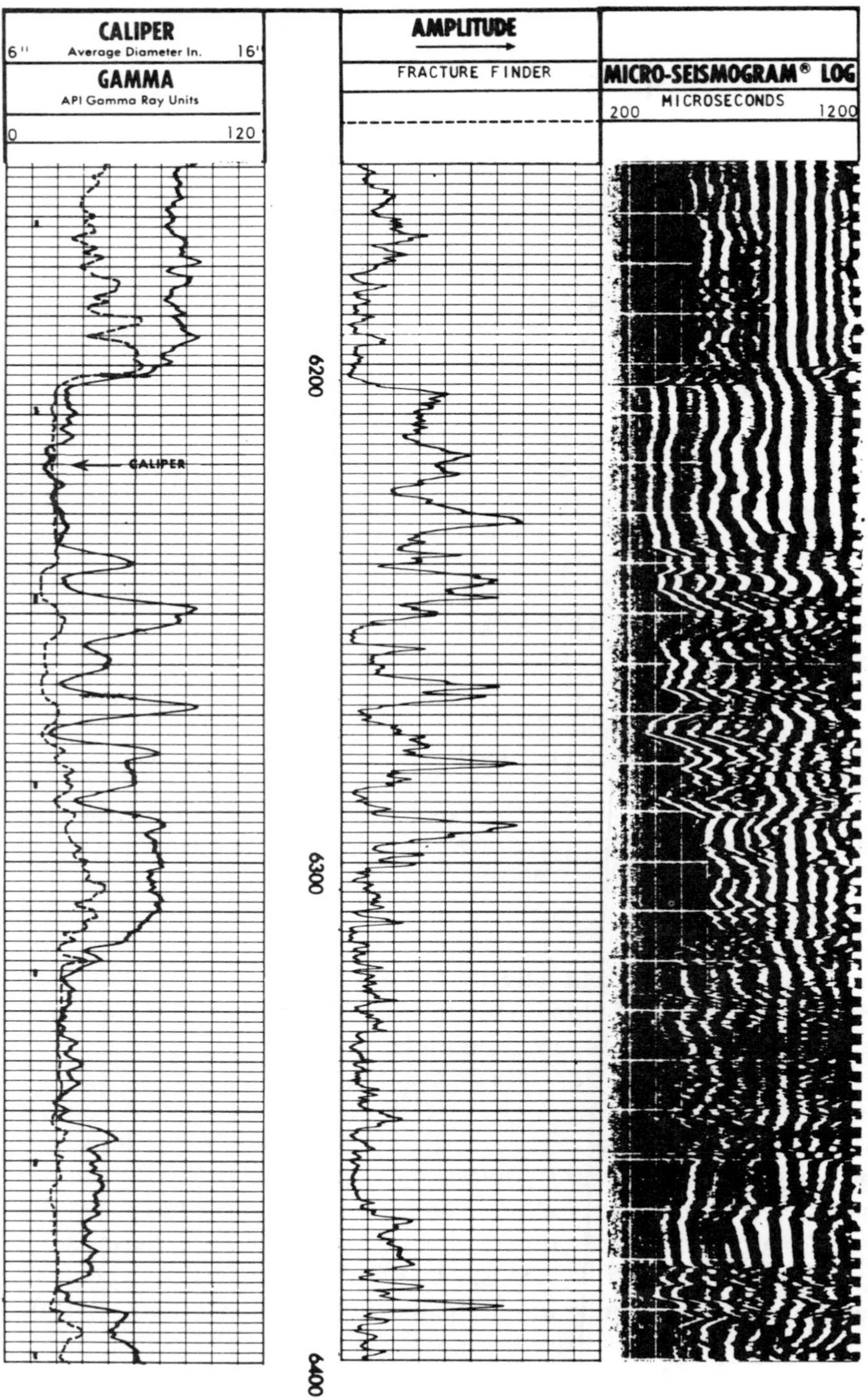

FIG. 10-39. Full waveform log: Welex's Micro-Seismogram (MSG) used in open hole.

should also be recorded, a single receiver Δt. If there is an amplitude decrease where Δt is constant, then the decrease is probably formation caused. However, if the Δt is wandering slightly, an amplitude decrease probably is the result of decentralization. The acoustic sonde should be equipped with adequate centralizers and, if hole conditions permit, some sort of hole size stand-offs.

In addition to amplitude measurements, fractures may be detected by disruptions of the wave train formation patterns. There are two effects: first, a slight phase shifting of the wave train pattern and second, a high-frequency "W"-shaped interference pattern. Synthetic model studies show these effects under noise-free conditions (Figure 10-40, upper), and they are also observable under borehole conditions (Figure 10-40, lower). As

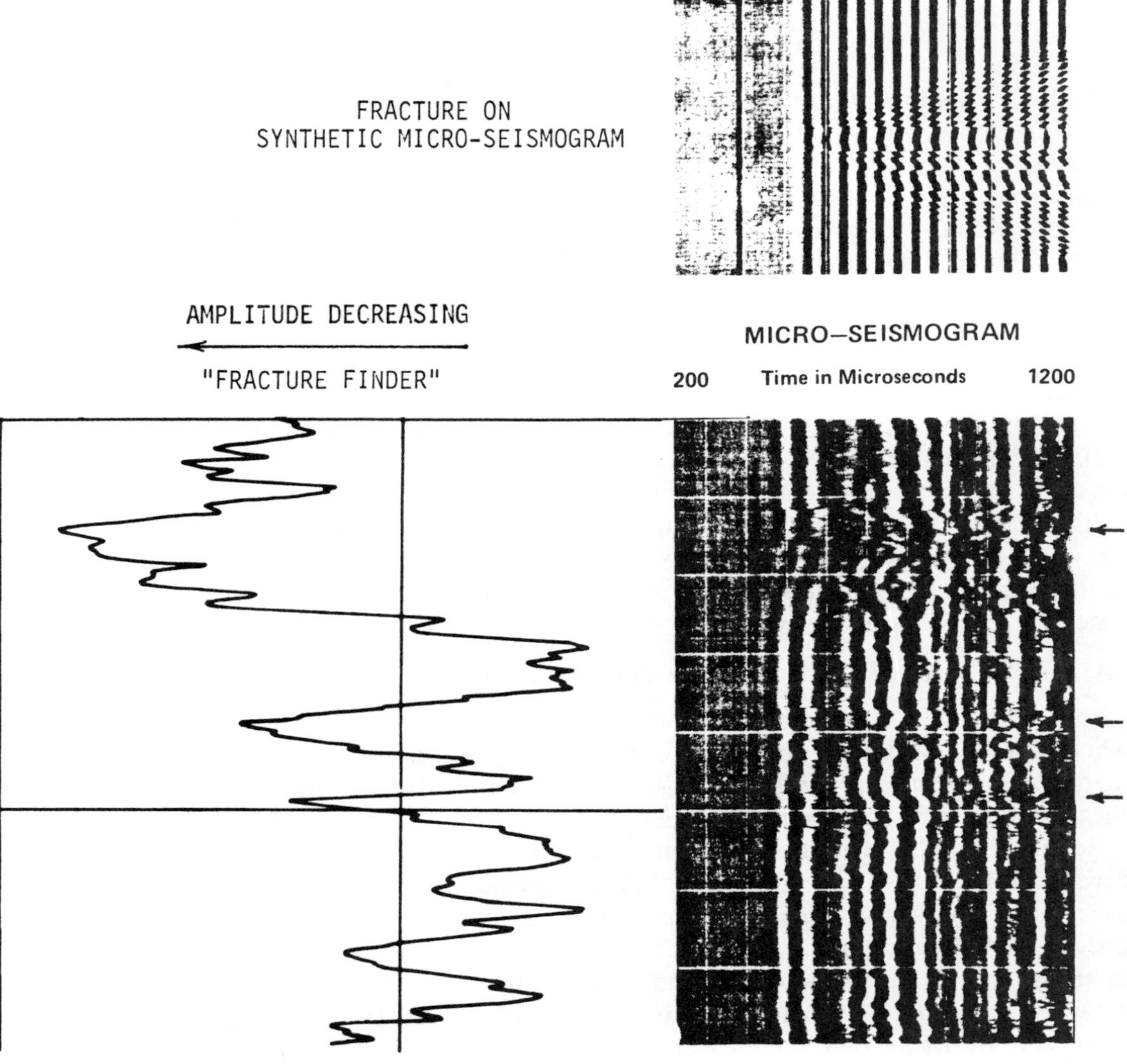

FIG. 10-40. Fractures on the acoustic full wave train recording. Courtesy of Welex.

with all fracture detection techniques, the acoustic indicators should be collaborated with other information.

S-WAVE LOGGING

By using the full wave train it is possible to measure both S-wave traveltime and the conventional P-wave traveltime. There are two basic approaches to obtain S-wave traveltime (Δt_s): (1) analysis of the waveform variable density recording or (2) direct measurement of shear Δt.

Determining Δt_s from the full wave train variable density recording first requires identification of S-wave arrivals. Four criteria are useful to help locate the S-wave arrivals:

1. The S-wave arrives after the P-wave (Figure 10-41, item A) and, based on probable formation properties,

$$1.4\ \Delta t_c \leq \Delta t_s \leq 1.8\ \Delta t_c. \qquad (10\text{-}3)$$

2. Higher S-wave amplitudes, as compared to P-wave amplitudes. On the variable density display higher S-wave amplitudes are indicated by darker bands (Figure 10-41, item C). This assumes the downhole receiver gain is not excessive. If it were large P-wave amplitudes and clipped S-waves input to the variable density plotter would be created which would yield a limited dynamic range amplitude recording.
3. A difference in slope of P- and S-wave arrivals, especially at changes in formations (Figure 10-41, item B).
4. Interference patterns of later S-wave arrivals with the P-wave arrivals (Figure 10-41, item D).

Once the S-wave is identified, the Δt_s must be calculated. Because the variable density display comes from a single transmitter-receiver system, the times picked from the display contain appreciable mud and borehole diameter time. If a conventional compressional borehole compensated Δt_c measurement were recorded, the S- and P-waves could be assumed to travel through the same fluid and formation. Thus, an approximate borehole compensated Δt_s would be:

$$\Delta t_s = \Delta t_c + (t_s - t_c)\ (\text{tool spacing})^{-1} \qquad (10\text{-}4)$$

where t_s and t_c would be shear and compressional arrival times from the variable density display. There is a small error in the difference in the shear and compressional critical angles.

Several direct Δt_s measurement techniques are possible or are being developed. Probably the simplest technique is detailed in Nations (1974). This method takes advantage of the common condition where the S-wave arrival is of greater amplitude than the P-wave arrival. The downhole cartridge gain is adjusted to lower the P-wave amplitude to below the detection threshold of the Δt measuring system (Figure 10-11), causing the system to trigger on the higher amplitude S-wave arrival. The main problem with this method occurs when the higher amplitude mud or tube

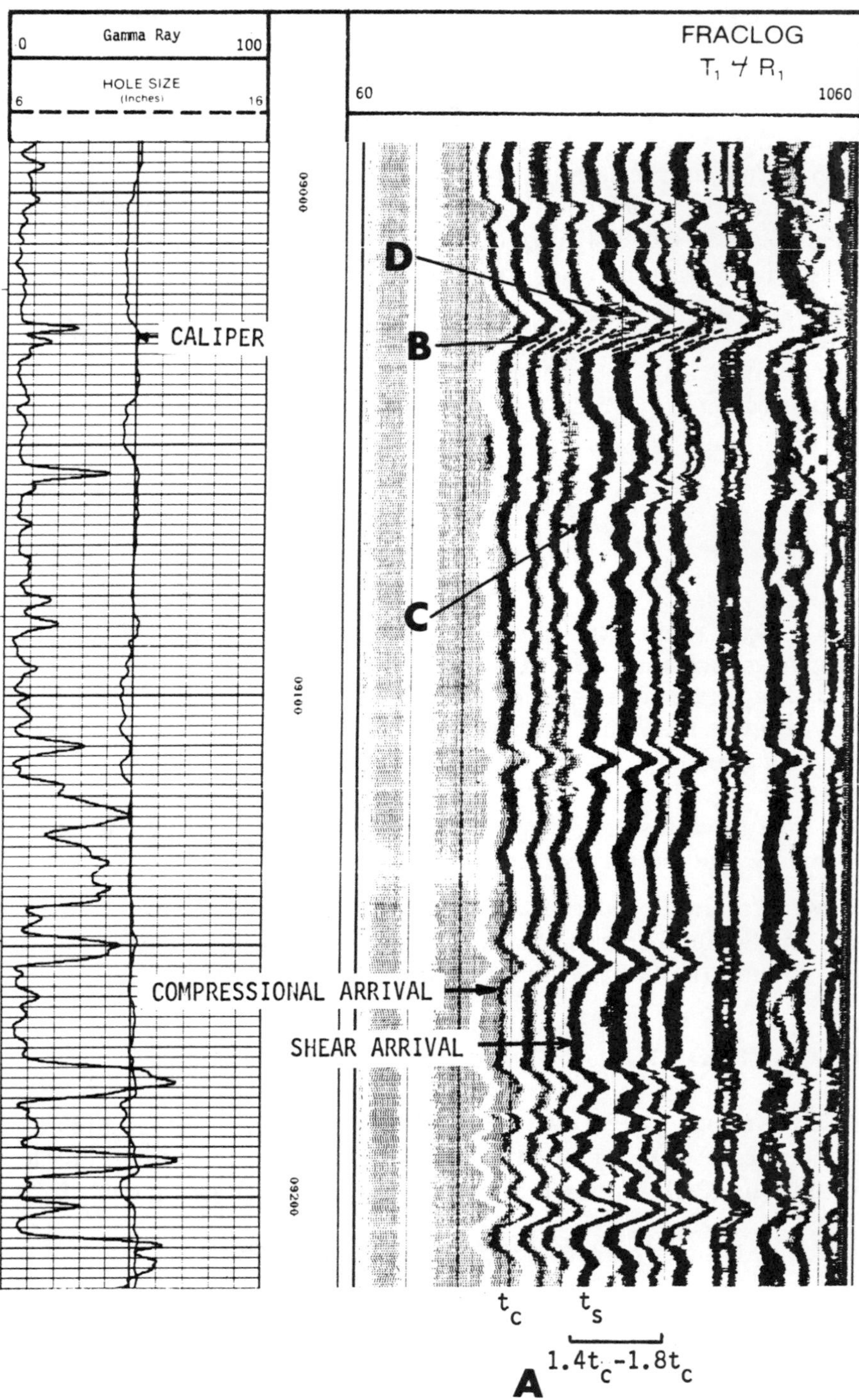

FIG. 10-41. S-wave identification (A) by time after *P*-wave, (B) difference in slope at formation changes, (C) amplitude o· darkness difference, or (D) overriding interference with *P*-waves.

wave arrivals are present at about the same time as the shear *S*-wave arrivals.

Another direct detection approach is detailed in Aron et al, (1978) for a long spaced single transmitter-multiple receiver tool. This system records the multiple receiver wave trains on digital tape and then uses a computer correlation technique to determine Δt_s and Δt_c.

A third direct detection approach is the use of a *S*-wave source, Kitsunezaki (1980). Again, this concept offers better methods for measuring *S*-wave arrival times, but has not yet reached commercial logging status.

With the measurement of *S*-wave and *P*-wave transit times and formation density, it is possible to determine several formation elastic properties:

$$(V_p/V_s)^2 = (0.5 - s)\ (1 - s)^{-1} \qquad (10\text{-}5)$$

$$(V_s)^2 = \mu\ \rho^{-1} = \mathrm{E}\ [2\varrho\ (1 - s)]^{-1} \qquad (10\text{-}6)$$

$$(V_p)^2 = (K + \frac{4}{3}\mu)\rho^{-1} \qquad (10\text{-}7)$$

where,

s is Poisson's ratio, the ratio of transverse strain to longitudinal strain,
μ is shear modulus, the stress-strain ratio for simple shear. Schlumberger uses the term *G*.
E is Young's modulus, the stress-strain ratio when a rod is pulled or or compressed,
and,
K is bulk modulus, the stress-strain ratio under simple hydrostatic pressure. Schlumberger uses the term c_b^{-1}.

These formation properties are derived for Birdwell's Elastic Properties Log and Schlumberger's Mechanical-Properties Log, discussed in Geyer and Myung (1970) and Tixier et al. (1975), respectively.

V_s/V_p LITHOLOGY AND GAS DETECTION

Observations by Picket (1963) and others indicate there is a relationship between V_p/V_s, or $\Delta t_s/\Delta t_c$, and lithology in "clean," single lithology formations. Picket's observations (Figure 10-42) show that:

V_p/V_s	"clean" lithology
1.6-1.7	Sandstones
1.8	Dolomites
1.9	Limestones

However, Tathum (1982) suggests the V_p/V_s ratio is controlled more by pore geometry than by lithology. More research is needed to determine if there may be a relationship between lithology and pore geometry.

The ratio does work in single lithologies but it is not deterministic enough to indicate component ratios in mixed lithologies. The ratio is also effected by the presence of gas (Kithas, 1976). Gas in the pore fluid tends to make Δt_c higher, or V_p slower, while the shear measurement is unaffected. There can be a 10 to 20 percent reduction in the $\Delta t_s/\Delta t_c$ ratio in gas-bearing formations. This technique also works in shaley sand formations of consistent lithology according to Leeth and Holmes (1978).

MICRO-ACOUSTIC TOOLS

Most acoustic logging tools have spacings of several feet, with current trends calling for even longer spacings. Small, pad-mounted acoustic systems, however, do have several advantages and applications. There are two current types: the vertical sidewall acoustic log and the horizontal circumferential acoustic log.

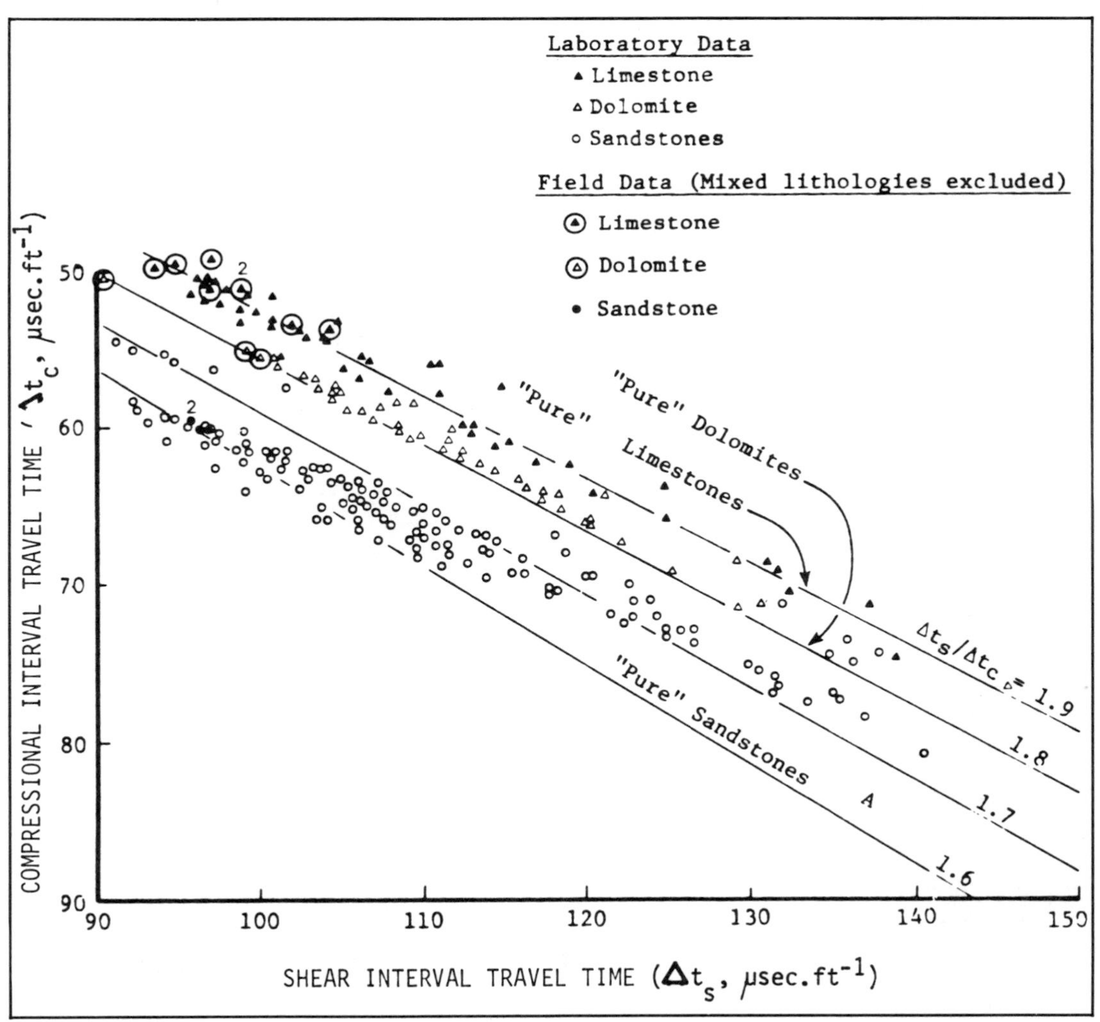

FIG. 10-42. Crossplot of shear and compressional interval traveltimes. For "pure" lithologies the $\Delta t_s/\Delta t_c$ or V_p/V_s ratio indicates lithology. The presence of gas reduces the ratio by 10 to 20 percent. (Adapted from Picket, 1963, © 1963 SPE-AIME.)

The sidewall acoustic log (SWA) used a very short, 6 and 9 inch spacing, pad-mounted transmitter-dual receiver configuration which was developed by Dresser-Atlas and was run simultaneously with a sidewall neutron log on a second pad mounted on the opposite side of the tool. The two

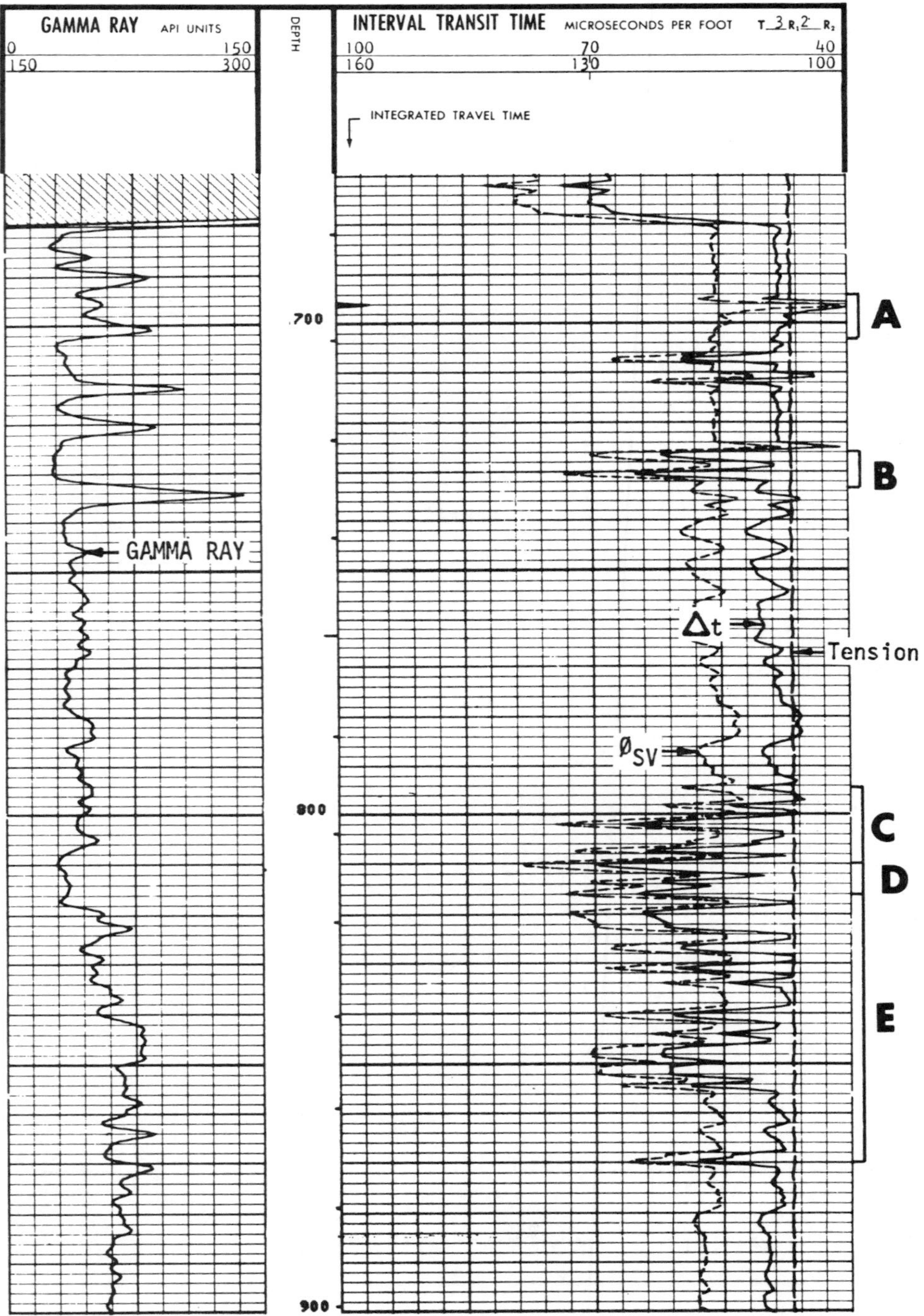

FIG. 10-43. Are the intervals A through E porosity breaks or cycle skips? Comparison with other log measurements (Figures 10-45 and 10-47) is needed to make the decision.

measurements compliment each other. The short acoustic spacing yields vertical resolution similar to the neutron measurement, though not the same depth of investigation. The acoustic-neutron combination allows good porosity, lithology, and gas identification but it is pessimistic in formations with significant secondary porosity.

The sidewall acoustic system offers several advantages over the more conventional spaced acoustic logs. Its prime advantage is negligible borehole distortion of the receiver waveforms. With borehole centered acoustic tools, the resultant receiver waveform is a combination of signals coming down all sides of the borehole wall. Time variations of the signals from the various paths result in constructive and destructive interference at the receivers. Thus the receiver waveform is effected not only by the desired information from the formation, but also borehole size and shape, sonde size and centering, borehole wall rugosity, borehole fluid, and mudcake characteristics. By moving the acoustic array to the borehole wall, most of these permutations are eliminated.

Because S-waves are attenuated at a greater rate with distance than P-waves, the shorter spacing allows a better S-wave measurement. In addition, a pad system allows recording of transit times in gas cut muds that would attenuate conventional spaced signals. Small anomalies, such as vugs or fractures, occupy a greater part of the measured volume and are emphasized. A pad system is also used in comparison with conventional acoustic tool measurements to provide information on near-borehole damage or alteration.

Acoustic measurements taken vertically do not detect vertical fractures because the sound waves bypass the fracture. However, if the measurement system is turned at right angles to the borehole, vertical fractures then lie across the path of the acoustic wave and are easier to detect. Shell Oil Company has developed the circumferential acoustic device (CAD), and Schlumberger the circumferential micro-sonic tool (CMT). The measurement system uses four acoustic transducer pads at right angles to the tool, with two consecutive pads acting as transmitter and receiver. With proper sequencing the borehole circumference is measured acoustically in four quadrants. Vertical fractures produce strong attenuation of the acoustic wave in that quadrant. Comparison of amplitudes in the different quadrants indicates anomalies produced by near-vertical fractures.

RELATIONSHIPS WITH OTHER LOG MEASUREMENTS

Acoustic logs tend to spike because of noise and attenuation cycle skips, making it good practice to compare the Δt measurement with other measurements before and during editing. With the newer digital processed acoustic logs, it is sometimes difficult to differentiate thin porosity streaks from cycle skips. Are the acoustic log breaks in intervals A through E in Figure 10-43 really porosity or are they cycle skips? Comparison with the other log measurements is helpful in reaching an accurate decision.

The shallow focused resistivity device of the dual induction log measures nearly the same formation as the acoustic log (Figure 10-44). This shallow device measures the consistent flushed zone that is not effected as greatly by hydrocarbons as the medium or deep devices, thus it makes a good comparison for the acoustic log. Because of borehole condition in this example a dual laterolog was run instead (Figure 10-45). With this tool the shallow laterolog measures deeper than the dual induction's shallow device. The micro-resistivity is useful but has much greater detail than the acoustic log. However, in this example the shallow device appears not to be working properly, because it never measures over 300 Ω•m which

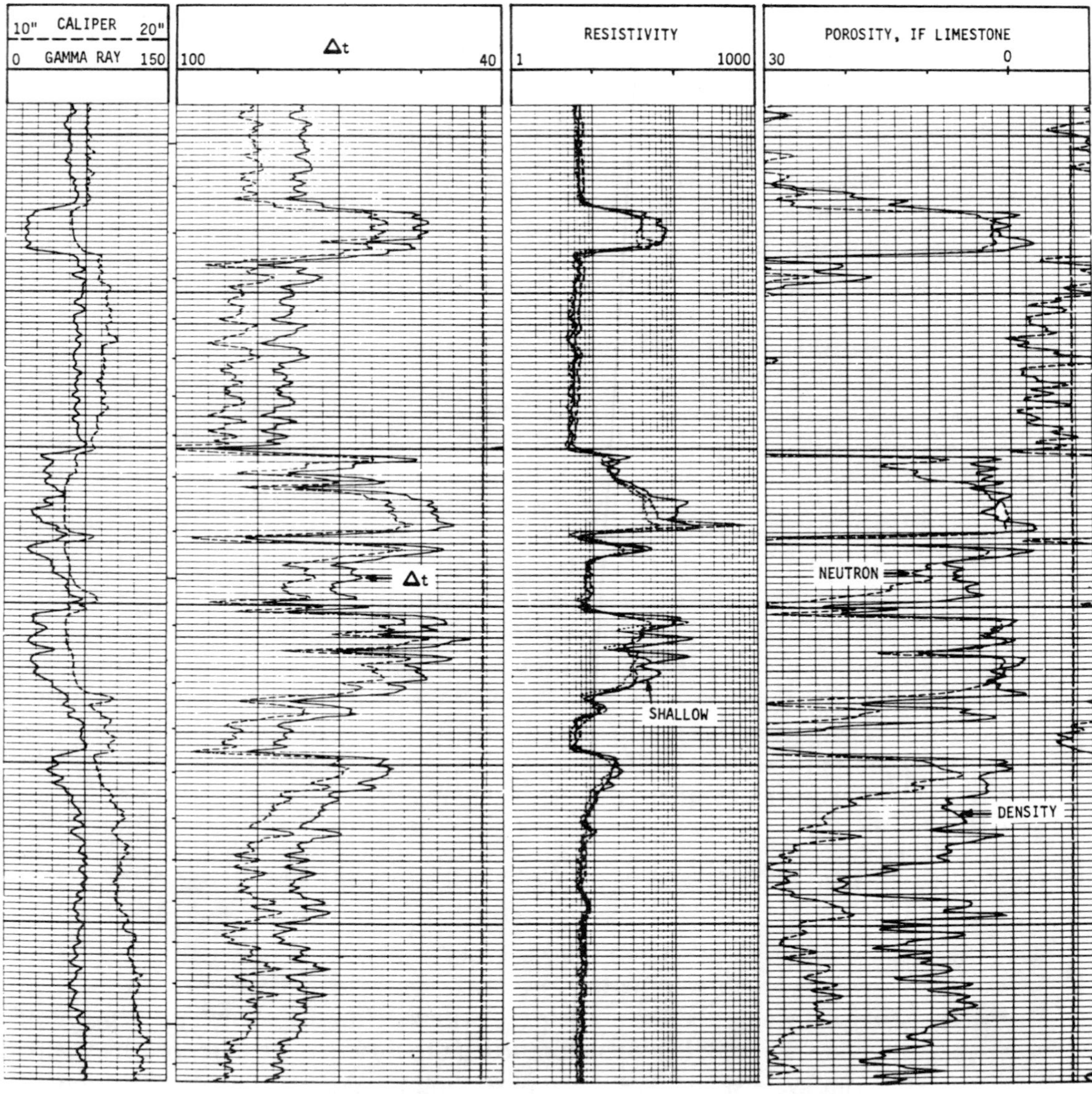

FIG. 10-44. Comparison of Δt — shallow resistivity — compensated neutron measurements for similarity of curve shapes.

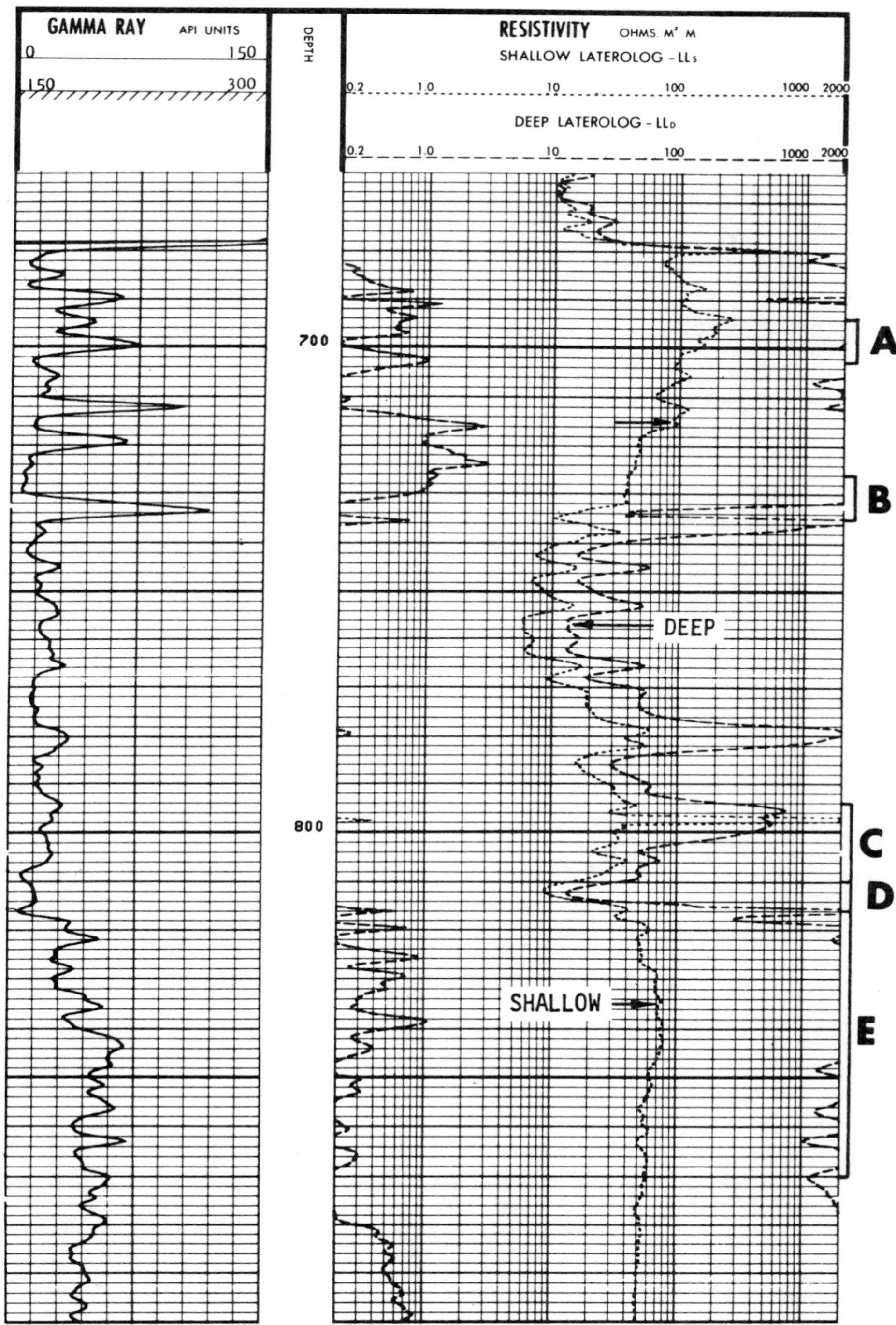

FIG. 10-45. Dual laterolog over the same interval as Figure 10-43. The shallow-focused device measures a deeper formation depth than the acoustic log. The shallow device also appears to be working improperly in high-resistivity intervals.

is a common fault indication of the shallow device. Thus the resistivity log provides no helpful comparison.

Another good choice is the compensated neutron measurement. While it has a deeper depth of investigation and a light hydrocarbon effect, it also seems to have a similar reaction to formations (Figure 10-44). The acoustic-neutron crossplot is also second only to the density-neutron crossplot for definition of porosity and lithology (Figure 10-46).

From the neutron measurement (Figure 10-47) it is obvious that all the Δt and ϕ_{SV} breaks of Figure 10-43, except interval D, are cycle skips. These breaks should be edited out before using the Δt measurement in a synthetic seismogram.

Another alternative measurement to use in editing acoustic curves,

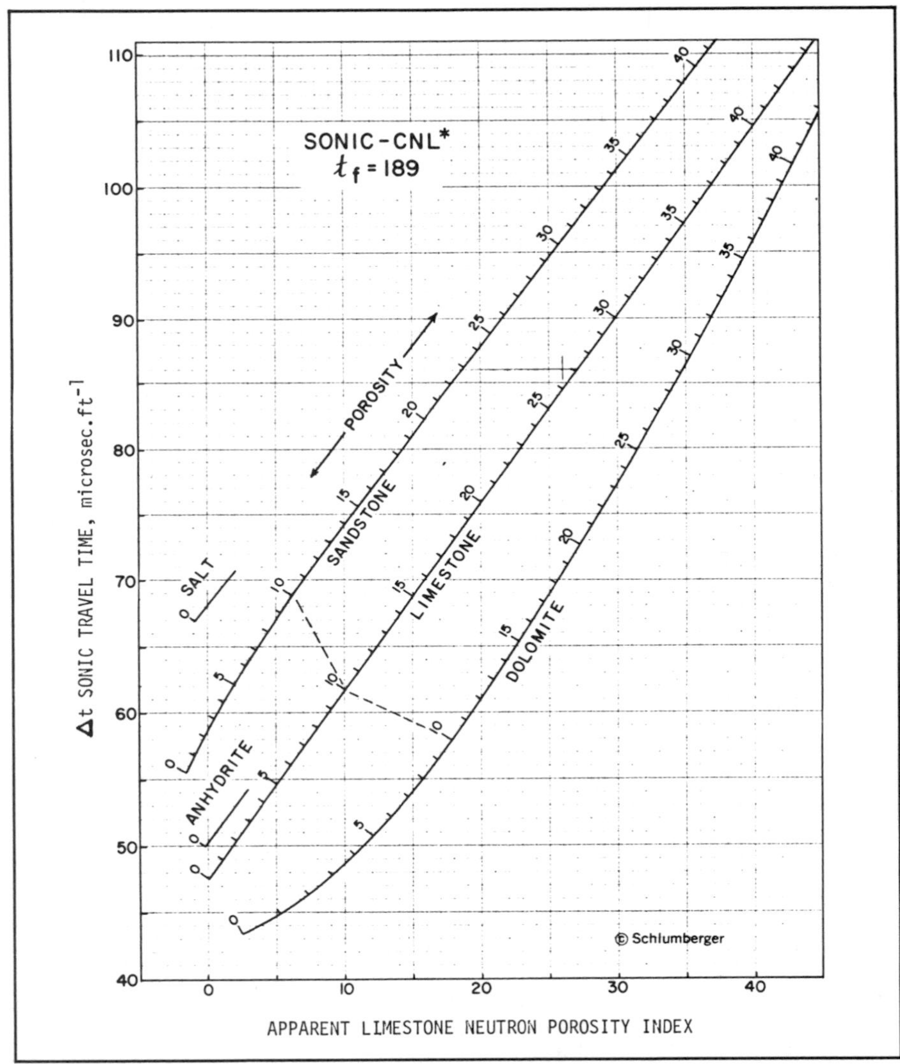

FIG. 10-46. Crossplot of Δt and neutron porosity: second only to the density-neutron crossplot for definition of porosity and lithology.

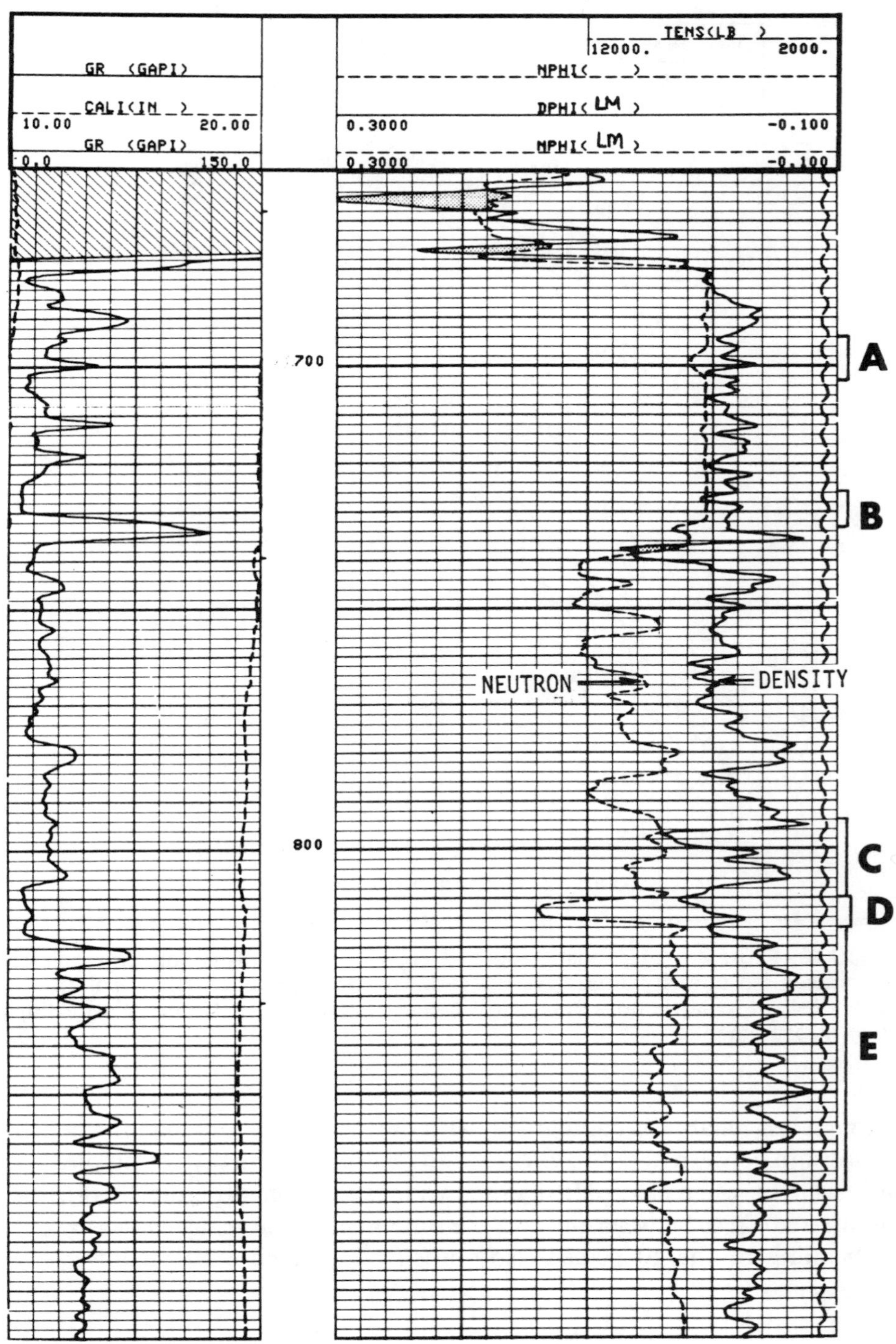

FIG. 10-47. Density-neutron log over the same interval as Figure 10-43 showing that the Δt breaks on all intervals except D are cycle skips.

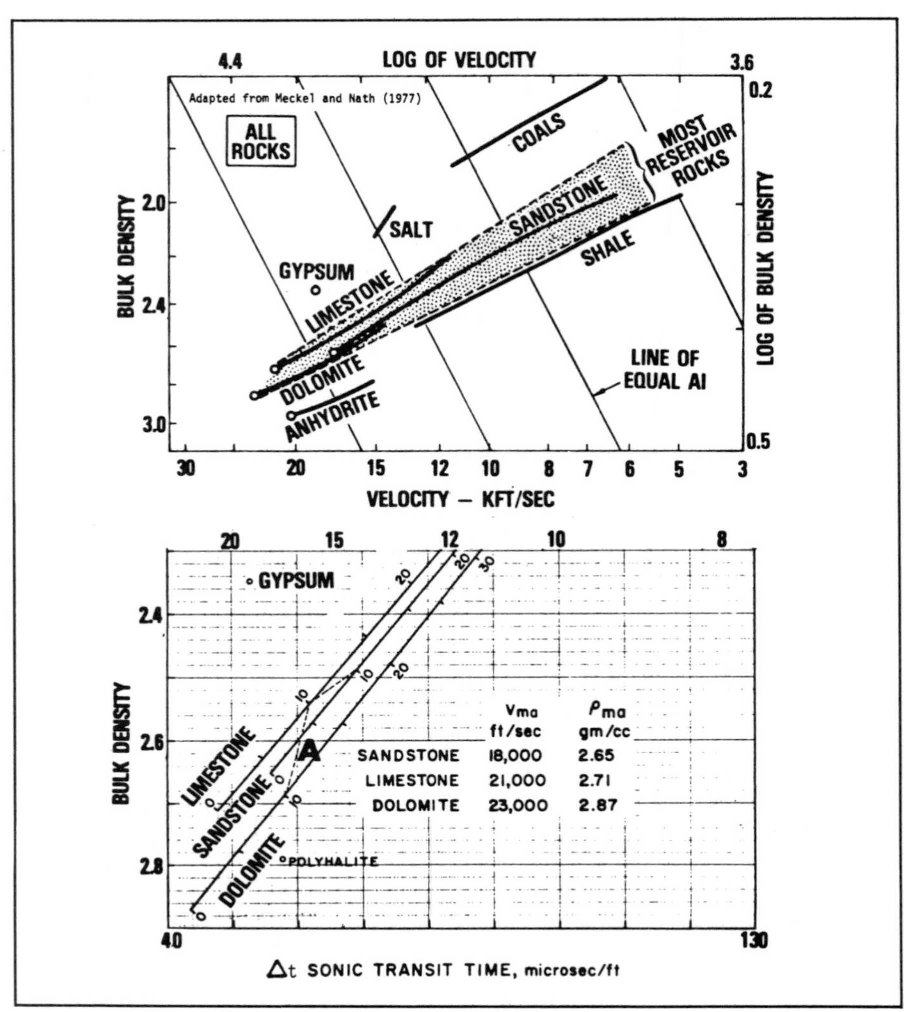

FIG. 10-48. Crossplot of acoustic log and compensated density log measurements: note the nonlinear iso-porosity lines (A) indicating a nonunique crossplot solution for porosity. Obviously, the matrix does make a difference. There is not a simple relationship between the two measurements.

although less reliable, is comparison with the density measurement. The relationship between acoustic velocity and density, however, is not simple. This is shown when crossplotting the two measurements (Figure 10-48). The matrix effect between the two measurements is large, as shown by the bending of the 10 percent iso-porosity line (Figure 10-48, lower, item A).

REFERENCES

Allaud, L. A., and Martin, M. H., 1977, Schlumberger, the history of a technique: John Wiley & Sons, Inc.

Anderson, W. L., and Walter, T., 1961, Application of open hole acoustic amplitude measurements: 36th Ann. Soc. Petr. Eng. Fall Conference, Dallas, SPE-122.

Aron, J., Murry, J., and Seeman, B., 1978, Formation compressional and shear interval-transit-time logging by means of long spacing and digital techniques: 53rd Ann. Soc. Petr. Eng. Fall Conference, Houston, SPE-7446.

Blakeman, E. R., 1982, A case study of the effect of shale alteration on sonic transit times: 23rd Ann. Soc. Prof. Well Logging Analysts Logging Symposium, Corpus Christi, paper II.

Doh, C. A., and Alger, R. P., ______________, Schlumberger Eng. Rept. No. 2 — Sonic logging: Schlumberger Well Survey Copmany.

Dzeban, I. P., 1970, Elastic wave propagation in fractured and vuggy media: IZV earth physics, translated from Russian by D. G. Fry.

Ferth, W. H., and Wichmann, P. A., 1977, Open hole porosity logs can be used in cased holes: Oil and Gas J., **75**, 84-86.

Fons, I., 1968, Acoustic logging through casing: Trans. 2nd Can. Well Logging Soc. 2nd Formation Evaluation Symposium, Calgary.

Geyer, R. L., and Myung, J. I., 1970, The 3-D velocity log; a tool for in-situ determination of the elastic moduli of rocks: 12th Ann. Symposium, Univ. of Missouri — Rolla.

Goetz, J. F., Dupal, L., and Bowler, J., 1971, An investigation into discrepancies between sonic log and seismic check shot velocities: Austral. Petr. Expl. Assn. J., **19**, 2, 131-141.

Kithas, B. A., 1976, Lithology, gas detection, and rock properties from acoustic logging systems: Trans. 17th Ann. Soc. Prof. Well Log Analysts Logging Symposium, Denver, paper R.

Kitsunezaki, C., 1980, A new method for shear wave logging: Geophysics, **45**, 1489-1506.

Leeth, R., and Holmes, M., 1978, Log interpretation of shaley formations using the velocity ratio plot: Trans. 19th Ann. Soc. Prof. Well Log Analysts Logging Symposium, El Paso, paper CC.

Misk, A., Mowat, G. R., Goetz, J., and Vivet, B., 1977, Effects of hole conditions on log measurements and formation evaluation: Trans. 5th European Logging Symposium, Paris.

Morris, R. L., Grine, D. R., and Arkfeld, T. E., 1963, Using compressional and shear acoustic amplitudes for the location of fractures: 38th Ann. Soc. Petr. Eng. Fall Conference, New Orleans, SPE 723.

Nations, J. F., 1974, Lithology and porosity from acoustic shear and compressional wave transit time relationships: Trans. 15th Ann. Soc. Prof. Well Logging Analysts Logging Symposium, McAllen, paper Q.

Pickett, G. R., 1963, Acoustic character logs and their applications in formation evaluation: J. Petr. Tech., SPE-452.

Seismograph Service Corp., Continuous velocity logging.

Tatham, R. H., 1982, Vp/Vs and lithology: Geophysics, **47**, 336-344.

Thomas, D. H., 1978, Seismic applications of sonic logs: the Log Analyst, 19, 1, 23-32.

Tixier, M. P., Alger, R. P., and Doh, C. A., 1959, Sonic Logging: J. Petr.

Tech., **11**, 5.

Tixier, M. P., Loveless, G. W. and Anderson, R. A., 1975, Estimation of formation strength from the mechanical-properties log: J. Petr. Tech., 27, 283-293.

Welex, Calibration principles and field calibration procedures for Welex logs: Publ. A-134.

Willis, M. E., and Toksoz, M. N., 1983, Automatic P and S velocity determination from full waveform digital acoustic logs: Geophysics, **48**, 1631-1644.

REFERENCES FOR GENERAL READING

Anderson, W. L., and Riddle, G. A., 1961b, Acoustic amplitude ratio logging: J. Petr. Tech., **13**, 1243-1248.

Ausburn, B. E., 1977, Well log editing in support of detailed seismic studies: Trans. 18th Ann. Soc. Prof. Well Log Analysts Logging symposium, Houston, paper F.

Botter, B. J., 1982, Circumferential acoustic waves in boreholes for the delineation of vertical fractures: Trans. 23rd Ann. Soc. Prof. Well Log Analysts Logging Symposium, Corpus Christi, Paper S.

Breck, H. R., Schoellhorn, S. W. and Baum, R. B., 1957, Velocity logging and its geologic and geophysical applications: Am. Assn. Petr. Geol. Bulletin, **41**, 8, 1667-1682.

Christensen, D. M., 1964, A theoretical analysis of wave proportion in fluid filled drill holes for the interpretation of three-dimensional velocity logs: Trans. 5th Ann. Soc. Prof. Well Log Analysts Logging Symposium, Midland, paper K.

Dupal, L., Gartner, J., and Vivet, B., 1977, Seismic application of well logs: Trans. 5th European Logging Symposium, Paris.

Gardner, G. H. F., Gardner, L. W., and Gregory, A. R., 1974, Formation velocity and density — the diagnostic basis for stratigraphic traps: Geophysics, **39**, 770-780.

Guy, J. O., Youmans, A. H. and Smith, W. D. M., 1971, The sidewall acoustic neutron log: Trans. 12th Ann. Soc. Prof. Well Log Analysts Logging Symposium, Dallas, paper X. Also, Soc. Prof. Well Log Analysts Acoustic Reprints, paper I.

Guyod, H., and Shane, L. E., 1969, Geophysical well logging — Volume I, Introduction to geophysical well logging and acoustic logging.

Hicks, W. G., 1959, Lateral velocity variations near boreholes: Geophysics, **24**, 451-464.

Koerperich, E. A., 1975, Evaluation of the circumferential microsonic log — a fracture detection device: Trans. 16th Ann. Soc. Prof. Well Log Analysts Logging Symposium, New Orleans, paper JJ.

Kokesh, F. P., Schwartz, R. J., Wall, W. B., and Morris, R. L., 1965, A new approach to sonic logging and other acoustic measurements: J. Petr. Tech., **17**, 282-286.

Meckel, L. D., and Nath, A. K., 1977, Geologic considerations for stratigraphic modeling and interpretation: *in* Seismic stratigraphy — applications to hydrocarbon exploration: Am. Assn. Petr. Geol., Memoir 26.

Purdy, C. C., 1982, Enhancement of long spaced sonic transit time data: Trans. 23rd Ann. Soc. Prof. Well Log Analysts Logging Symposium, Corpus Christi.

Raymer, L. L., Hunt, E. R., and Gardner, J. S., 1980, An improved sonic transit time-to-porosity transform: Trans. 21st Ann. Soc. Prof. Well Log Analysts Logging Symposium. Lafayette, paper P.

Setser, G. G., 1981, Fracture detection by circumferential propagation of acoustic energy: 56th Ann. Soc. Petr. Eng. Fall Conference, San Antonio, SPE-10204.

Stripling, A. A., 1958, Velocity log characteristics: J. Petr. Tech., **10**, 207-212.

Summers, G. C., and Broding, R. A., 1952, Continuous velocity logging: Geophysics, **17**, 598-614.

Timur, A. (Editor), 1978, Soc. Prof. Well Log Analysts Reprint Volume — Acoustic Logging: Soc. Prof. Well Log Analysts.

Vogel, C. B., and Herolz, R. A., 1977, The CAD, a circumferential acoustic device for well logging; 52nd Ann. Soc. Petr. Eng. Fall Conference, Denver.

Waller, W. C., Cram, M. E., and Hall, J. E., 1975, Mechanics of log calibration: Trans. 16th Ann. Soc. Prof. Well Log Analysts Logging Symposium, New Orleans, paper GG.

Welex, ________, Basic acoustics and fracture finder/Micro-Seismogram Logs: Welex, a Halliburton Company, Publ. EL-1009.

Youmans, A. H., Guy, J. O., and Engle, A. W., 1970, Field tests with an experimental sidewall acoustic logging device: 3rd Can. Well Logging Soc. Formation Evaluation Symposium, Calgary, paper 7060.

Chapter 11

BOREHOLE SEISMIC PRINCIPLES

Borehole seismic is the placement of a geophone within the wellbore to relate borehole measurements to surface derived seismic measurements. Two different uses of geophone measurements are (1) using just the first arrival times at a relatively wide sampling interval to compute vertical traveltime measurements (a check-shot survey), or (2) using the entire signal wave train at close intervals to obtain a vertical seismic profile (VSP).

Seismograph Service Corporation, through first their Birdwell Division and later their Velocity and Special Projects Division, was the prime developer of borehole seismic. They offered check-shot surveys beginning in 1967 and vertical seismic profiles in 1977.

MEASUREMENT TECHNIQUE

The borehole seismic measurement system has three components: surface source, depth measurement, and a downhole geophone (Figure 11-1). The surface pulse can be anything used as a seismic source: vibrator, dynamite, air-gun, thumper, etc. However, as the number of recorded levels increases, dynamite becomes less suitable.

Ideally, like the surface seismic method, it would be better to use a single-source pulse and record simultaneously at different consecutive geophone stations. However, present technology limits require use of a single repositioning geophone used as often as necessary to obtain the desired downhole coverage.

For vertical traveltime determination, where only the first arrival time is measured, repeat uniformity of the source waveform is not critical, provided sufficient first break energy reaches the geophone. However, for

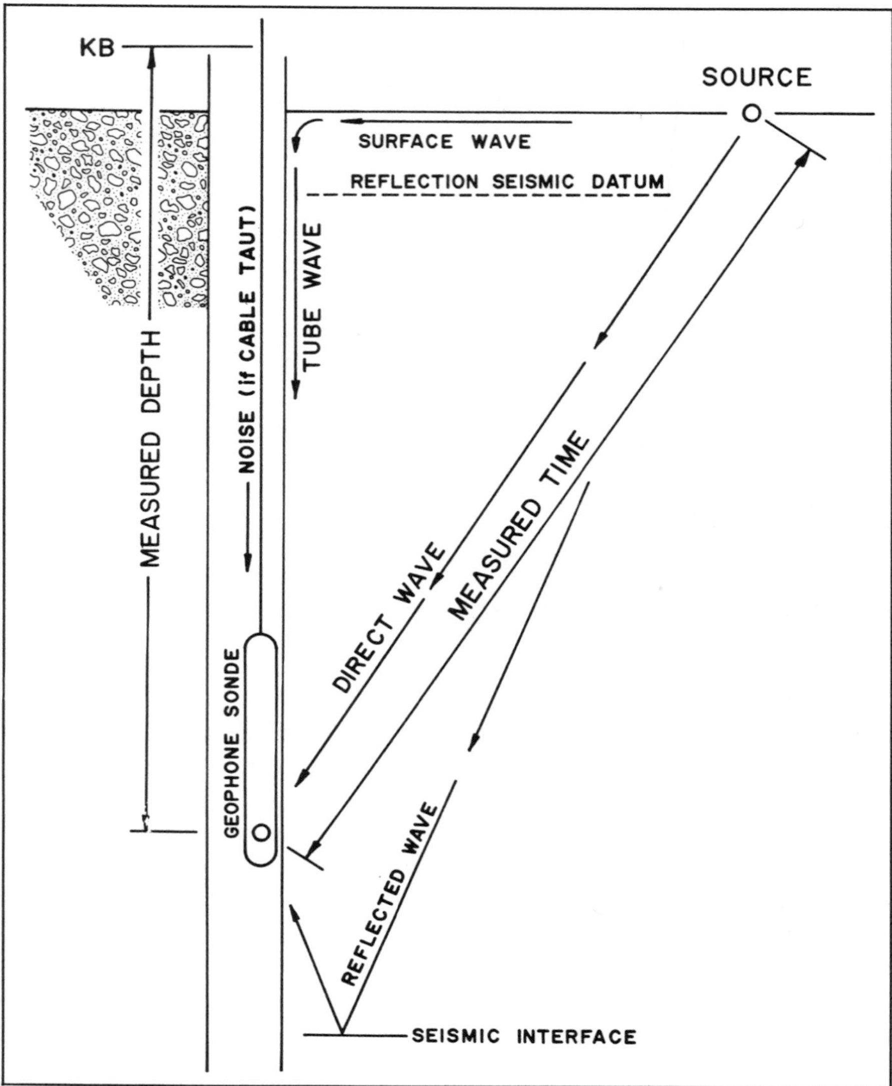

FIG. 11-1. The basic borehole seismic system.

a VSP shot pulse variations effect the downhole waveform and its interpretation. Are the resultant downhole recorded trace variations then a geologic or a source waveform change? Because achieving source uniformity can be difficult, some form of waveshaping may be necessary after recording. An inverse filter operator is designed to shape the near-source geophone recording into the desired-source waveform. This same operator is then applied to the downhole recorded trace.

Cable movement measurements are used for depth control of the downhole geophone. At present no service companies use a method to tie-in geophone stations to other logging measurements.

The downhole detector sonde contains several geophones, a downhole amplifier, and a wellbore clamping mechanism. Figure 11-2 shows

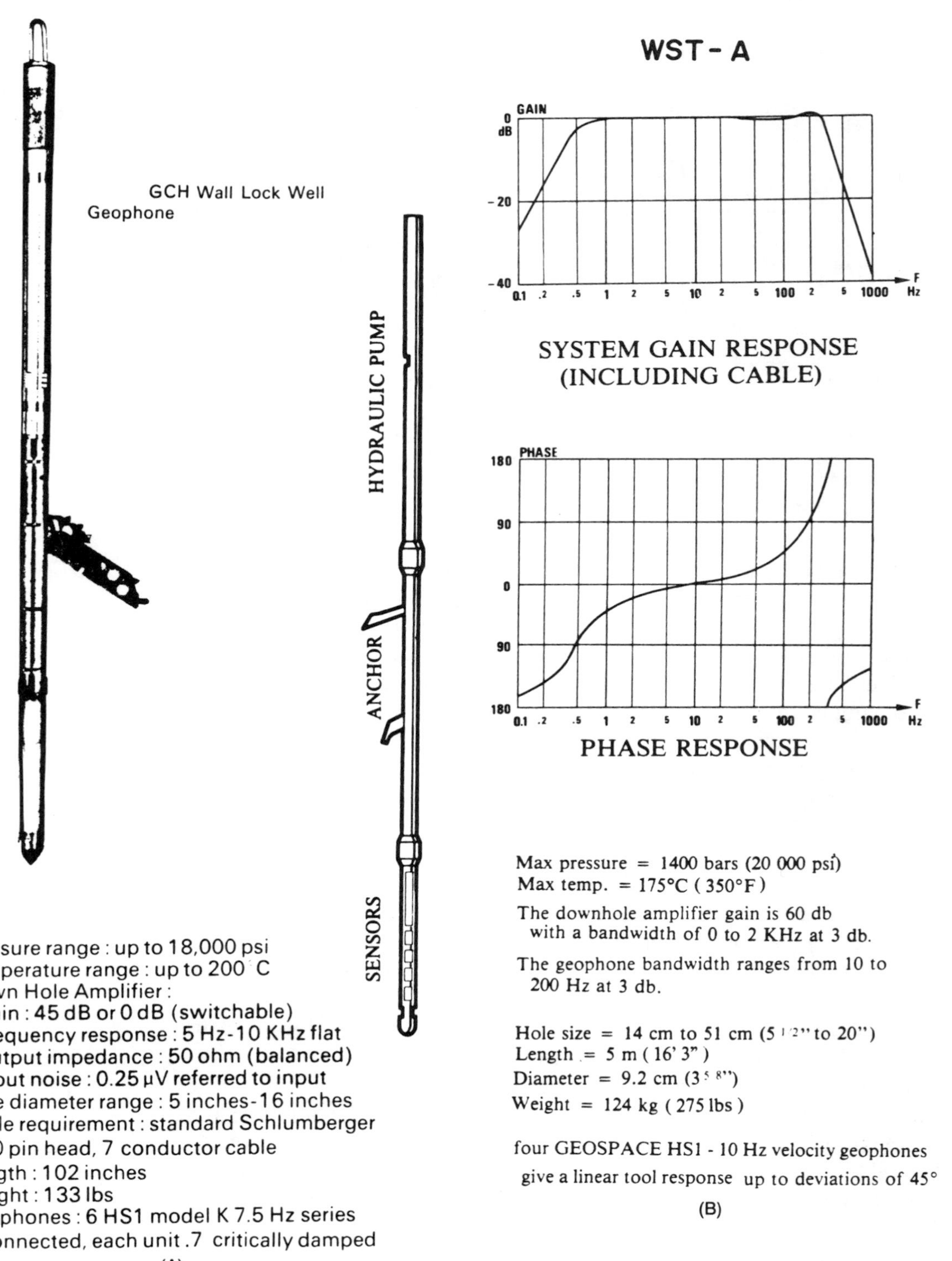

FIG. 11-2. Borehole geophone tool specifications (A) Birdwell/SSC, and (B) Schlumberger. Additional details are given in Appendix E.

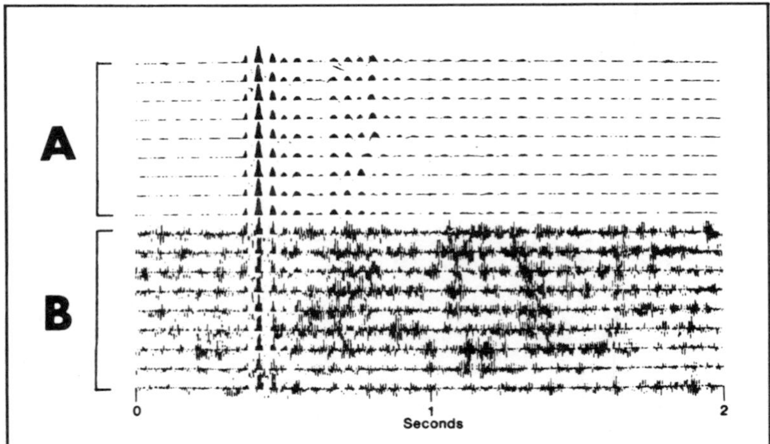

FIG. 11-3. Geophone record showing traces (A) recorded when the sonde was stationary due to adequate clamping and traces (B) due to noise from the sonde creeping down the wellbore. When the sonde was slipping (B), the first direct arrival was detectable, but not later arrivals. (Balch, et al., 1982.)

specifications for the two major downhole geophones. The geophone could be used suspended from the derrick like other logging sondes. This method suffers, however, from poor coupling to the formation and from surface noise being guided through the taut cable to the geophone. Instead, an extendable arm clamps the geophone sonde to the formation or casing. Once clamped the cable is slacked off which significantly reduces the cable as a noise source.

If the arm cannot hold the tool solidly to the wellbore, however, noise of considerable amplitude is generated as the sonde creeps downward (Figure 11-3). With check shots, tool creep is sometimes tolerated because the direct downward arrival is usually higher in amplitude than the creep noise. However with a VSP, the important lower-amplitude secondary arrivals can easily be buried in the tool creep noise. Tool creep can be eliminated by carefully choosing anchoring locations in consolidated, smooth, in-gauge hole sections of the wellbore. But for best interpretation results, uniform spacing between VSP stations is necessary. If the noise is not excessive, stacking of multiple shots as the geophone slowly creeps through the desired station is a solution.

Another operational problem is tube wave or tube noise. Tube noise is caused by the source ground wave striking the top of the borehole and propagating a strong amplitude wave down the borehole (Figure 11-1). The tube wave arrives downhole after the direct arrival and thus is not a problem in vertical traveltime determination. However, it interferes with the secondary arrivals on a VSP recording. The solution is to isolate the strong surface wave from the borehole top by either moving the source a sufficient distance from the wellbore top or by lowering the borehole fluid level or by both.

The analog recording obtained from each check-shot level (Figure 11-4)

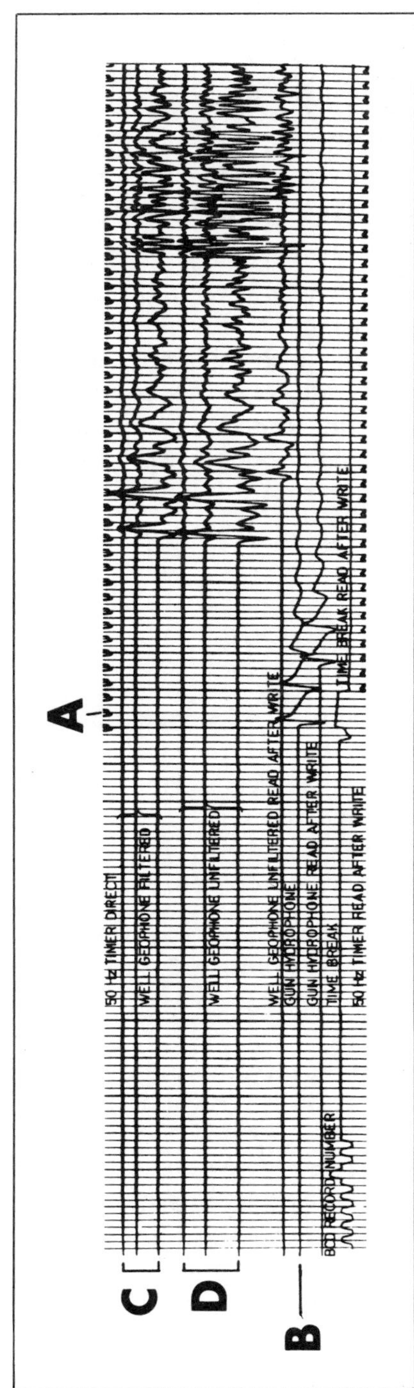

FIG. 11-4. Analog field record obtained from a borehole geophone tool: (A) timing reference, (B) near source geophone, (C) filtered borehole geophone, and (D) unfiltered filtered borehole geophone.

typically contains a timing signal (item A), near-source geophone trace (item B), and several different amplitude traces of filtered and unfiltered downhole geophone signal (items C and D). Items B, C, and D are also recorded digitally. Often more than one recording is made at each borehole station to allow stacking of the best traces to improve the signal-to-noise ratio. With digital recording, digital filtering and deconvolution can be applied after the data are gathered.

VERTICAL TRAVELTIME DETERMINATION

The first use of the borehole geophone was to measure the time a surface pulse arrives at points along the wellbore and thereby accurately determine depth-vertical time relationships which aid in reflection seismic interpretation. Using the field analog recordings or the processed, enhanced digital playbacks the elapsed time from source to geophone is measured. This measurement must be corrected to a vertical time, from a desired seismic datum, using simple geometry (Figure 11-5):

$$T_{gd} = (T \cos i) - \Delta sd \, V_e^{-1} \qquad \text{(11-1)}$$

and

$$i = \arctan (H D_{gs}^{-1}) \qquad \text{(11-2)}$$

where,

T_{gd} is vertical traveltime from seismic datum to geophone depth,
T is observed record time from source to geophone,
Δsd is difference between source elevation and seismic datum,
V_e is surface correctional velocity,
H is horizontal offset of source from borehole,
and D_{gs} is difference between source elevation and geophone depth.

A vertical borehole is assumed. Figure 11-6 shows examples of the data reduction calculations for a Vibroseis® survey. The correction to vertical time becomes less than 1 ms at an offset to depth (H:D_{gs}) ratio of greater than 1:25.

If dynamite is used, the source elevations and offsets will change for each station, making the correction to vertical time more complex (Figure 11-7).

If the borehole is deviated more than 8 degrees, the geometrical reduction to vertical traveltime will need true vertical depths of the geophone stations and their spacial location with reference to the wellhead and seismic source (Figure 11-8). A borehole directional survey, discussed in Chapter 12, should be run to provide accurate spacial locations of the geophone stations.

The conventional traveltime determination has a limited number of control points tied to specific depths, which yield a gross depth-time curve.

®Trade and service mark of Conoco, Inc.

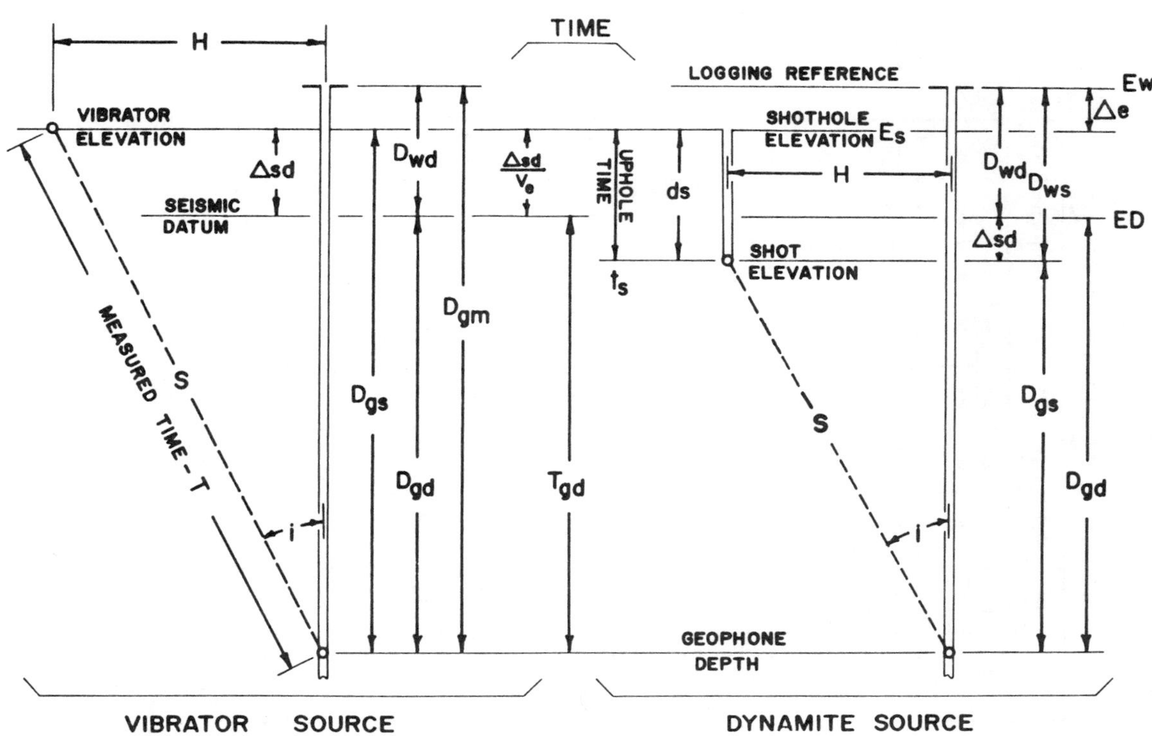

$$T_{gd} = (T \cdot \cos i) - \frac{\Delta sd}{V_e}$$

D_{gm} = Geophone depth below well elevation
D_{wd} = Difference between well elevation and elevation datum = Ew − ED
D_{gd} = Geophone depth below elevation datum = $D_{gm} - D_{wd}$
t_s = Uphole time in shothole
t_R = Refraction time from reference geophone
ds = Depth of shot
H = Horizontal distance from well to shothole
Δsd = Difference between shot elevation and elevation datum = E_s − ds − ED
D_{gs} = Geophone depth below shot elevation = $D_{gd} + \Delta sd$
cos i = $D_{gs}/\sqrt{H^2 + D_{gs}^2}$
T = Observed traveltime from shot to well geophone
Gr = Quality grade of well geophone "break"
T_{gs} = Traveltime for D_{gs} distance = T cos i
$\Delta sd/V_e$ = Time correction from shot to elevation datum
T_{gd} = Traveltime for D_{gd} distance = $T_{gs} - \Delta sd/V_e$
V_a = Average velocity to depth D_{gd} = D_{gd}/T_{gd}
ΔD_{gd} = Interval distance = $D_{gd_n} - D_{gd_m}$
ΔT_{gd} = Interval time for ΔD_{gd} distance = $T_{gd_n} - T_{gd_m}$
V_i = Interval velocity = $\Delta D_{gd}/\Delta T_{gd}$
S = Direct diagonal distance from shot to geophone = D_{gs}/cos i
ED = Elevation or reference datum
V_e = Elevation correction velocity
Δ_e = Difference between well elevation and shothole elevation = $E_w - E_s$

FIG. 11-5. Reduction of borehole geophone data to vertical traveltime assuming a vertical wellbore.

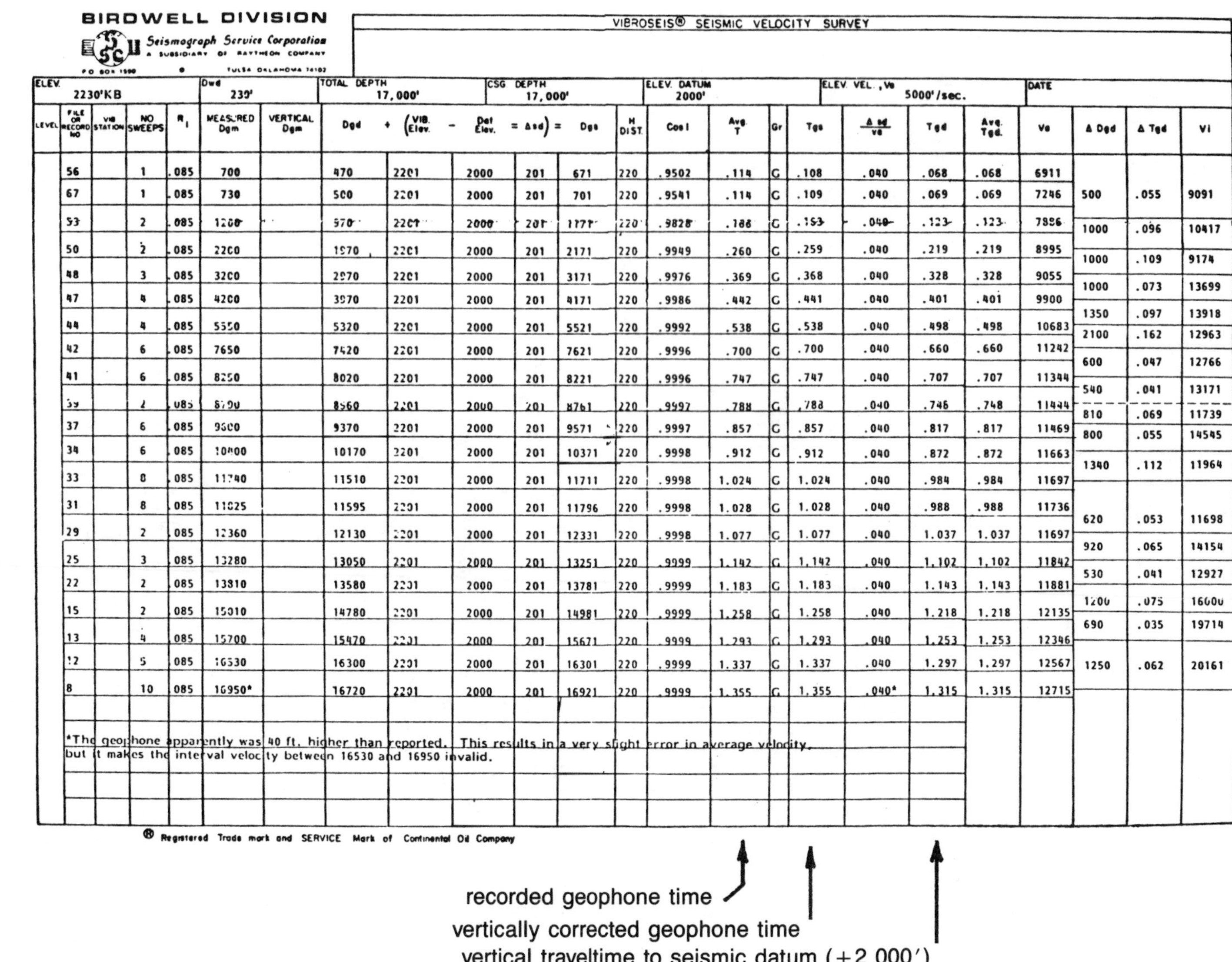

BIRDWELL DIVISION
Seismograph Service Corporation
A SUBSIDIARY OF RAYTHEON COMPANY
P.O. BOX 1590 • TULSA, OKLAHOMA 74102

VIBROSEIS® SEISMIC VELOCITY SURVEY

ELEV. 2230'KB | Dwd 239' | TOTAL DEPTH 17,000' | CSG DEPTH 17,000' | ELEV. DATUM 2000' | ELEV. VEL., Ve 5000'/sec. | DATE

LEVEL	FILE OR RECORD NO	VIB STATION	NO SWEEPS	R_1	MEASURED Dgm	VERTICAL Dgm	Dgd +	(VIB. Elev. −	Dat Elev.	= Δsd) =	Dgs	H DIST	Cos I	Avg. T	Gr	Tgs	Δsd/Ve	Tgd	Avg. Tgd	Ve	Δ Dgd	Δ Tgd	Vi
	56		1	.085	700		470	2201	2000	201	671	220	.9502	.114	G	.108	.040	.068	.068	6911			
	67		1	.085	730		500	2201	2000	201	701	220	.9541	.114	G	.109	.040	.069	.069	7246	500	.055	9091
	53		2	.085	1200		970	2201	2000	201	1171	220	.9828	.186	G	.153	.040	.123	.123	7886			
	50		2	.085	2200		1970	2201	2000	201	2171	220	.9949	.260	G	.259	.040	.219	.219	8995	1000	.096	10417
	48		3	.085	3200		2970	2201	2000	201	3171	220	.9976	.369	G	.368	.040	.328	.328	9055	1000	.109	9174
	47		4	.085	4200		3970	2201	2000	201	4171	220	.9986	.442	G	.441	.040	.401	.401	9900	1000	.073	13699
	44		4	.085	5550		5320	2201	2000	201	5521	220	.9992	.538	G	.538	.040	.498	.498	10683	1350	.097	13918
	42		6	.085	7650		7420	2201	2000	201	7621	220	.9996	.700	G	.700	.040	.660	.660	11242	2100	.162	12963
	41		6	.085	8250		8020	2201	2000	201	8221	220	.9996	.747	G	.747	.040	.707	.707	11344	600	.047	12766
	39		4	.085	8790		8560	2201	2000	201	8761	220	.9997	.788	G	.788	.040	.748	.748	11444	540	.041	13171
	37		6	.085	9600		9370	2201	2000	201	9571	220	.9997	.857	G	.857	.040	.817	.817	11469	810	.069	11739
	34		6	.085	10400		10170	2201	2000	201	10371	220	.9998	.912	G	.912	.040	.872	.872	11663	800	.055	14545
	33		8	.085	11740		11510	2201	2000	201	11711	220	.9998	1.024	G	1.024	.040	.984	.984	11697	1340	.112	11964
	31		8	.085	11825		11595	2201	2000	201	11796	220	.9998	1.028	G	1.028	.040	.988	.988	11736			
	29		2	.085	12360		12130	2201	2000	201	12331	220	.9998	1.077	G	1.077	.040	1.037	1.037	11697	620	.053	11698
	25		3	.085	13280		13050	2201	2000	201	13251	220	.9999	1.142	G	1.142	.040	1.102	1.102	11842	920	.065	14154
	22		2	.085	13810		13580	2201	2000	201	13781	220	.9999	1.183	G	1.183	.040	1.143	1.143	11881	530	.041	12927
	15		2	.085	15010		14780	2201	2000	201	14981	220	.9999	1.258	G	1.258	.040	1.218	1.218	12135	1200	.075	16600
	13		4	.085	15700		15470	2201	2000	201	15671	220	.9999	1.293	G	1.293	.040	1.253	1.253	12346	690	.035	19714
	12		5	.085	16530		16300	2201	2000	201	16301	220	.9999	1.337	G	1.337	.040	1.297	1.297	12567	1250	.062	20161
	8		10	.085	16950*		16720	2201	2000	201	16921	220	.9999	1.355	G	1.355	.040*	1.315	1.315	12715			

*The geophone apparently was 40 ft. higher than reported. This results in a very slight error in average velocity, but it makes the interval velocity between 16530 and 16950 invalid.

® Registered Trade mark and SERVICE Mark of Continental Oil Company

FIG. 11-6. Reduction of borehole geophone data to vertical traveltime using the terms and definitions of Figure 11-5.

SEISMIC REFERENCE SERVICE

Company:
Well:
Location:

Elev.: +1315' K.B.
Elev. Datum: +1169'
Elev. Velocity (ve): *
Dwd: 146'
Csg. Depth: 20460'
Total Depth
Date:

Rec. No.	Dgm	Dgd	Formation	S.P. No.	Shot Seq.	Dyn. Chg.	t_S	t_R	Elev.	d_S	H Dist.	Δsd	Dgs	cos i	T	Gr.	Tgs	$-\frac{\Delta sd}{ve}$	Tgd	Average Tgd	Average Velocity (va)	Δ Dgd	Δ Tgd	Interval Velocity (vi)
2	1800	1654		A3	A	10	.020	.086	1281	96	740	+16	1670	.91426	.168	G	.154	0	.154					
1	1800	1654		B2	A	10	.023	.102	1282	96	595	+17	1671	.94206	.169	G	.159	-.004	.155	.1545	10,706			
3	3000	2854		A4	A	10	.020	.086	1281	96	740	+16	2870	.96833	.258	G	.250	0	.250	.2485	11,485	1200	.0940	12766
5	4200	4054		A2	A	10	.021	.088	1281	96	735	+16	4070	.98408	.329	F	.324	0	.324					
4	4200	4054		B3	A	10	.017	.106	1282	96	590	+17	4071	.98966	.334	G	.331	-.010	.321	.3225	12,571	1200	.0740	16216
6	5300	5154		B4	A	25	.023	.102	1282	90	590	+23	5177	.99357	.413	G	.410	-.004	.406	.4080	12,632	1100	.0855	12865
7	6500	6354		A1	A	25	.019	.079	1281	90	735	+22	6376	.99342	.498	F	.495	-.001	.494	.4910	12,941	1200	.0830	14458
19	8100	7954		A5	C	100	.024	.103	1281	50	750	+62	8016	.99563	.586	G	.583	0	.583	.5795		3300	.1585	20820
8	9800	9654		B5	A	50	.027	.115	1282	130	585	-17	9637	.99816	.646	G	.645	0	.645	.6495	14,864			
18	11400	11254		A7	B	100	.010	.106	1283	10	815	+104	11358	.99744	.745	G	.743	-.010	.733	.7280	15,459	1600	.0785	20382
9	13000	12854		B5	B	100	.027	.116	1282	110	585	+ 3	12857	.99897	.804	F	.803	0	.803	.8090	15,889	1600	.0810	19753
17	14600	14454		A8	B	200	.025	.104	1282	45	825	+68	14522	.99839	.920	G	.919	0	.919	.9125	15,840	1600	.1035	15459
10	16250	16104		B6	A	200	.023	.115	1285	70	855	+46	16150	.99860	1.000	G	.999	-.004	.995	1.0025	16,064	1650	.0900	18333
16	18350	18204		A8	A	200	.019	.104	1282	70	825	+43	18247	.99898	1.149	G	1.148	-.001	1.147	1.1385	15,989	2100	.1360	15441
11	20450	20304		A5	A	200	.021	.100	1281	70	750	+42	20346	.99932								2100	.1230	17073
12	20450	20304		B6	B	200	.024	.113	1285	95	855	+21	20325	.99912	1.256	P	1.255	-.003	1.252	1.2615	16,095			
15	21800	21654		A6	B	300	.024	.100	1284	35	805	+80	21734	.99931	1.352	F	1.351	0	1.351	1.3410	16,148	1350	.0795	16981
14	24600	24454		A6,7A		400	.021	.101	1283	50	805	+64	24518	.99946	1.512	P	1.511	0	1.511			2800	.1585	17666
13	24600	24454		B789A		400	.026	.121	1286	85	875	+32	24486	.99936	1.490	F	1.489	-.001	1.488	1.4995	16,308			

The uphole times of A3-A and B5-A were used as control to correct "A" and "B" side check shots, respectively to datum. The average shot elevation of A3-A and B5-A was used as seismic datum. Corrections were made by subtracting the difference between the uphole time of the control record and the uphole time of the record being corrected from Tgs, as follows: Tgd = Tgs - (Ts - Ts)
A3-A or B5-A

Useable data was not recorded from shot locations on both sides of the well at all check levels; so an adjustment was necessary to provide comparable average vertical times at each level. The single values were corrected by adding, or subtracting, as necessary, one-half the differential, as shown on the Graphic Comparison of Sides.

The 8100' check shot level was not used in final results or adjustments to the Sonic Log. The time indicates the well geophone obviously did not reach the logged depth.

FIG. 11-7. Correction to vertical time for dynamite data.

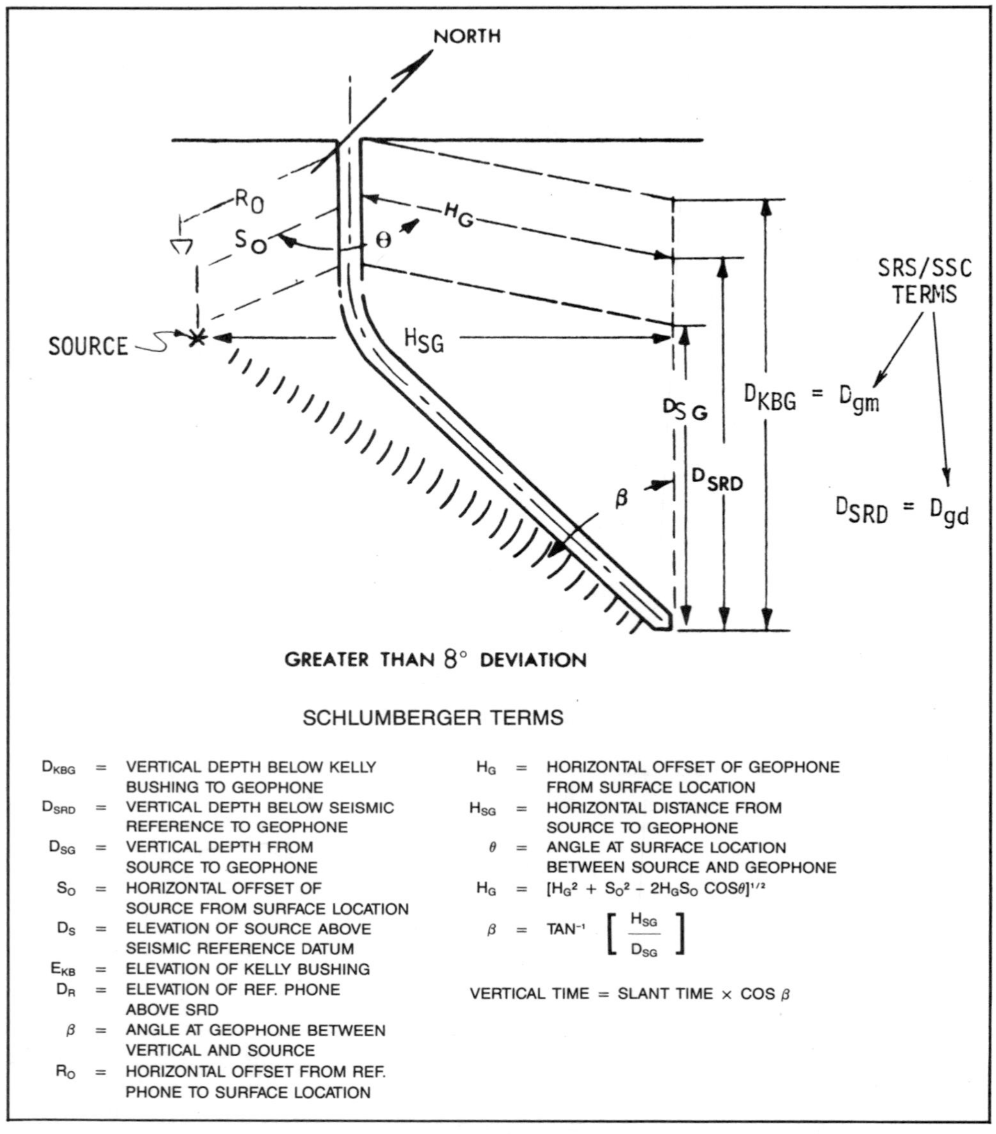

FIG. 11-8. Geometry for reduction of geophone data to vertical traveltime for a deviated borehole.

A linear interpolation between geophone points is rarely exact. For a more accurate and detailed depth-to-time conversion, the acoustic log is used to interpolate between geophone stations. First the acoustic log is integrated and adjusted to the seismic datum by using the time from the shallowest geophone station (Table 11-1). A comparison is made between the two "running" times by subtracting the geophone times from the datum adjusted and integrated acoustic log times. The resultant deviations are plotted for each geophone station depth (Figure 11-9). The inverted term is drift; geophone time minus integrated acoustic log time.

The deviation plot shows the intervals where the acoustic log differs

Table 11-1. Conversion of integrated acoustic log times to running time from the seismic datum, and calculation of deviation.

Depth (ft)	Geophone (ms) from datum)	BHC Integration	Adjusted BHC	Deviation (Acoustic-Geophone)
5 040	446.0	1 380.8	446.0	0.0
6 810	573.0	1 253.8	573.0	0.0
8 580	722.0	1 105.0	721.8	−0.2
10 110	844.0	982.7	844.1	+0.1
11 840	972.0	843.0	983.8	+11.8
13 750	1 124.0	691.8	1 135.0	+11.0
15 115	1 239.0	576.0	1 250.8	+11.8
16 890	1 382.0	419.5	1 407.3	+25.3
18 000	1 472.0	328.4	1 498.4	+26.4
18 540	1 510.5	289.9	1 536.9	+26.4
19 175	1 567.5	233.5	1 593.3	+25.8
21 650	1 707.0	92.9	1 733.9	+26.9
22 560	1 756.0	43.3	1 783.5	+27.5
23 437	—	BHC FR	1 826.8	

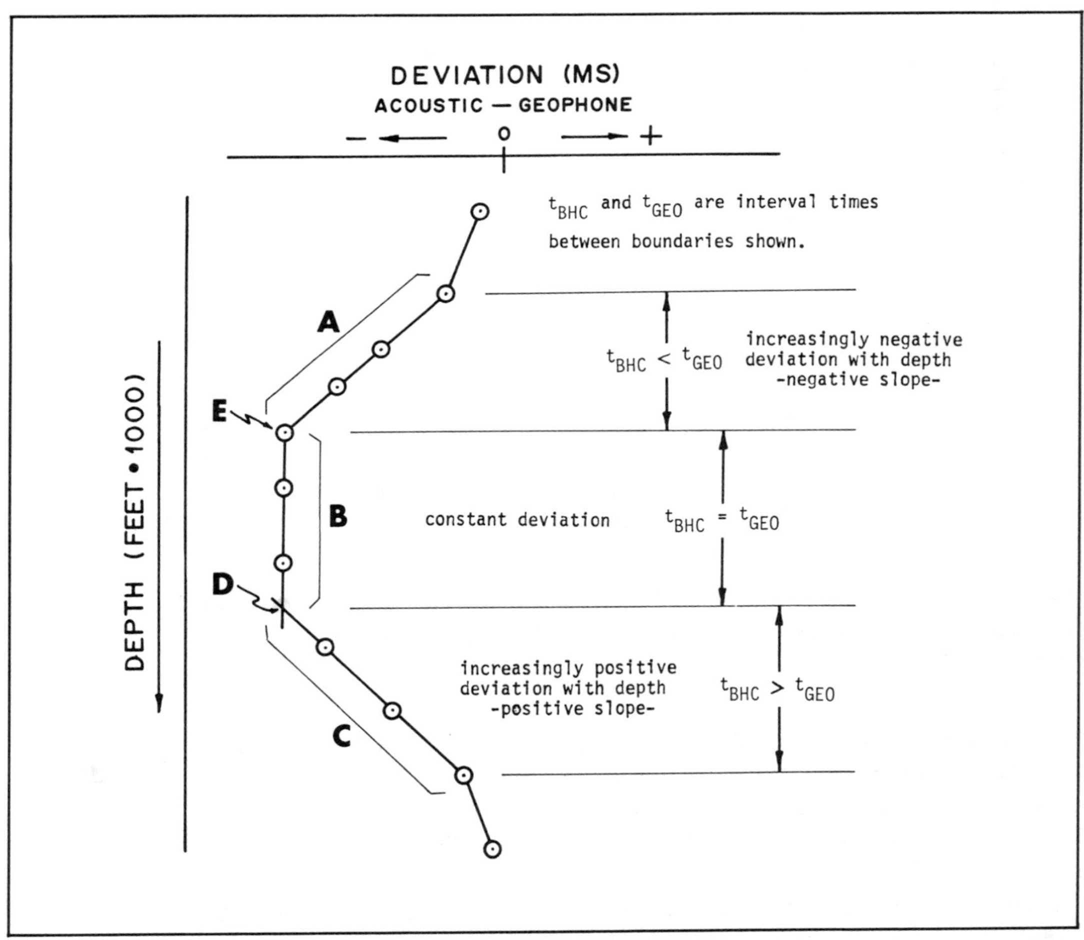

FIG. 11-9. Deviation plot showing a comparison of integrated acoustic log times minus borehole geophone times at geophone depths. Note that interval A is bounded (“kneed”) by geophone stations, but at D there is no geophone station. Thus some correction technique must be available that is not directly dependent on the presence of a geophone station.

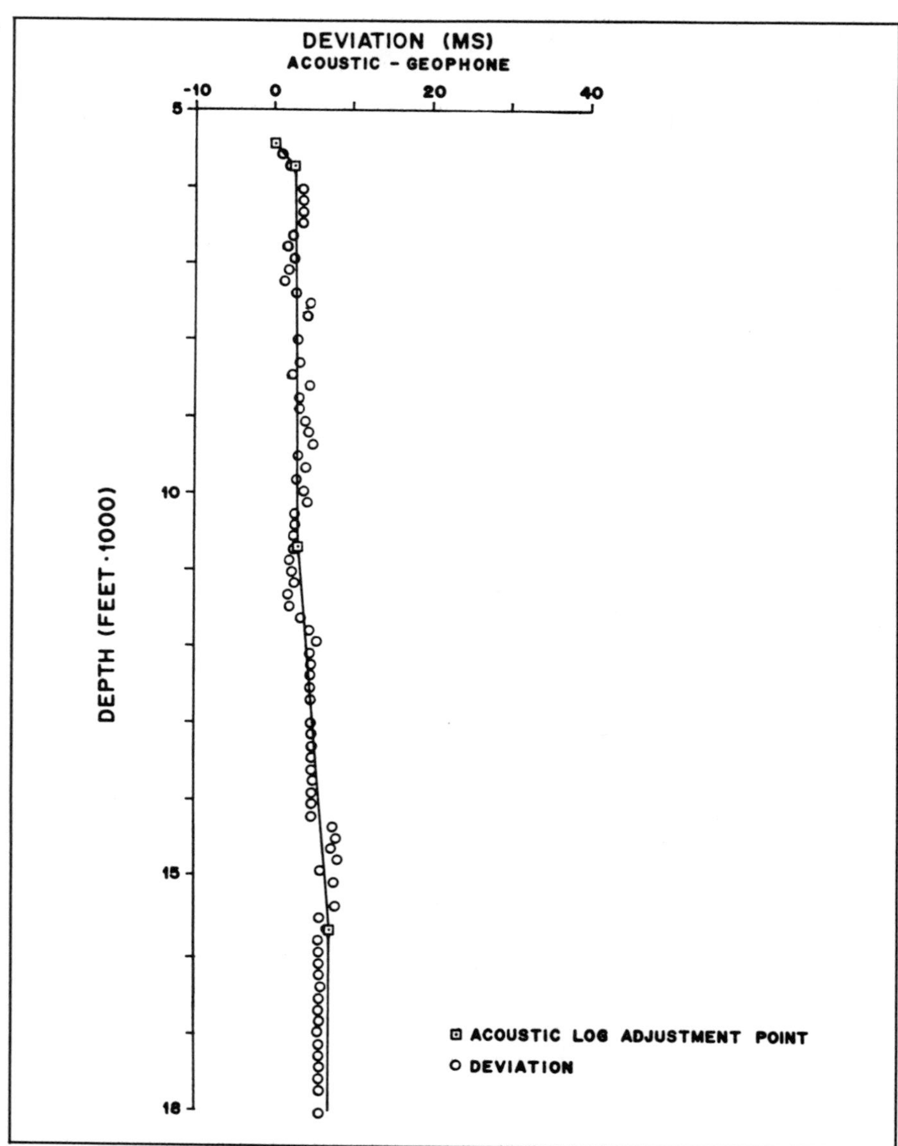

FIG. 11-10. VSP deviation plot showing greater detail for determining deviation trends.

from the interval geophone times. In intervals where the deviation becomes increasingly negative with increasing depth (Figure 11-7, interval A), or negative slope, the integrated acoustic log time is less than the geophone time for that interval. Here the acoustic log Δt needs to be increased to match interval geophone time.

Because a continuous running time is used in the comparison, intervals where geophone and integrated-acoustic times are equal are shown by constant, vertical deviation (interval B in Figure 11-9). In this case the deviation difference can be a positive, zero, or negative constant. In these intervals the integrated acoustic log can be used as is for interpolation between geophone stations.

In intervals where the deviation is increasingly positive with depth, or a positive slope, the geophone interval time is greater than integrated acoustic-log interval time. Here the acoustic log Δt values are decreased for the interval-integrated time to match the geophone interval time.

With the VSP a considerably more detailed deviation plot is obtained (Figure 11-10) which greatly aids the choice of correction intervals.

DIFFERENCES BETWEEN ACOUSTIC LOG AND GEOPHONE TIME

Differences between the integrated acoustic log times and geophone times result from many causes. Goetz, et. al. (1979) gives a complete discussion of error sources of the two techniques.

One error source is the accuracy of the two techniques' basic measurements. The acoustic log uses a time base measuring system that yields traveltime accuracies of 0.4 μs/ft for each individual measurement of the compensated sum. A geophone record is typically picked to the nearest millisecond.

For the geophone survey measurements, the determined times can be too long because of the electronic circuitry or because of the method used to pick the geophone records. Electronic errors are a result of either delays in the filtering circuits or signal amplifier DC offsets.

Geophone survey times can be too short under conditions of relative borehole dip and lateral formation changes, as well as through operational errors. For relative borehole dip either the formation can be dipping or the borehole deviated. In either, an increasing velocity with depth creates a faster refracted path as compared to the borehole path (Figure 11-11). Thus geophone time is less than acoustic log time ($t_{BHC} > t_{GEO}$). This difference for shallow depths can be greater than 5 ms one-way time (Figure 11-11, item B) and for relative bed angles greater than 25 degrees (Figure 11-11, item C).

Wells drilled close to but not through high-velocity formations such as near salt domes, volcanic dikes, or massive reefs can present a geophone first arrival path of considerably shorter time than that measured by the acoustic log.

Operational problems, such as the source being at a different depth, or the geophone being at a different depth than desired can contribute to an incorrect geophone time. This is especially true if the source is deeper than expected.

Some difference is possible from the frequency content of the seismic source. Experiments in Ward and Hewitt (1977) show a difference in deviation for a dynamite, a 55 Hz, and a 35 Hz source (Figure 11-12). As the mono-frequency becomes lower, the difference compared to the acoustic log increases.

Errors are just as likely to occur during acoustic log measurements. Noise and cycle skip can result in integrated acoustic log times either too

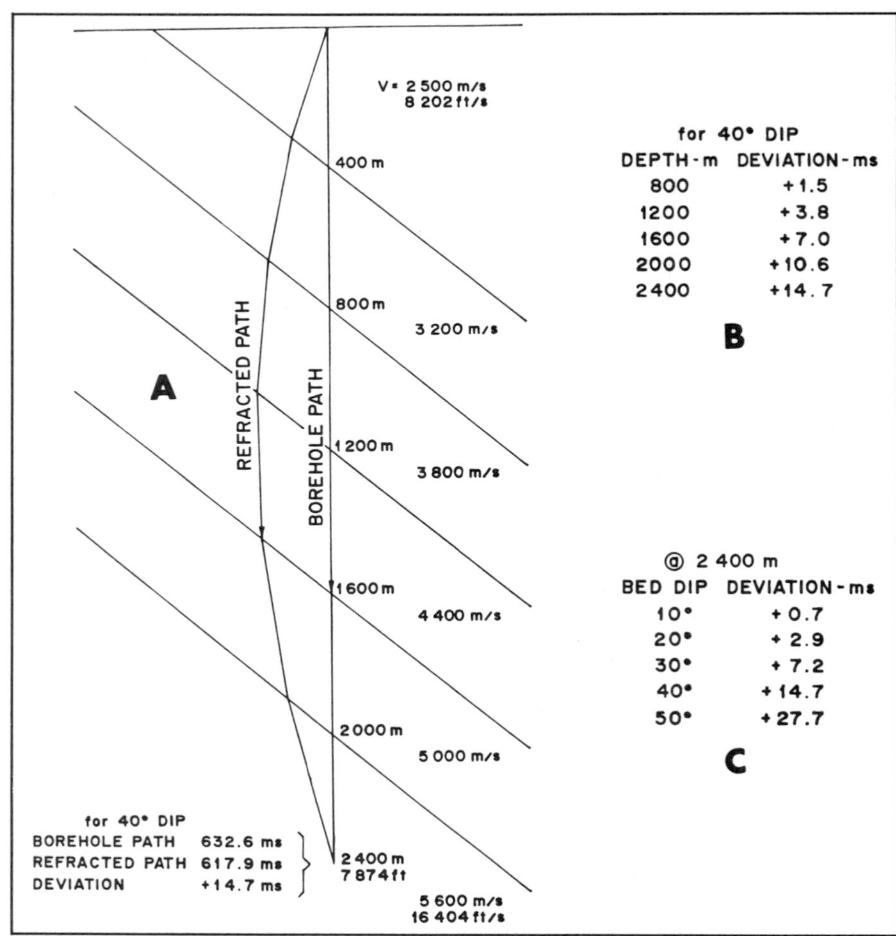

FIG. 11-11. Large relative bed dipping due to formation dip or deviated borehole. Analysis of a six layer model (A) shows the borehole path time is greater than the refracted path time ($t_{BHC} > t_{GEO}$). Further, the difference can be significant (greater than 5 ms) at shallow depths (B) and for beds dipping greater than 25 degrees (C).

long or too short depending upon which receiver is affected. Measurement detection "stretch," large hole conditions, and formation alteration can produce acoustic logs with too long an integrated time. Formations with traveltimes greater than mud traveltime, near-borehole velocity inversions, the presence of gas in the pore volume, and high-formation dip can all produce integrated times that are too short. These were all discussed in Chapter 10.

ACOUSTIC LOG CORRECTION TO GEOPHONE TIME

Using the information from the deviation plot, a correction can be obtained for the acoustic log to achieve agreement with the vertical traveltime survey and to provide fine detail in time-depth conversion.

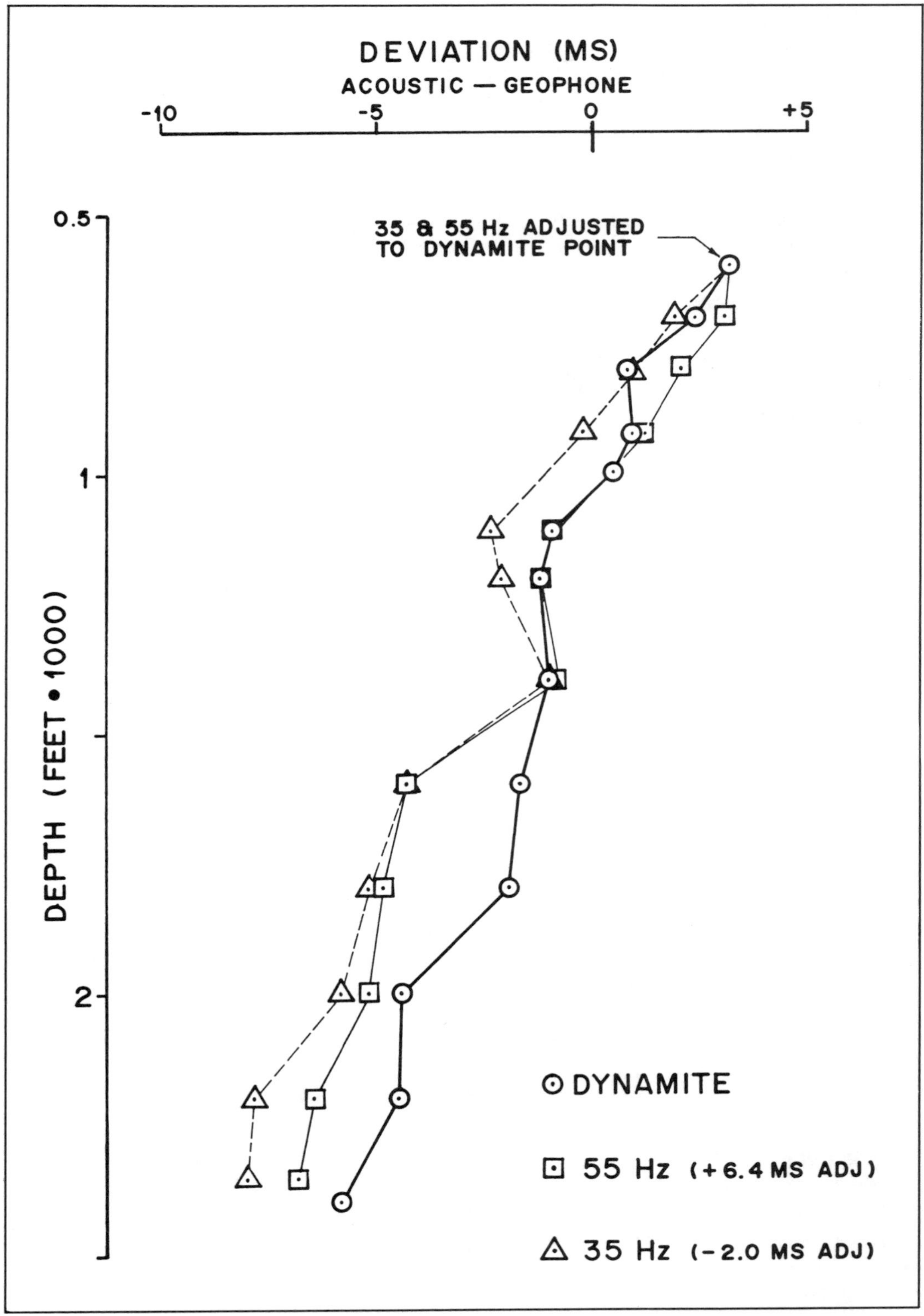

FIG. 11-12. Deviation plot comparison of seismic source frequency (Ward and Hewitt, 1977).

The deviation plot is used to pick interval boundaries for applying corrections to the acoustic log. These boundaries are the intersections of linear deviation segments (Figure 11-9, points D and E). Correctional boundaries may or may not correspond to geophone stations. When corrections are applied to the acoustic log a tare may result that can produce a change in reflectivity for a synthetic seismogram from the corrected acoustic log. Thus, correction boundaries should be very carefully chosen, preferably at distinct geologic and acoustic log changes.

The simplest correction techniques are either a linear or a differential adjustment. The linear correction is used when the interval integrated acoustic time is less than the interval geophone time, or for a negative slope region of the deviation plot (Figure 11-13 above 6 500 ft). This method adds a fixed time to every Δt measurement for that interval. In effect the Δt measurement is shifted a positive amount. If there are geophone stations at the deviation curve slope changes, the linear correction factor is:

$$\Delta t \text{ correction} = (t_{GEO} - t_{BHC})\, \Delta Z^{-1} \qquad (11\text{-}3)$$

where t_{GEO} is the interval geophone time, t_{BHC} is the interval integrated acoustic log time, and ΔZ is the interval thickness. For example, in Figure 11-13:

Depth (ft)	BHC time	Geophone time	Deviation (ms)
3 000	249.7	248.5	+1.2
6 500	488.0	491.0	−3.0
Interval time	238.3	242.5	−4.2

$$\Delta t \text{ correction} = \frac{242.5\text{-}238.3}{3\ 500} = +1.2\ \mu\text{s/ft}.$$

Thus for the interval 3 000 to 6 500 ft the acoustic log must be shifted to increase Δt by 1.2 μs/ft.

If the correction boundaries are not at a geophone station then equation (11-3) must be generalized using deviation curve projected values at the boundaries:

$$\Delta t \text{ correction} = (\text{Dev}_{upper} - \text{Dev}_{lower})\, \Delta Z^{-1} \qquad (11\text{-}4)$$

where Dev_{upper} is projected deviation curve value at the upper boundary, and Dev_{lower} is projected deviation curve value at the lower boundary. It can be seen that the same results are obtainable in the example using the deviation values.

A fractional correction is used when the interval integrated acoustic log time is greater than the interval geophone time, or when the deviation plot has a positive slope. This correction method assumes that the highest transit times caused by borehole conditions, such as caving, mud filtrate invasion, etc., are the major contributors to differences in the time measurements. This method then applies a fractional multiplier to each acoustic log Δt reading for that interval. Thus higher transit times, or lower velocities, are decreased more than lower transit times. If there

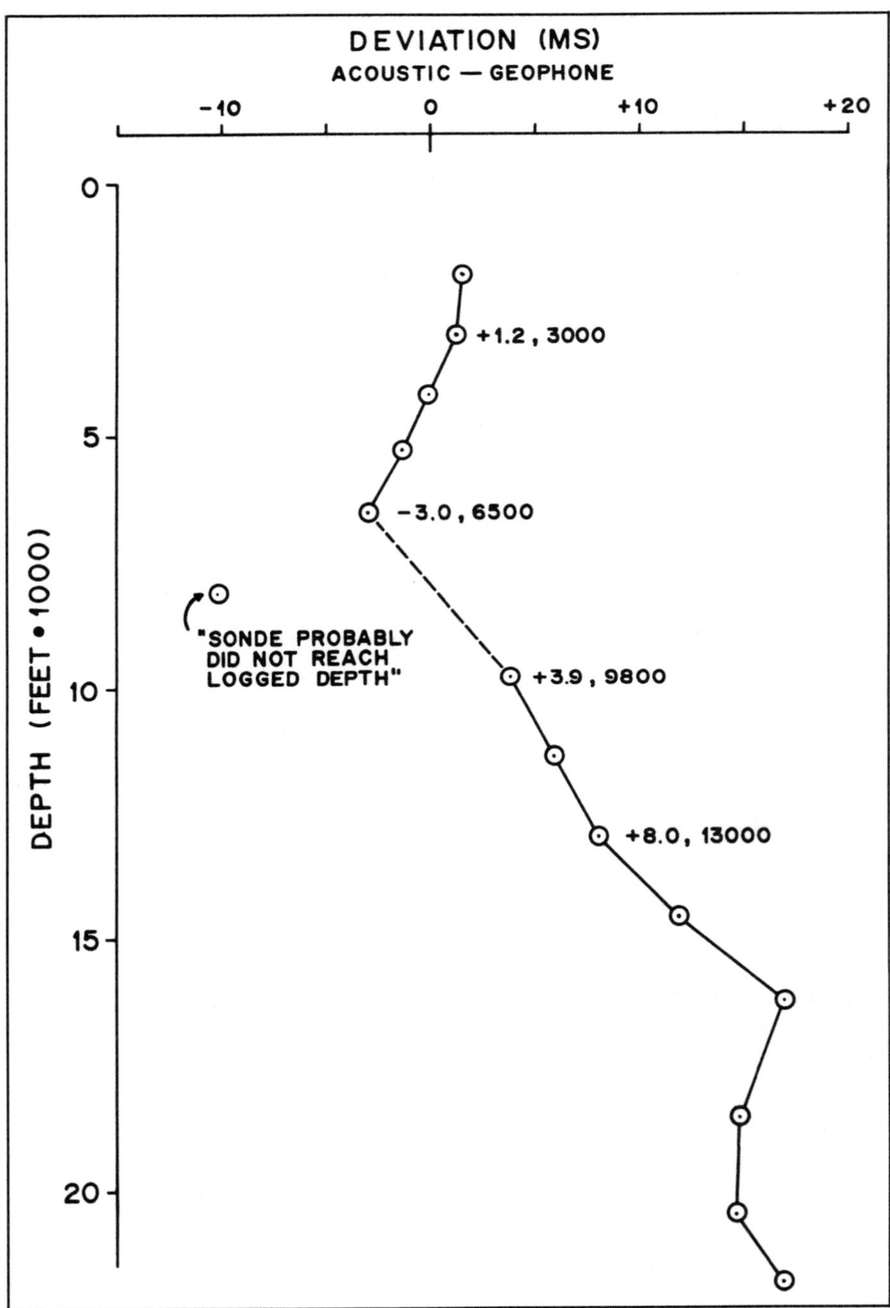

FIG. 11-13. Deviation plot. The interval from 3 000 to 6 500 ft shows $t_{BHC} < t_{GEO}$ or increasing negative deviation with depth, indicating linear correction is needed. The interval from 9 800 to 13 000 ft shows $t_{BHC} > t_{GEO}$ or increasingly positive deviation, indicating differential fraction correction is needed. The geophone time at 8 100 ft appears to be an incorrect measurement.

are geophone stations at the deviation plot changes, the differential fraction is:

$$\text{fraction} = t_{GEO}\, t_{BHC}^{-1} \tag{11-5}$$

For example, in Figure 11-13:

Depth (ft)	BHC time	Geophone time	Deviation (ms)
9 800	653.4	649.5	+3.9
13 000	817.0	809.0	+8.0
Interval time	163.6	159.5	+4.1

$$\text{fraction} = \frac{159.5}{163.6} = 0.974\,9\,.$$

If the correction point boundaries are not at geophone stations, then equation (11-5) is generalized using deviation plot projected values at the boundaries:

$$\text{fraction} = \frac{t_{BHC} - \text{Dev}_{lower} + \text{Dev}_{upper}}{t_{BHC}}\,. \tag{11-6}$$

A refinement of the fractional correction method is to apply a correction only above a defined traveltime. The problem is how to properly determine the traveltime value to begin applying the correction.

The above correction methods sometimes use the general correction formula

$$\Delta t_{corrected} = (\alpha \cdot \Delta t_{log}) + \beta \tag{11-7}$$

where α is the fractional correction term determined by equation (11-5) or (11-6), and β is the linear correction term determined by equation (11-3) or (11-4).

Once the corrections are calculated, a geophone calibrated velocity log (Figure 11-14) can be plotted from the correct acoustic log. This presentation normally contains the adjusted traveltime (item A of Figure 11-14), the original acoustic log integration (item B), the corrected acoustic log integration (item C), and annotated geophone stations (item D). These are normally presented in both linear depth and linear two-way time scales (Figure 11-15).

Also included will be the data used to determine the original deviation plot (Figure 11-16, item A) and the residual deviation after correction of the acoustic log (item B).

VERTICAL SEISMIC PROFILE

The prime event on a geophone recorded trace is the first arrival of the downgoing, direct-source wave. Closer examination of the trace shows that there are many other arrivals. These later arrivals have origins from both above and below the geophone station. If many stations are recorded closely along the wellbore, these later secondary arrivals can be traced

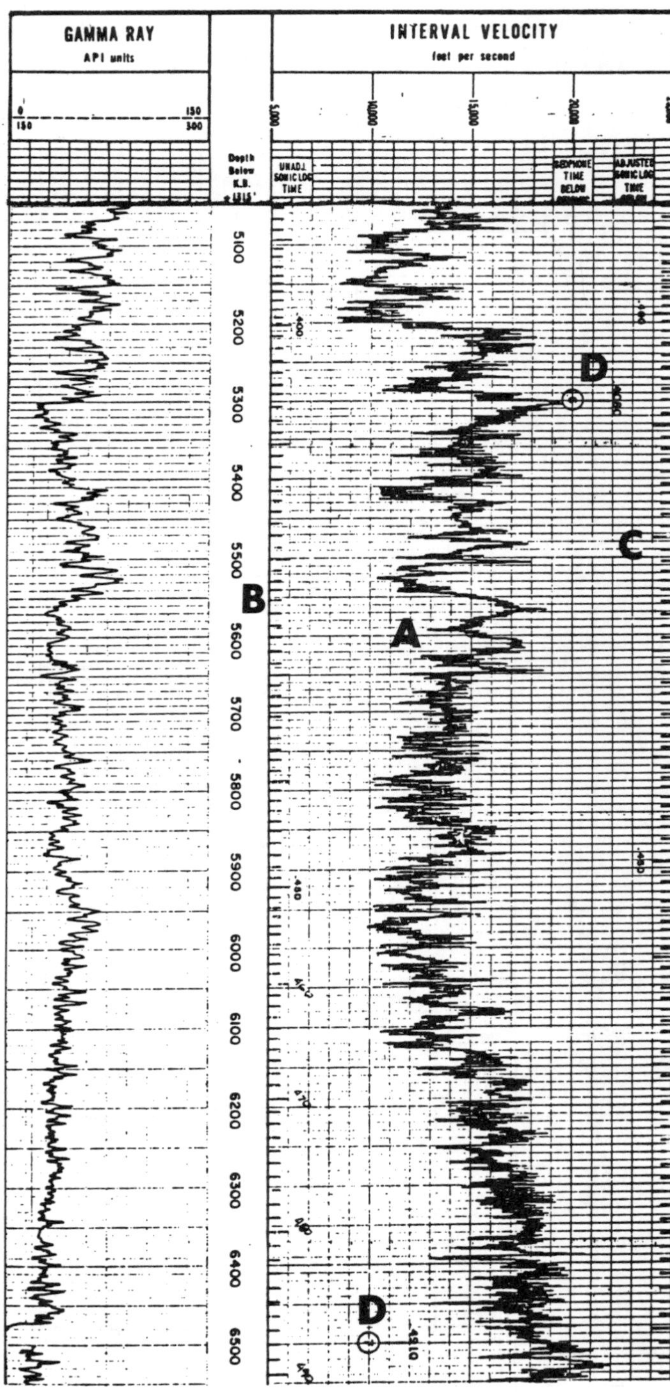

FIG. 11-14. Calibrated velocity log in a linear-with-depth format with (A) adjusted acoustic transit time, (B) original acoustic log integration, (C) adjusted acoustic log integration, and (D) geophone stations and times.

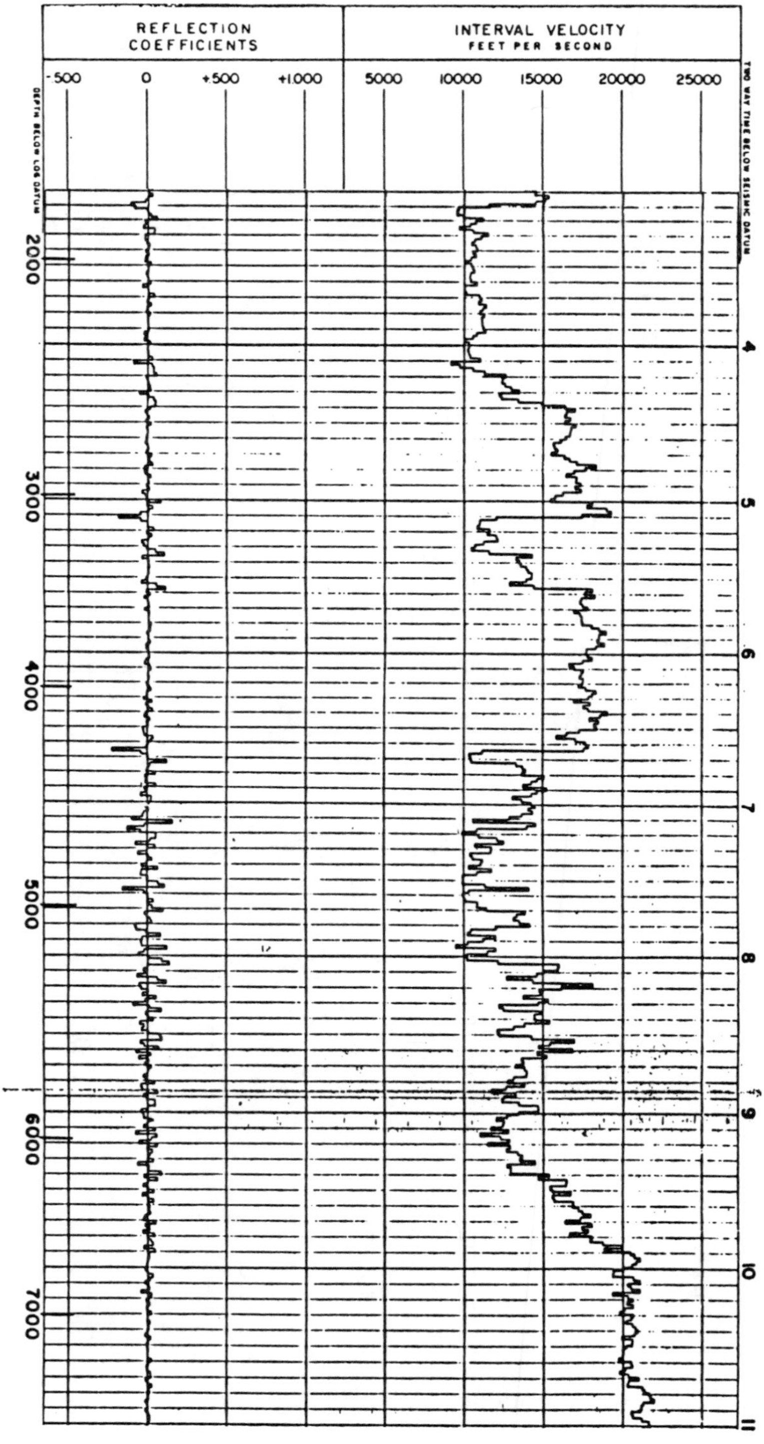

FIG. 11-15. Calibrated velocity log in a linear two-way time format.

TIME COMPARISON

DEPTHS BELOW		SRS	SONIC LOG - UNADJUSTED						SONIC LOG - ADJUSTED					
Log Datum K.B.	Seismic Datum +1169'	VERTICAL TIME	VERTICAL TIME	AVERAGE VELOCITY	INTERVAL DEPTH	INTERVAL TIME	INTERVAL VELOCITY	DIFFERENCE	DIFFERENCE	VERTICAL TIME	AVERAGE VELOCITY	INTERVAL DEPTH	INTERVAL TIME	INTERVAL VELOCITY
* 146	0	.0000	-.0002					-.0002	-.0002	-.0002				
1,800	1,654	.1545	.1560	10,603	1654	.1562	10,589	+.0015	'-.0001	.1544	10,712	1654	.1546	10,699
3,000	2,854	.2485	.2497	11,430	1200	.0937	12,807	+.0012	-.0001	.2484	11,490	1200	.0940	12,766
4,200	4,054	.3225	.3224	12,574	1200	.0727	16,506	-.0001	+.0001	.3226	12,567	1200	.0742	16,173
5,300	5,154	.4080	.4066	12,676	1100	.0842	13,064	-.0014	±.0000	.4080	12,632	1100	.0854	12,881
6,500	6,354	.4910	.4880	13,020	1200	.0814	14,742	-.0030	-.0002	.4908	12,946	1200	.0828	14,493
8,100	7,954	.5795	.5693	13,972	1600	.0813	19,680	-.0102		.5687	13,986	1600	.0779	20,539
9,800	9,654	.6495	.6534	14,775	1700	.0841	20,214	+.0039	±.0000	.6495	14,864	1700	.0808	21,040
11,400	11,254	.7280	.7339	15,335	1600	.0805	19,876	+.0059	+.0002	.7282	15,455	1600	.0787	20,330
13,000	12,854	.8090	.8170	15,733	1600	.0831	19,254	+.0080	+.0002	.8092	15,885	1600	.0810	19,753
14,600	14,454	.9125	.9244	15,636	1600	.1074	14,898	+.0119	-.0001	.9124	15,842	1600	.1032	15,504
16,250	16,104	1.0025	1.0194	15,798	1650	.0950	17,368	+.0169	+.0001	1.0026	16,062	1650	.0902	18,293
18,350	18,204	1.1385	1.1532	15,786	2100	.1338	15,695	+.0147	+.0001	1.1386	15,988	2100	.1360	15,441
20,450	20,304	1.2615	1.2760	15,912	2100	.1228	17,101	+.0145	-.0002	1.2613	16,098	2100	.1227	17,115
21,800	21,654	1.3410	1.3579	15,947	1350	.0819	16,484	+.0169	-.0006	1.3404	16,155	1350	.0791	17,067
24,600	24,454	1.4995	1.5169	16,121	2800	.1590	17,610	+.0174	+.0001	1.4996	16,307	2800	.1592	17,588

*Seismic Datum

A B

FIG. 11-16. Tabulation showing deviation plot information (A) before correction to geophone times and, (B) after correction.

to their origins (Figure 11-17). In effect a seismic section is created along the borehole using a borehole geophone and a surface seismic source — a VSP. The ability to trace reflections to their origins and determine the subsurface character *and* its surface expression are the prime features of the VSP. The problem of changing source waveform for the many geophone stations was discussed earlier in this chapter.

Two operational problems discussed earlier that effect VSP quality and usability are geophone sonde creep and tube waves. Because the tube wave, shown in Figure 11-18, is in the same frequency bandpass as the downgoing direct wave, frequency filtering cannot be used to remove this problem. The best approach is to avoid generating a tube wave. Some success can be achieved in reducing the tube wave after data gathering by building downhole arrays in the computer or a velocity filter.

Because the VSP recording (Figure 11-19) actually contains two sets of information, two different steps are needed in processing the data. First, remove the first break time from each trace causing alignment of the first arrival pulse (Figure 11-20), allowing observation of the downgoing

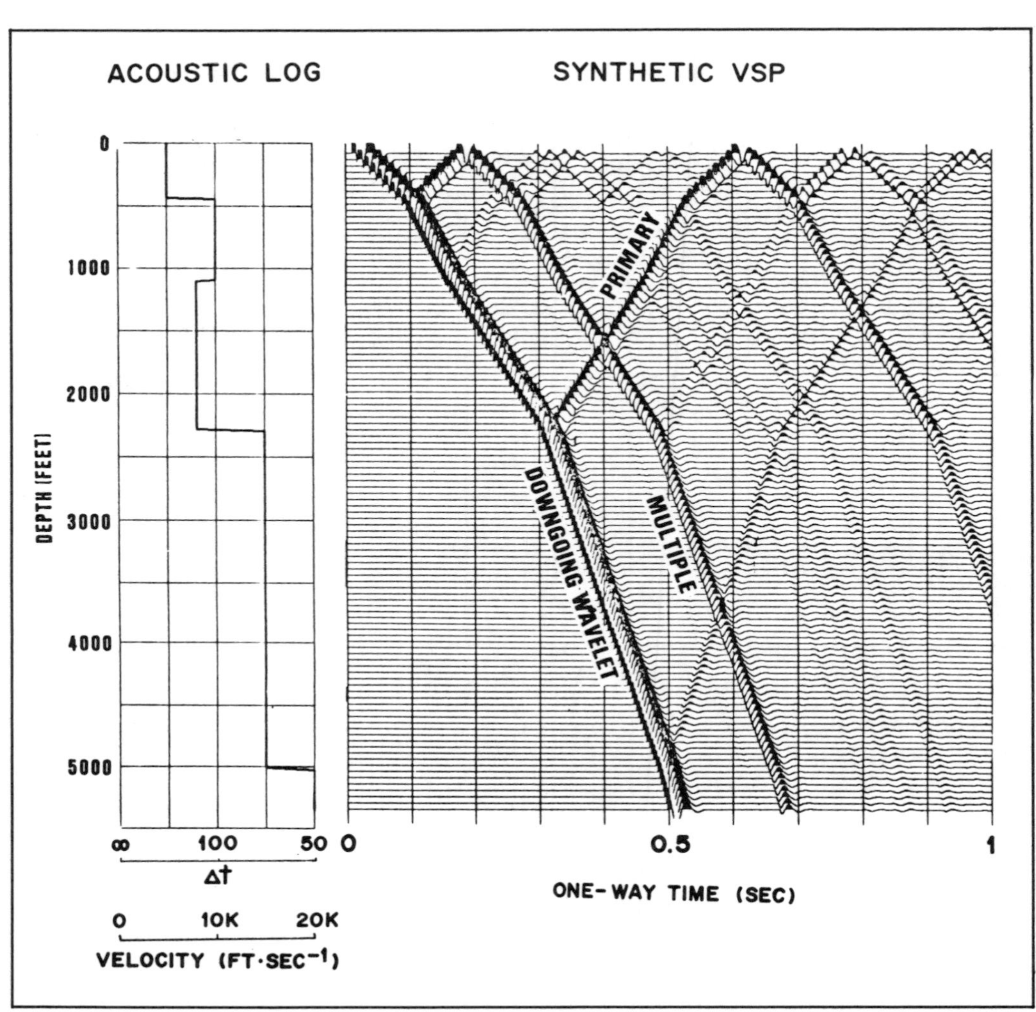

FIG. 11-17. A simplified vertical seismic profile (VSP). (After Wyatt, 1982.)

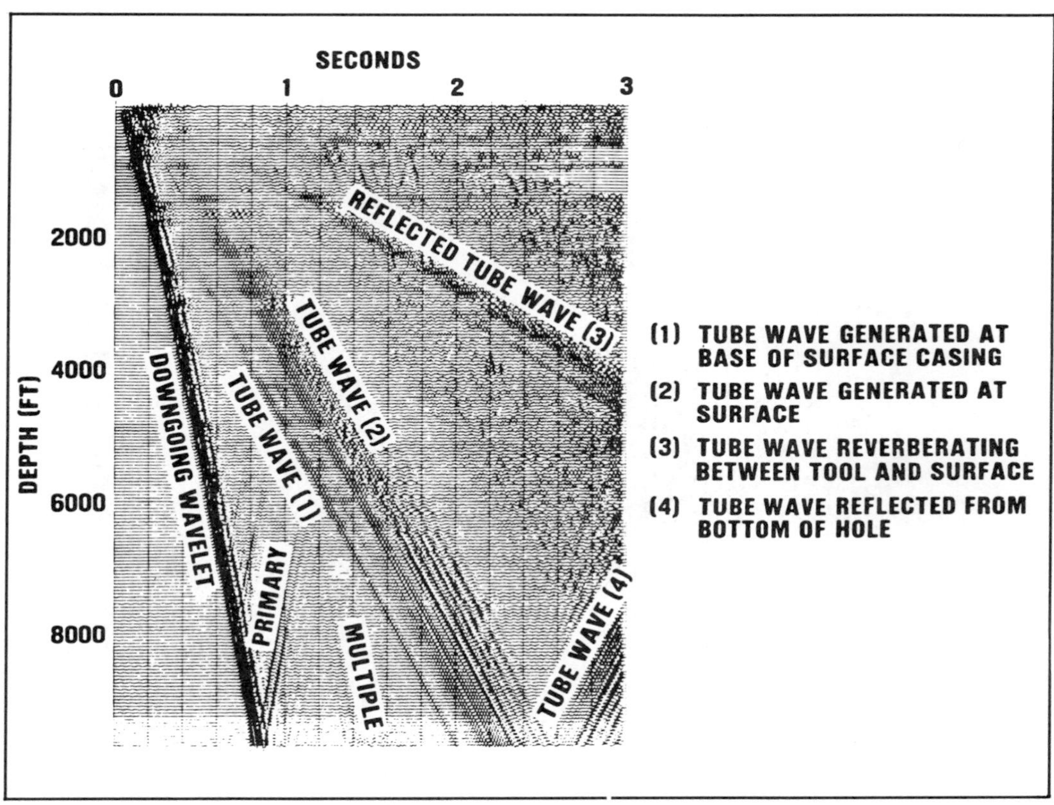

FIG. 11-18. VSP with strong tube wave. Note that the frequency of the tube wave is nearly identical with the downgoing wave. (Hardage, 1981.)

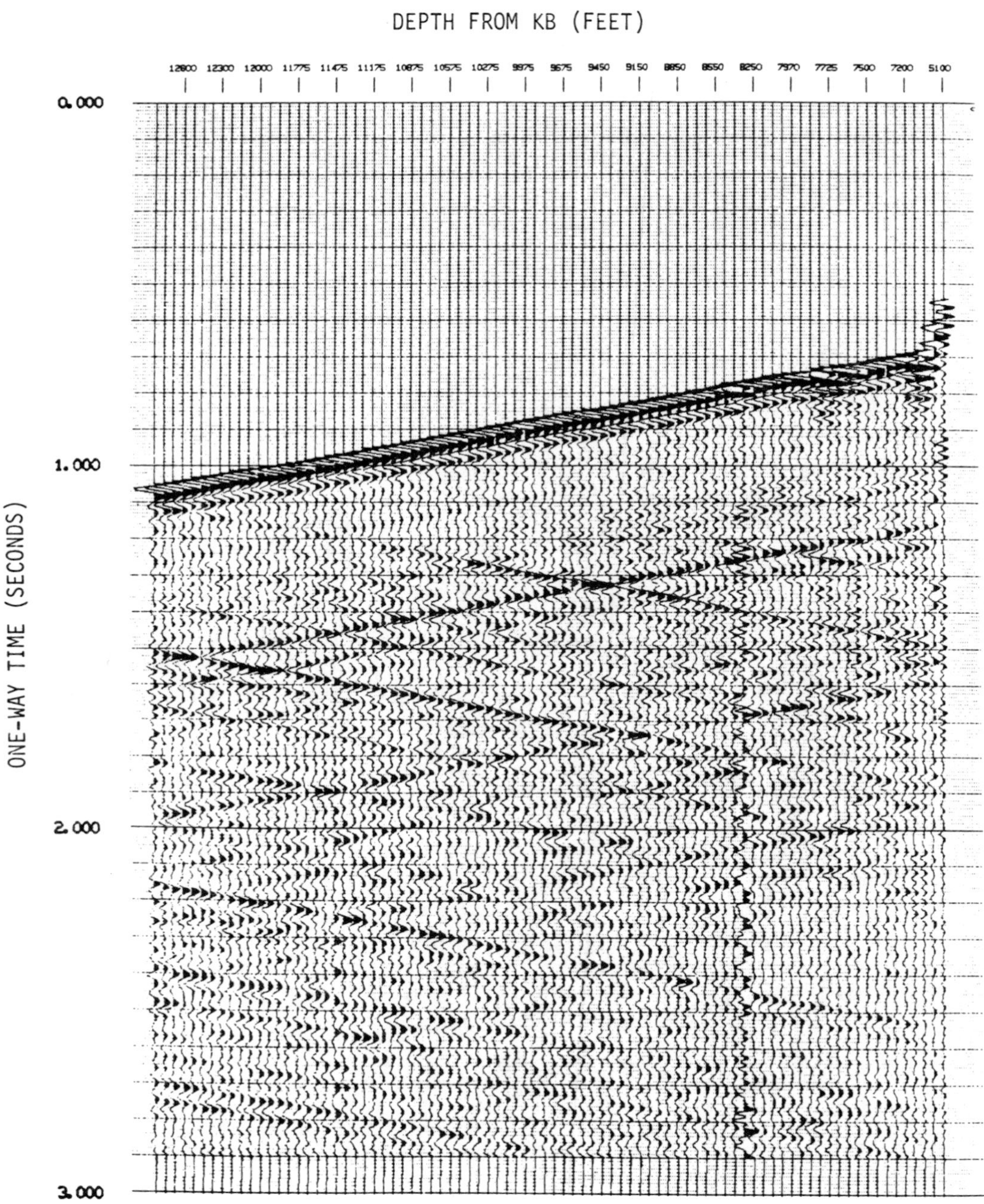

FIG. 11-19. Field recording using a 10-60 Hz linear sweep and a geophone and trace spacing of 75 ft.

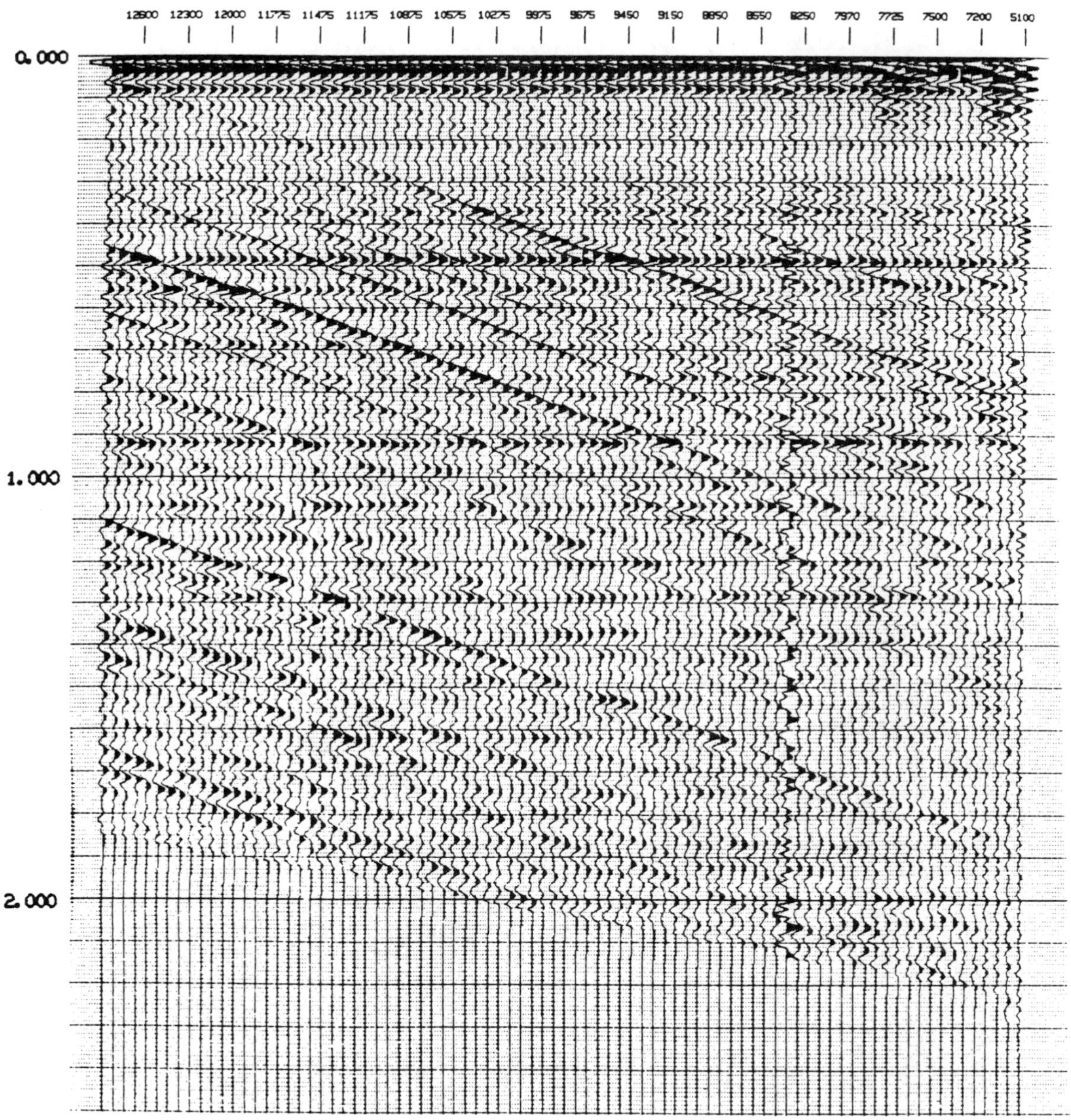

FIG. 11-20. Alignment of downgoing waves (Figure 11-16) by subtracting the first arrival times.

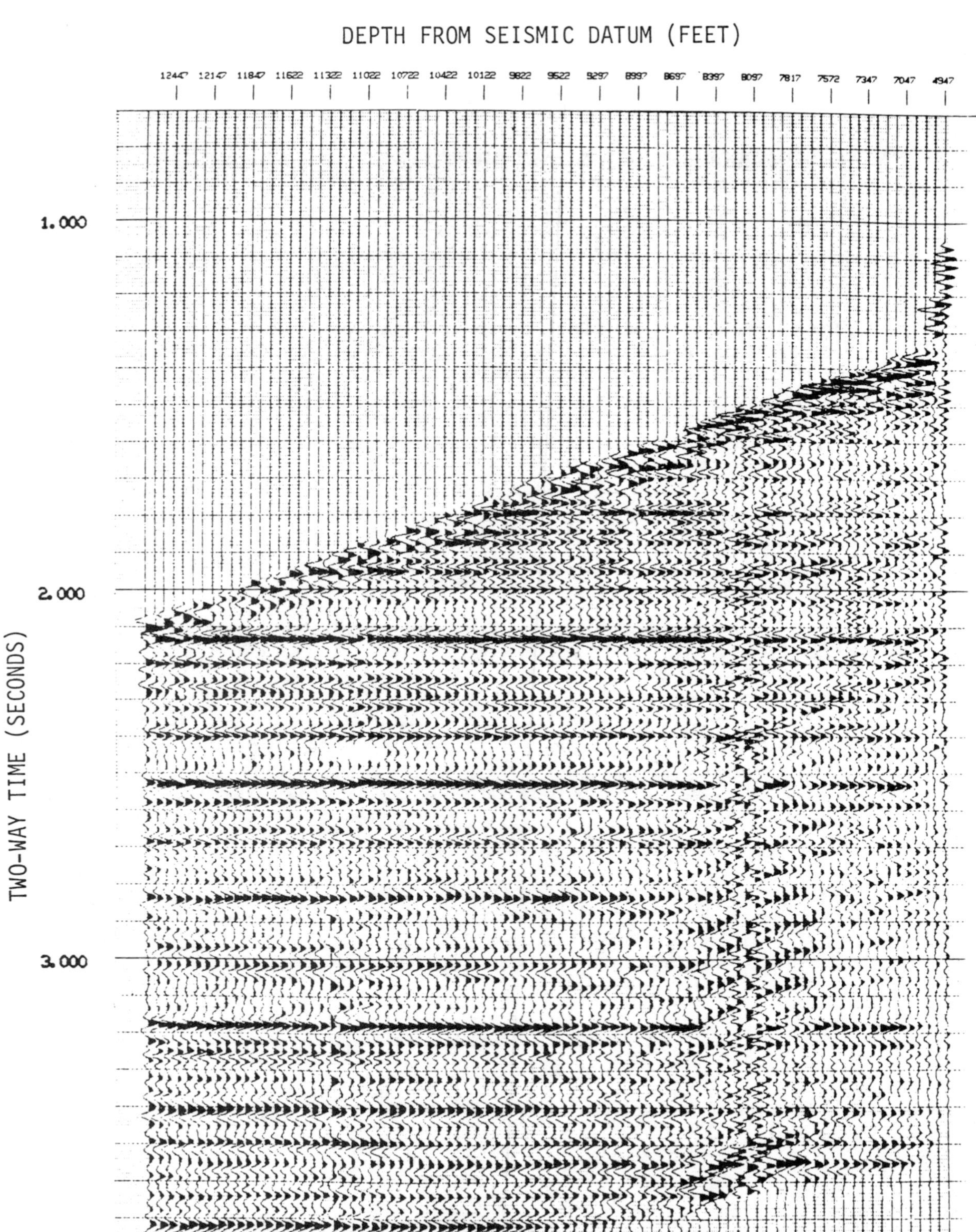
DEPTH FROM SEISMIC DATUM (FEET)
12447 12147 11847 11622 11322 11022 10722 10422 10122 9822 9522 9297 8997 8697 8397 8097 7817 7572 7347 7047 4947
1.000
2.000
3.000
TWO-WAY TIME (SECONDS)

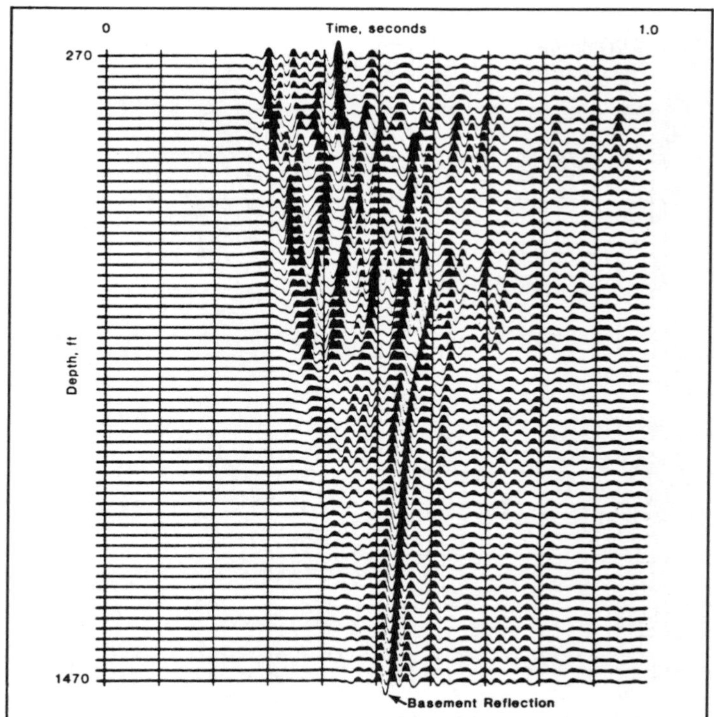

FIG. 11-22. An example of an upgoing wave being attenuated before it reaches the surface. (Balch et al., 1982.)

direct wave in detail and determine any changes with depth.

To emphasize the upgoing arrivals, that become what is recorded with the reflection seismic technique, an additional time equal to the first break time is added to each trace (Figure 11-21). This addition converts the traces into two-way time and causes alignment of the upgoing reflections. Thus, it is possible to follow a particular reflection, whose origin depth is accurately known, to the surface and also to observe how it changes shape as it travels to the surface. This observation shows that some reflections do not reach the surface, but the depth they are attenuated (Figure 11-22) can be accurately determined.

The upgoing alignment can be stacked and used as a "synthetic seismogram" (Figure 11-23) that ties depth and borehole events to the surface gathered seismic section (Figure 11-24). Because the same method is used to create both, the methods can tie extremely well. With a stacked upgoing VSP, events below the bottom of the well are also available to tie the seismic data. This further removes ambiguity that is often present when using the limited time section of a conventional acoustic log derived synthetic seismogram.

FIG. 11-21. Alignment of upgoing waves (Figure 11-16) by adding the first arrival times determined in the downgoing alignment step of Figure 11-20.

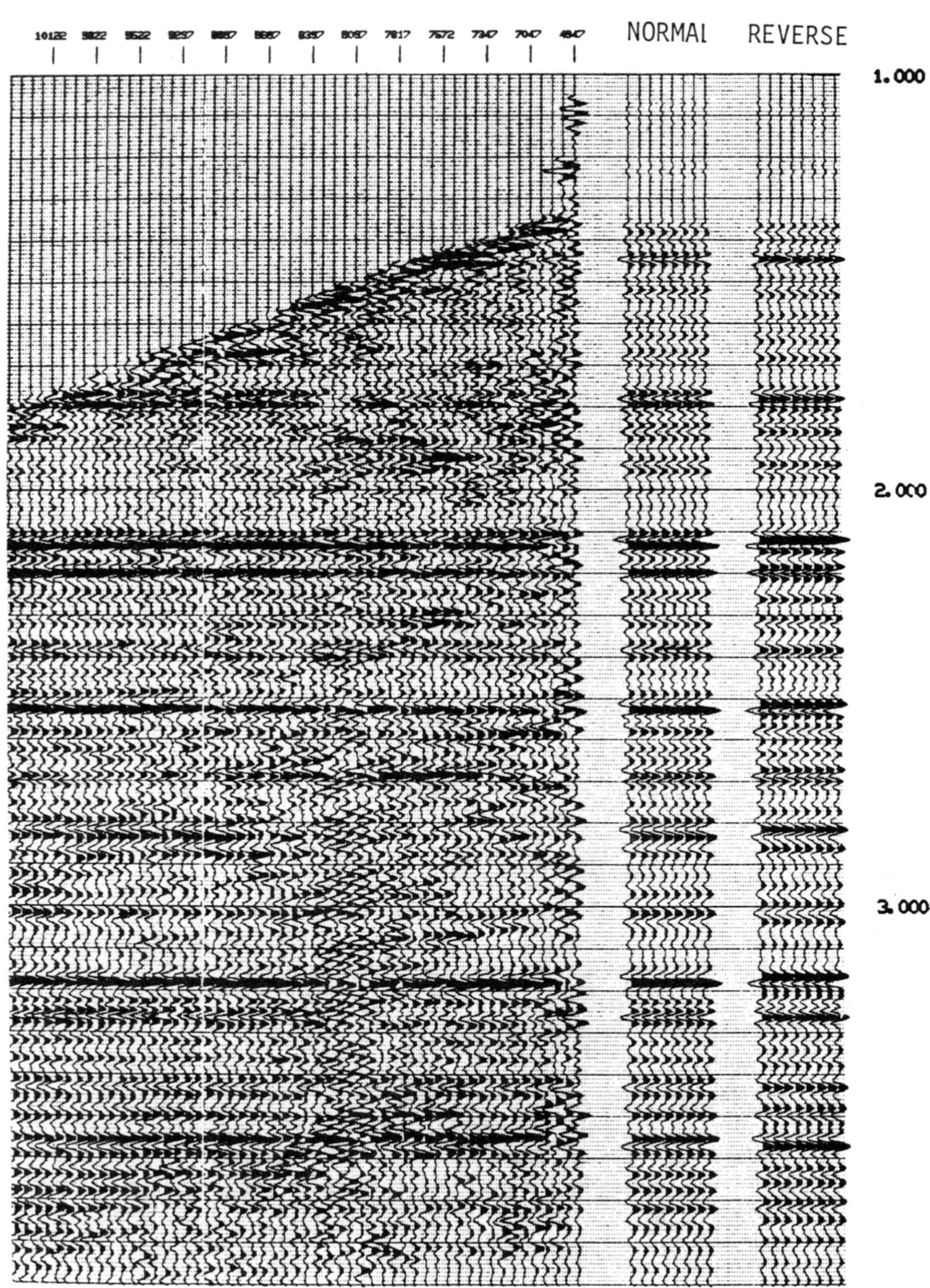

FIG. 11-23. Upgoing VSP alignment stacked and placed in synthetic seismogram format.

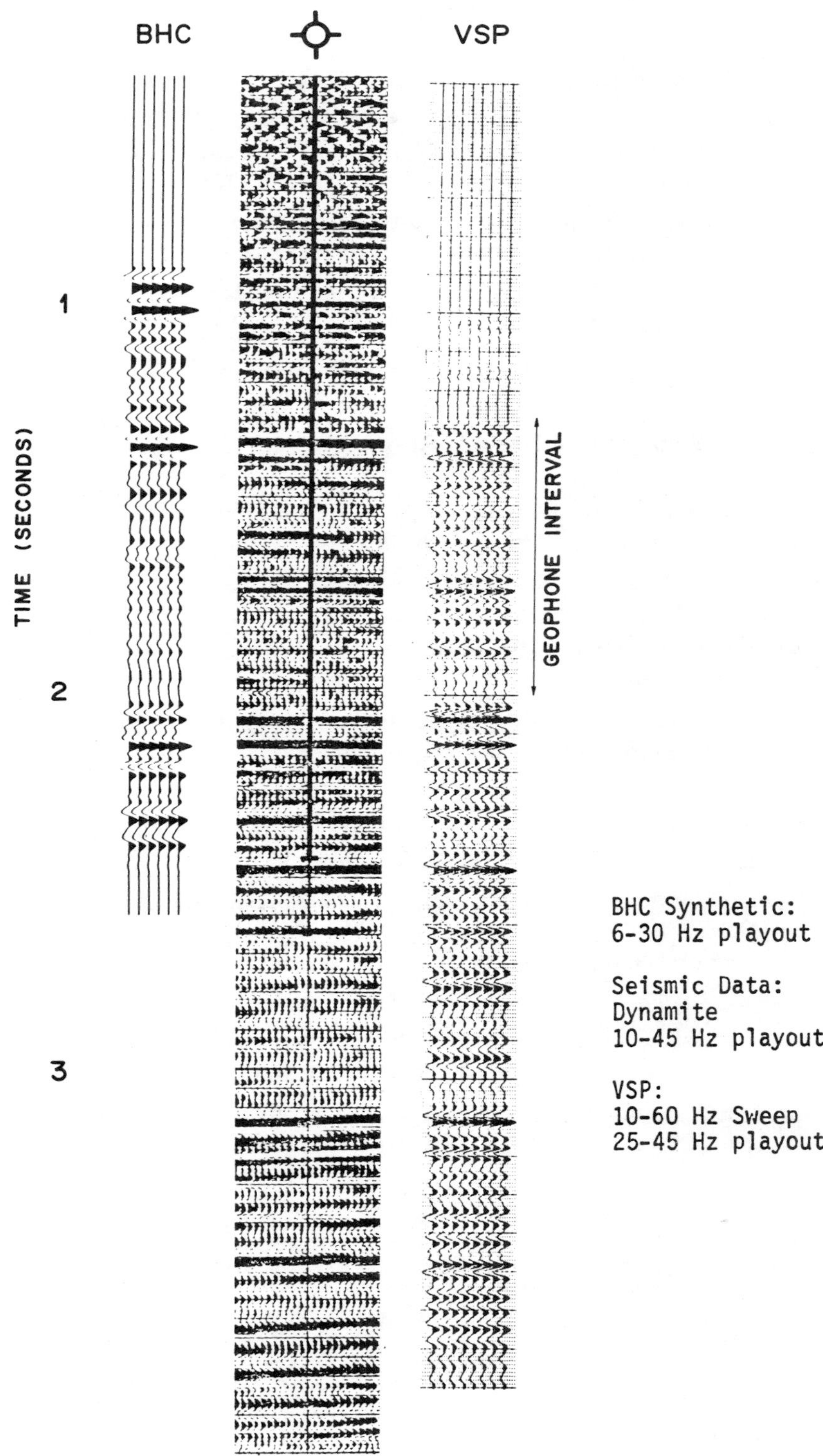

FIG. 11-24. Comparison of VSP stack, acoustic log derived synthetic, and surface gathered seismic data.

REFERENCES

Balch, A. H., Lee, M. W., Miller, J. J., and Taylor, R. T., 1982, The use of vertical seismic profiles in seismic investigations: Geophysics, **47**, 906-918.

Goetz, J. R., Dupal, L., and Bowler, J., 1979, An investigation into discrepancies between sonic log and seismic check shot velocities: Austral. Petr. Expl. Assn. J., **19**, 2, 131-141.

Hardage, B. A., 1981, An example of tube wave noise in vertical seismic profiles: 49th Ann. Soc. Expl. Geophys. Convention, New Orleans.

Ward, R. W., and Hewitt, M. R., 1977, Monofrequency borehole traveltime survey: Geophysics, **42**, 1137-1145.

Wyatt, K. D., 1982, Synthetic vertical seismic profile: Geophysics, **46**, 880-891.

REFERENCES FOR GENERAL READING

Babur, K., and Mons, F., 1981, Vertical seismic log: recording, processing and uses: Trans. 7th European Logging Symposium, Paris.

Boss, F. E., 1970, How the sonic log is used to enhance the seismic reference service velocity survey: Can. Well Logging Soc. J., **3**, 17-31.

Deeming, T., 1979, Synthetic seismograms and vertical seismic profiles: 49th Ann. Soc. Expl. Geophys. Convention, New Orleans.

Dupal, L., Gartner, J., and Vivet, B., 1977, Seismic applications of well logs: Trans. 5th European Logging Symposium, Paris.

Gal'perin, E. I., 1973, Vertical seismic profiling: Soc. Expl. Geophys., 270.

Hubbard, T. P., 1979, Deconvolution of surface recorded data using vertical seismic profiles: 49th Ann. Soc. Expl. Geophys. Convention, New Orleans.

Kennett, P., Ireson, R. L., and Conn, P. J., 1979, Vertical seismic profiles, their application in exploration geophysics: 41st Ann. Eur. Assn. Expl. Geophys. Convention, Hamburg.

Poster, C. K., 1982, Comparison of seismic sources for VSP's in a cased well: Can. Soc. Expl. Geophys. Ann. Convention, Calgary.

Seismograph Services Limited, 1980, The VSP modeling atlas: Holwood.

Wuenschel, P. C., 1976, The vertical array in reflection siesmology — some experimental studies: Geophysics, **41**, 219-232.

Chapter 12

DIPMETER PRINCIPLES

The resistivity dipmeter tool uses three or four horizontal, equally-spaced, microresistivity measurements to determine formation dip magnitude and direction. The accurate determination of in-place formation dip is a two-part problem. First, the relative dip of formation beds, expressed in bed displacement along the borehole wall, is measured and then orientation of the borehole wall is measured.

Dipmeter tool experimentation was begun in 1927 by the Schlumberger brothers. In 1932, the first commercial anisotropic resistivity dipmeter was run in Russia.

How to measure borehole orientation was solved first. The teleclinometer used an induction magnet and a series of vertically oriented electromagnets. By stopping the tool in the borehole and making four consecutive voltage measurements, it was possible to calculate sonde orientation to magnetic north and the angle and azimuth of the tool's inclination from vertical. This system was replaced with the photoclinometer in 1940. The photoclinometer was considerably simpler because it contained a magnetic compass beneath a concave graduated glass dish which held a free moving ball. A picture of the two devices was taken after the tool was stopped in the borehole. The compass indicated magnetic north for a reference side of the sonde, with the ball position indicating sonde deviation from vertical and the direction of the downward side of the sonde. About 1957 the photoclinometer was replaced by the poteclinometer. Resistivity potentiometers were driven effortlessly by a compass for reference azimuth, and by pendulums for tool deviation from vertical and relative bearing of the sonde high side. These readings were sent to the surface periodically along with the resistivity measurements.

The detection system to measure the dipping formation beds took longer to develop. In some formations, primarily shales, apparent resistivity differs depending on whether the measurement is made parallel or perpendicular to the bedding. The anisotropic or "electromagnetic" dipmeter was a downhole adaptation of surface methods to measure resistivity anisotropy. The voltage difference between two electrodes that were equidistant from a source electrode, but on opposite sides of the borehole, was measured. If the formation beds are perpendicular to the borehole, then there is no voltage difference between the measure electrodes (Figure 12-1, A). For dipping beds (Figure 12-1, B) there is a potential difference caused by the anisotropy, with the magnitude of the difference related to the formation dip. However, core dip information was required to accurately determine the true dip angle. For orientation this dipmeter used the teleclinometer.

The correlation dipmeter concept was introduced about 1940. Three SP curves were recorded and correlation offset displacements determined

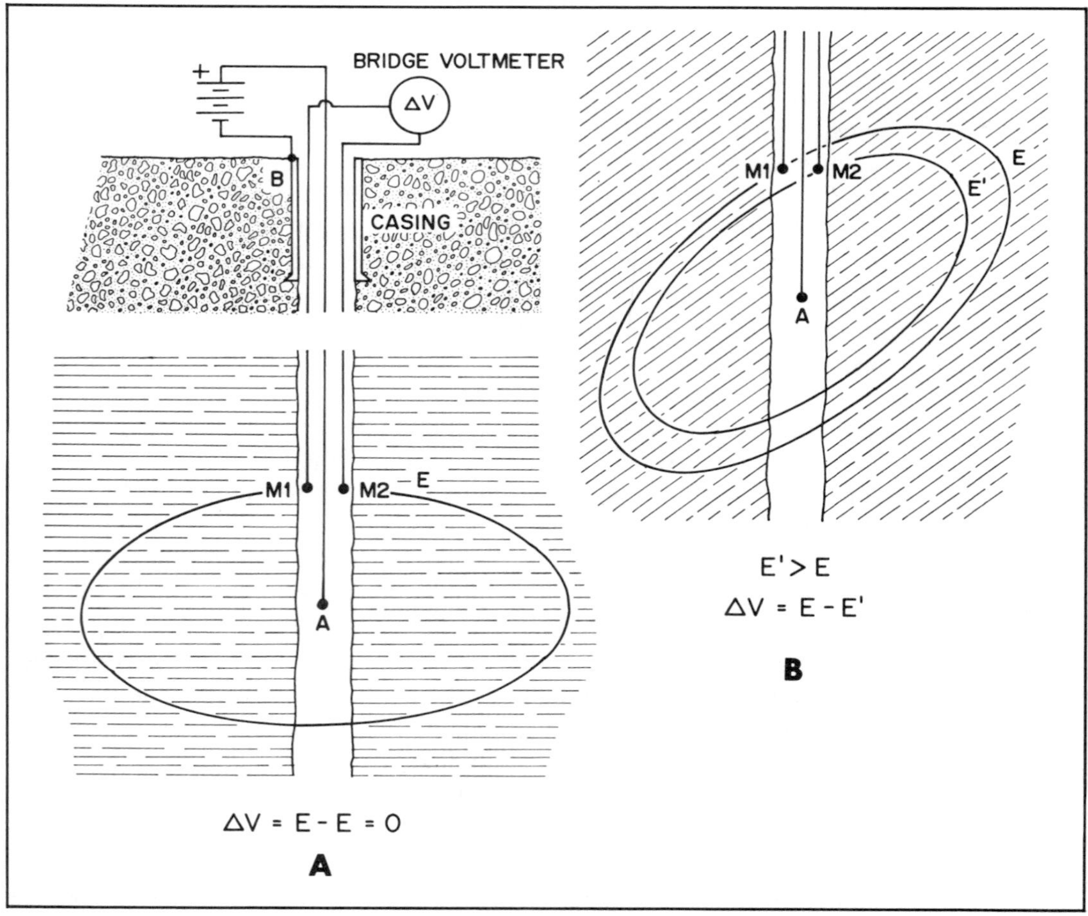

FIG. 12-1. The anisotropic resistivity dipmeter. The difference in potential measured between the two electrodes (M1 and M2) is a function of the anisotropic resistivity, and indirectly, of the relative-to-wellbore formation dip.

for SP deflections at bed boundaries. For maximum accuracy sharp boundary deflections were needed. The disadvantage of this method was that the dip calculation could only be made at a permeable-impermeable formation boundary. If this surface was also a bedding plane, then formation dip was calculated. But if it were an unconformity, then it did not necessarily apply to the permeable formation. For this reason, the anisotropic dipmeter was often preferred and was still used long after the introduction of the SP correlation dipmeter.

The first resistivity correlation dipmeter was introduced in the middle 1940s. This tool used a micro lateral resistivity device, MN about 1 inch and AO about 3 inches, and the photoclinometer orientation system. This advance allowed correlations to be made within a bedded formation. In 1952 the tool was further improved with use of microlog resistivity devices and the introduction of a continuous reading teleclinometer using a flux-gate compass and two right-angle pendulums.

In 1955 the resistivity device was changed to a microlaterolog configuration for still finer resistivity detail and the potelinometer was introduced in 1957 as the orientation device. The tool was completely redesigned in 1966 for digital recording, allowing computer correlation of the microresistivity measurements. A fourth resistivity arm was added for increased accuracy and dip quality determination.

MICRORESISTIVITY — FORMATION DIP RELATIONSHIP

If resistivity measurements are made that emphasize intraformation bedding surfaces, it becomes possible to measure true formation dip. For a vertical borehole with a conductive horizontal bed, if four right-angle, oriented resistivity measurements are made (Figure 12-2, item A), the conductivity breaks will occur at the same depth. However, if the bed were at 10 degrees down to the east dip, the breaks would have a maximum displacement of 0.70 inches in an 8-inch diameter borehole (Figure 12-2, item B). Obviously, the microresistivity devices would have to be sampling considerably finer than 0.70 inches to measure this dip accurately. From the relative displacements and the borehole diameters, the dip angle and direction can be calculated.

In reality the four resistivity pads are not located at fixed directions within the borehole. The natural torque of the wireline causes the tool slowly to rotate as it is pulled uphole making it necessary to measure the orientation of at least one of the pads and to use that measured direction in calculating the dip angle and dip direction (Figure 12-2, item C).

Also, because the borehole is not always vertical, its angle of deviation from vertical and the direction of borehole dip must be measured. Thus the displacement of dipping conductive breaks becomes complex in real conditions (Figure 12-2, item D and Figure 12-3).

If the borehole diameter, vertical displacement of at least three resistivity curves, orientation of the resistivity curves, sonde deviation from vertical,

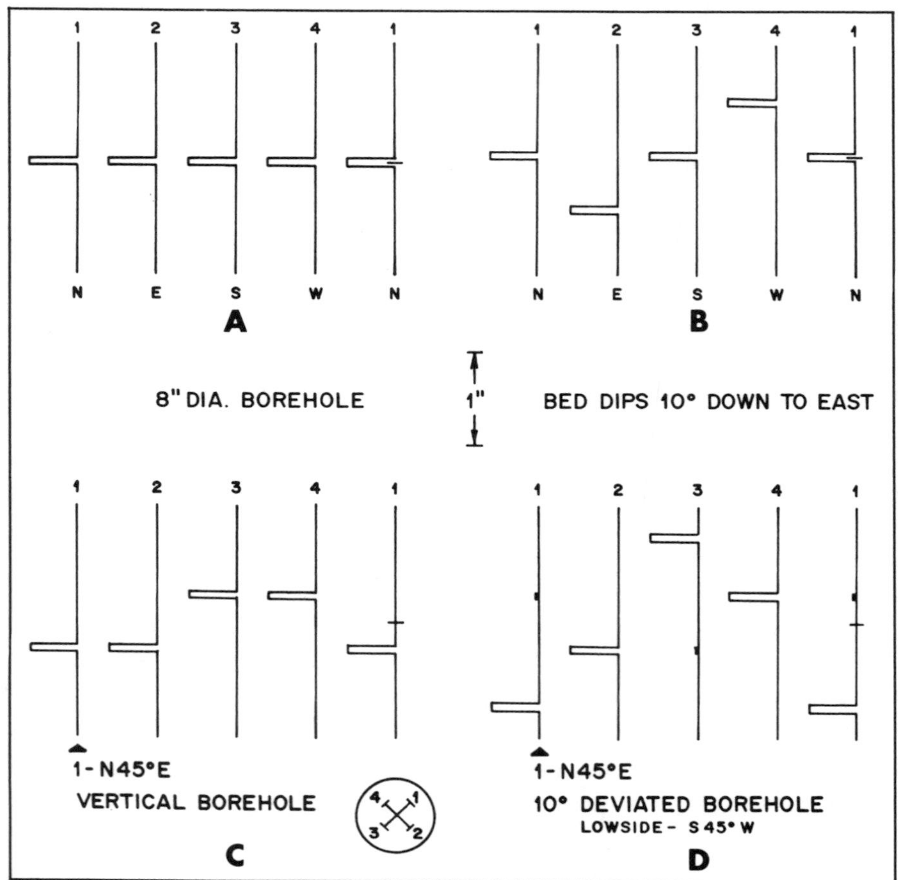

FIG. 12-2. Formation dip shown on microresistivity measurements: (A) flat bed measured with fixed orientation electrodes, (B) dipping bed measured with fixed orientation electrodes, (C) dipping bed with freely orienting vertical tool, and (D) dipping bed with freely orienting tilted tool.

and down direction of the sonde axis are known, then the bed dip direction and magnitude can be calculated.

DIGITAL HIGH-RESOLUTION DIPMETER TOOL

Schlumberger in 1966 introduced their digital dipmeter tool which has an improved correlation resistivity measurement system with four microresistivity measurement pads, and which digitally records all data.

To provide maximum formation bedding detail the tool uses a focused microresistivity system that primarily measures minor resistivity changes, as opposed to absolute resistivity values. The system (Figure 12-4) measures the focused horizontal button current, which varies with the conductivity of the formation. The system has a current return very close to the button, providing for a shallow depth of investigation. The button is elongated

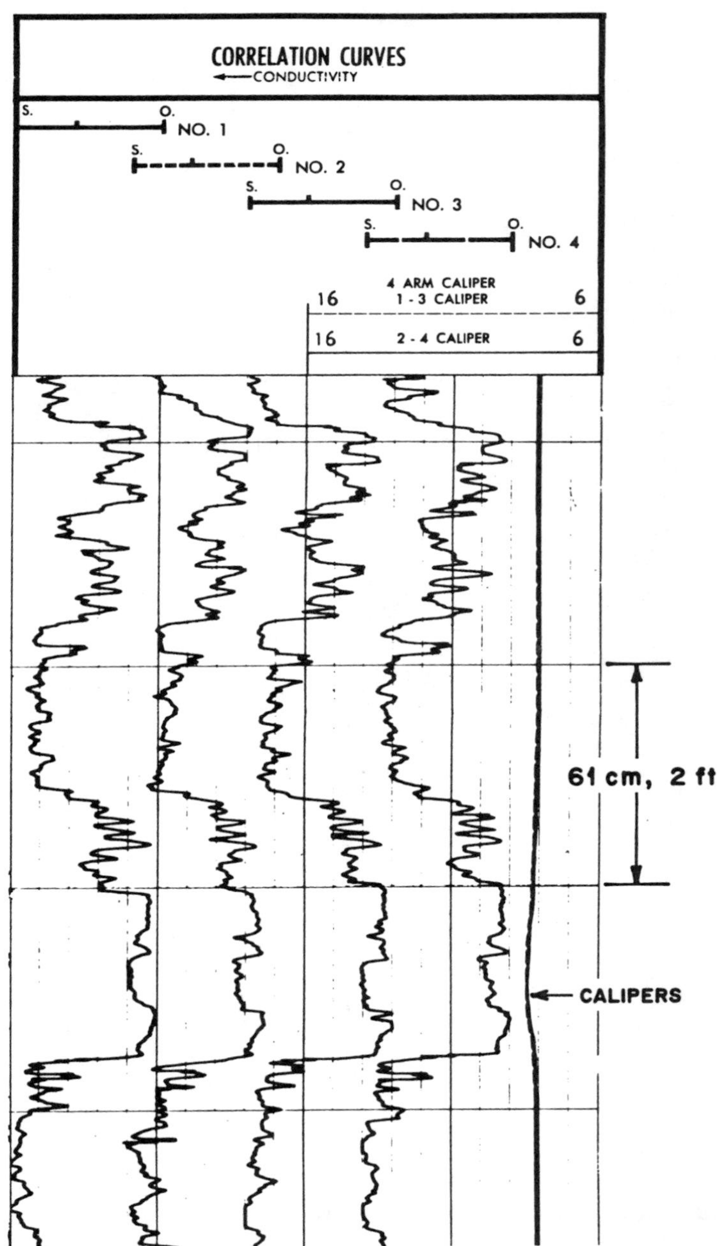

FIG. 12-3. Dipmeter microresistivity measurements. Note that the conductivity breaks occur at different depths around the borehole, defining relative-to-tool formation dip planes.

horizontally for more formation response without sacrificing vertical resolution. This system emphasizes resistivity variation or contrast, providing the multitude of fine detail necessary for good vertical correlations. The current is sampled digitally every 0.2 inches of borehole. The button voltage is measured periodically and used with curve averaging of the current measurement to computer generate a correlation curve

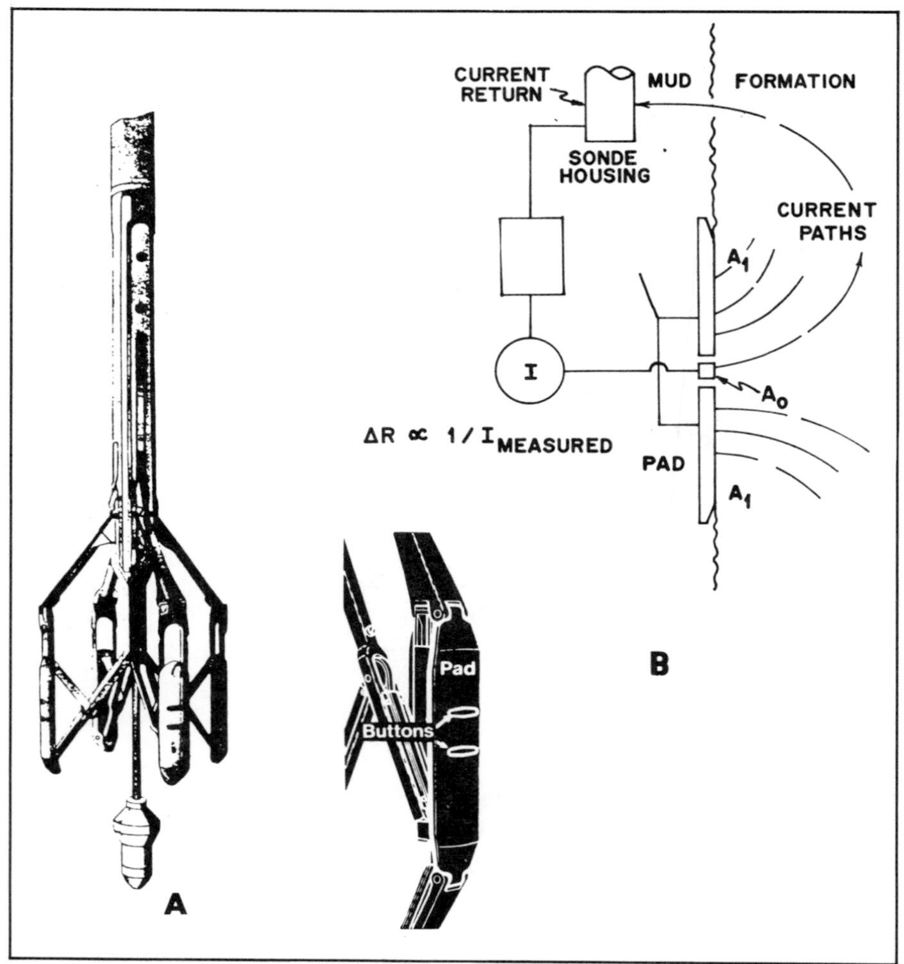

FIG. 12-4. The dipmeter microresistivity system: (A) the four resistivity measuring pads, and (B) the measurement principles. By only measuring current changes, differences in the formation resistivity are emphasized and more formation information is given to calculate formation dips.

(Figure 12-5, item C) for interlog depth control.

The addition of the fourth microresistivity measurement allows for improved dipmeter quality determination. With just three measurements only a single dip calculation can be obtained at each depth. With the addition of the fourth measurement, it is possible to make four three-point dip calculations and by comparing the various calculated dips, an indication of the dip quality is made (Figure 12-5, item B). With the three-arm system if one arm loses contact with the formation no dip can be calculated. With four correlation measurements, if one arm loses contact with the formation one dip calculation is still obtainable.

In reality there are five correlation measurements made. The fifth is on the same pad and just above one of the four measure buttons. The sonde is not always moving at the same velocity as the cable at the surface

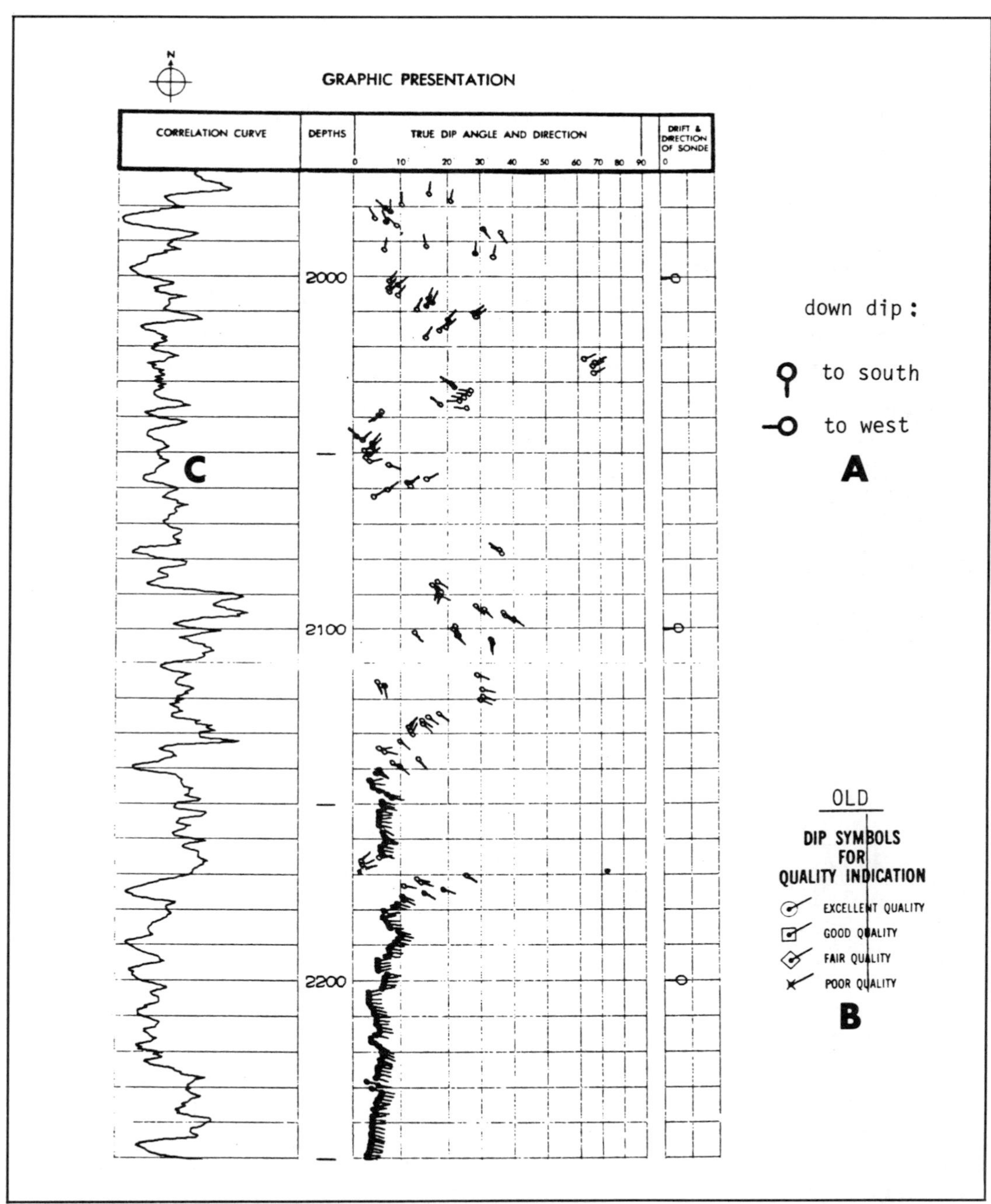

FIG. 12-5. Calculated dipmeter output using the tadpole plot format. The plot shows dip magnitude on the horizontal scale and direction down-dip by the compass direction of the tail (A). The quality of the dips (B) is based upon the amount of agreement of the different three-arm dip calculations at each level. The resistivity correlation curve (C) is created using periodic voltage measurements and averaging the microresistivity current measurements.

due to variable tool drag and cable stretch. Comparing two correlation curves made a fixed short distance apart allows calculation of slight variations of tool velocity, which are used to obtain more accurate displacements in the correlation process.

Five geometric measurements are made and transmitted to the surface for use in data reduction of the correlated microresistivity measurements (Figure 12-6):

Azimuth of the No. 1 microresistivity electrode with respect to magnetic north,

Relative bearing of the No. 1 microresistivity electrode to the high side of the sonde — this is for the "dip direction" of the borehole.

deviation angle of the sonde from vertical, and

two perpendicular borehole diameter *caliper* measurements

From these measurements it is possible to locate the electrodes spacially,

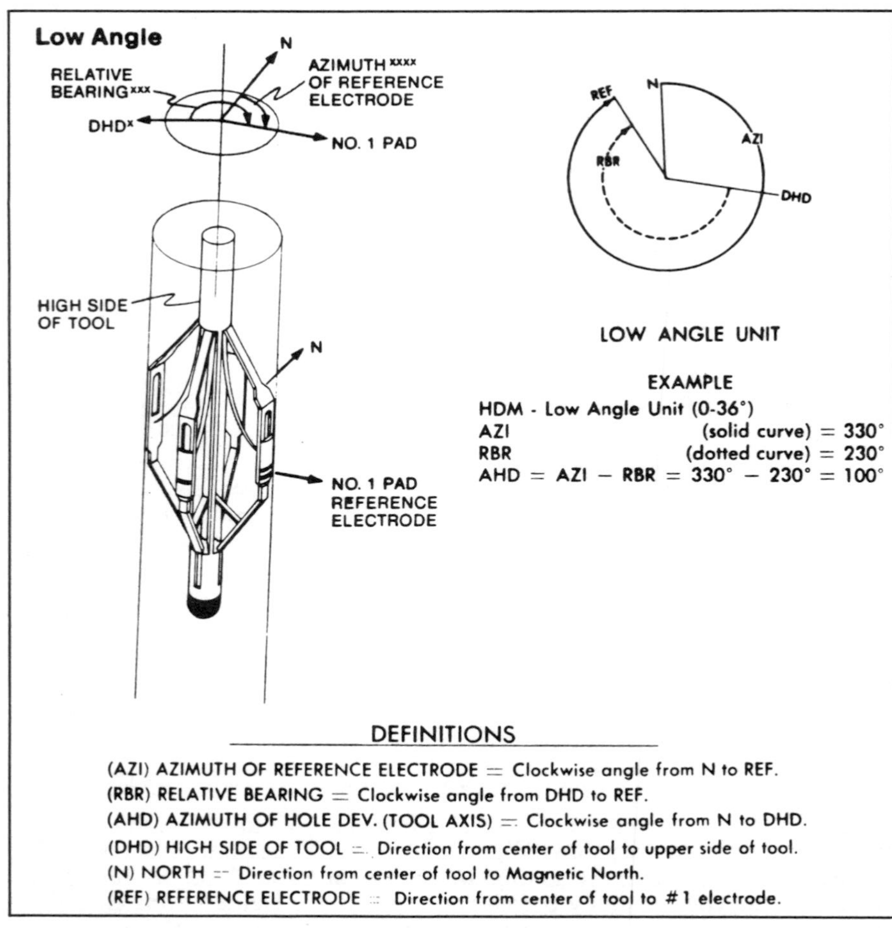

FIG. 12-6. Definition of dipmeter geometric orientation measurements of the tool and the borehole: azimuth, relative bearing, and deviation — for the low angle (36 degrees and less deviation) measurement system.

relative to magnetic north and vertical, and to properly relate the microresistivity correlated measurements to true bed dip.

FIELD PROCEDURES

Surface field calibrations (Figure 12-7) are of three functions: (1) a check of the geometric orientation devices (steps 10-12), (2) a calibration of the two caliper measurements (steps 4-5), and (3) a check of the resistivity measurement system (Steps 6-9). For the last function it is important that all four measurements exhibit the same deflection response.

The field recording consists of the primary digital tape containing all survey information, a 60 inch/100 ft optical back-up paper recording (Figure 12-3), and a 5 inch/100 ft survey record (Figure 12-8).

The conventional dipmeter resistivity pads are designed to operate in conductive muds, up to the super-saturated salt muds encountered in the Michigan Basin. The tool can be run in nonconductive oil-based muds by using special knife blade pads.

COMPUTER PROCESSING

Originally computer processing techniques used a correlation method. Each curve was correlated against an equal interval length of every other resistivity curve, and a correlation coefficient calculated for each sample shift — the maximum coefficient value determined the correlation displacement. The computer calculations of dip magnitude and direction (Figure 12-9) were determined by the primary dip computation parameters: correlation length, search angle, and step.

Correlation length or search segment is the length of each curve used in making the correlation calculations. Search angle is the maximum amount of curve displacement allowed in the correlation calculations. For example, in an 8-inch diameter borehole, for dips up to 45 degrees, the search angle maximum displacement would be 4 inches. Step, the amount to move up in depth to begin making the next set of correlations, is normally one-half the correlation length. Typical parameters in Michigan reef exploration are 4 ft, ± 45 degrees, and 1 or 2 ft for correlation length, search angle, and step, respectively.

MICHIGAN REEF STRATIGRAPHIC EXAMPLES

An excellent example of the use of the dipmeter is in detection of Silurian-Niagaran pinnacle reefs in the Michigan Basin. The reefs are overlain by several draping and infilling formations, which the dipmeter can use to determine the direction and distance to the reef crest. Only some of the formations are usable in this determination. The ideal situation is that of draping finely bedded formation such as shales. However,

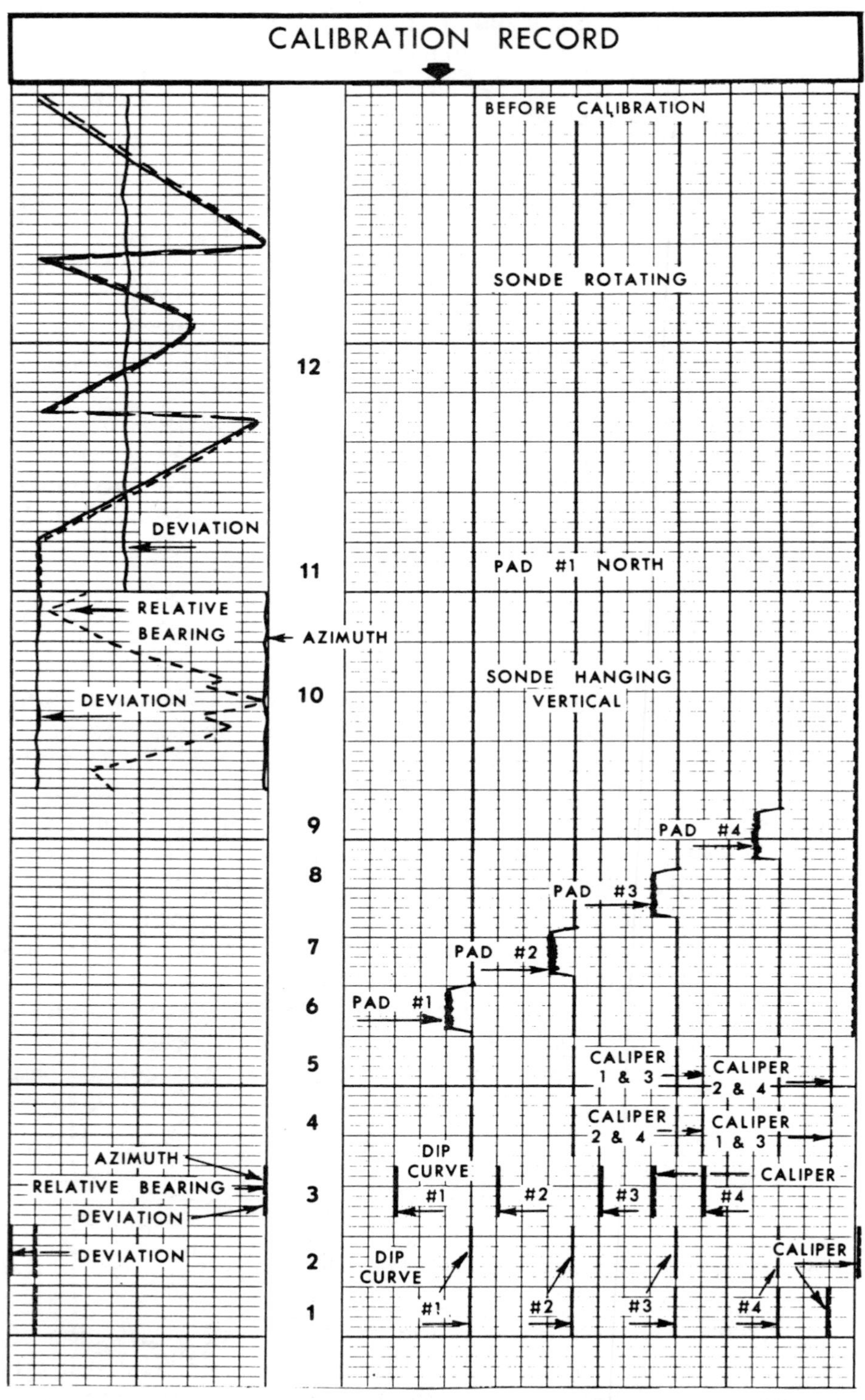

FIG. 12-7. Surface calibration for Schlumberger's digital dipmeter tool. Steps 4 and 5 are calibration of the two-caliper measurements. Steps 6 through 9 check the resistivity measurement response and should all have equal deflection. Steps 10 through 12 check the orientation measurements. (Schlumberger, 1974).

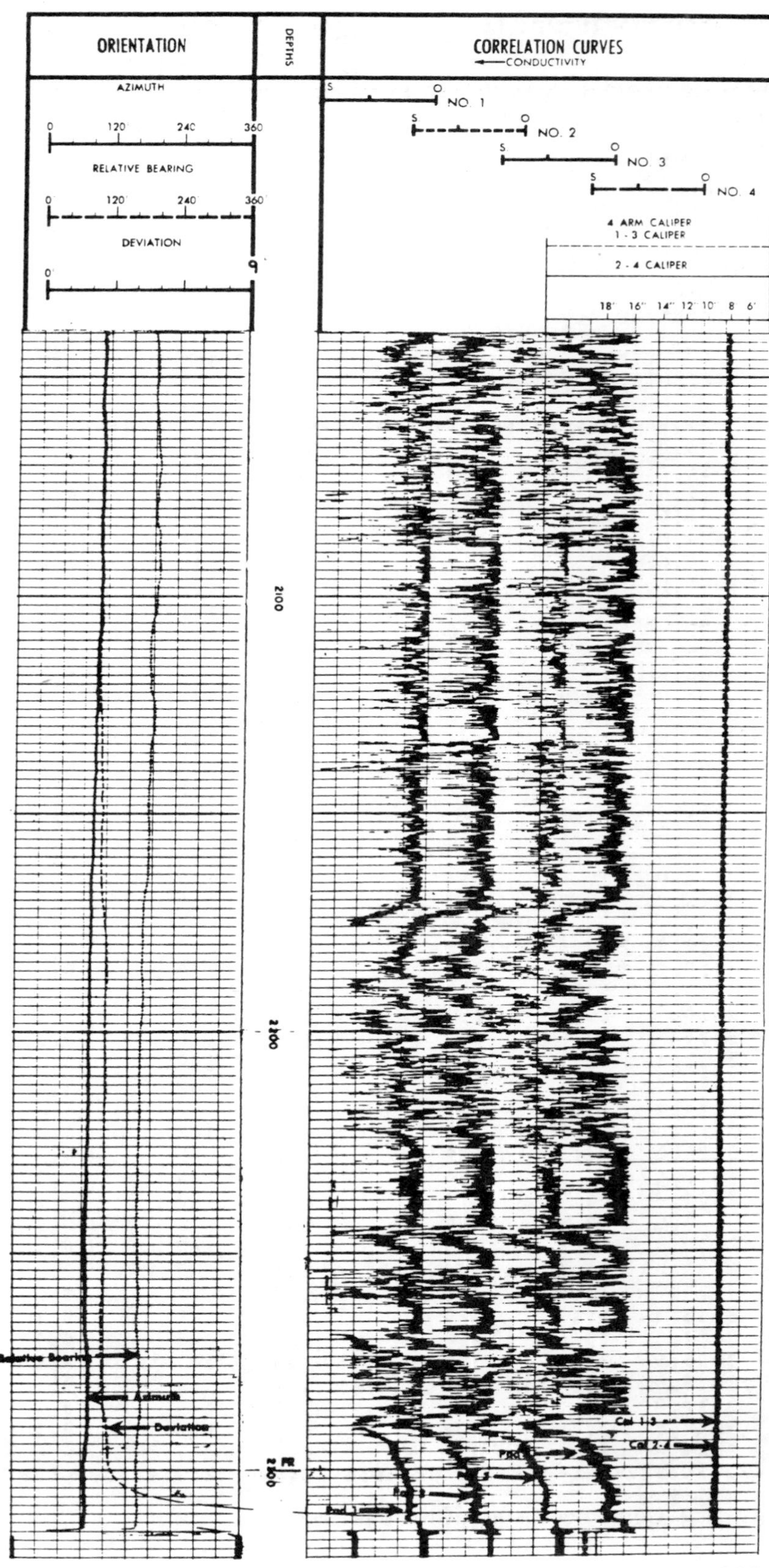

FIG. 12-8. The field recording is a 5 inch per 100 ft monitor log of the dipmeter measurements.

by *C* shale time (Figure 12-10 item C), the interreef areas have been nearly infilled and this formation exhibits predominately regional dip. Salt and anhydrites contain no consistent bedding and exhibit only random dips. The best drape indications are in the A1 and A2 carbonate formations (items A and B).

Figure 12-11 is the computed dipmeter for a well drilled on the extreme edge of a reef. The prime geologic indications of an offreef position are

	FORMATION		BOREHOLE				QUAL. INDEX
DEPTH	DIP	DIP AZI.	DEV.	DEV. AZI.	DIAM 1-3	DIAM 2-4	BEST =4
3595.0	36.1	347	3,9	153	8.1	8,1	2
3596.0			4,0	153	8.1	8,1	
3597.0	19,7	331	4,0	154	8,2	8,1	2
3598.0	21,6	335	4.1	154	8.2	8,1	2
3599.0	22,9	345	4,2	154	8.1	8,1	4
3600.0	24,9	1	4,2	154	8.2	8,1	2
3601.0	24.6	342	4.3	153	8.2	8,1	2
3602.0	24,3	334	4.3	153	8.2	8,1	4
3603.0	22.8	336	4,3	152	8.1	8,1	4
3604.0	22,8	331	4,3	152	8.1	8,1	4
3605.0	23.6	334	4,4	153	8.1	8,1	4
3606.0	25,0	337	4,4	153	8.1	8,2	4
3607.0	25.0	336	4,5	153	8.2	8,2	4
3608.0	24.6	336	4.6	153	8.3	8,2	4
3609.0	25.1	335	4,7	152	8,3	8,1	4
3610.0	27.1	336	4.8	152	8.3	8,1	4
3611.0	27.6	336	4.8	152	8.3	8,1	4
3612.0	28,3	336	4.8	152	8.2	8,1	4
3613.0	26,2	337	4.9	152	8.2	8,1	4
3614.0	27.8	334	5.0	152	8.2	8,1	4
3615.0	25.2	337	5.0	151	8.2	8,1	2
3616.0	25.7	336	5.0	151	8.2	8,1	2
3617.0	27.6	332	5.1	149	8.2	8,1	2
3618.0	23.3	336	5.1	149	8.3	8,1	1
3619.0	17.2	329	5.1	149	8.3	8,1	3
3620.0	16.2	337	5,1	149	8.3	8,1	3
3621.0	33,6	346	5.1	149	8.3	8,1	1
3622.0	33.2	348	5,1	149	8.3	8,1	1
3623.0	27,9	341	5,1	149	8.3	8,1	1
3624.0	37.8	335	5.2	149	8.3	8,1	1
3625.0	22.9	329	5,2	149	8.2	8,1	1
3626.0	24,2	329	5,2	149	8.2	8,1	1
3627.0	22.1	330	5.3	148	8.2	8,1	1
3628.0	29.3	339	5.3	147	8.2	8,1	4
3629.0	32,6	335	5.3	146	8.2	8,1	4
3630.0	29.6	334	5.3	147	8.2	8,1	4
3631.0	28,1	330	5,3	147	8.2	8,1	4
3632.0	30.2	331	5.4	147	8.2	8,1	4
3633.0	31,9	329	5,5	146	8.1	8,2	4
3634.0	32,8	326	5,6	146	8.1	8,2	4
3635.0	31.9	329	5,6	147	8.1	8,1	4
3636.0	32,3	332	5,7	147	8.1	8,1	4
3637.0	32,4	334	5,7	148	8.2	8,1	4
3638.0	32.4	333	5,8	148	8.3	8,1	4
3639.0	32.4	335	5.8	147	8.5	8.1	4

FIG. 12-9. Computer listing of the dip calculation results.

the presence of an A1 evaporite and a thin Brown Niagaran formation. The A1 carbonate shows the best indication of actual direction to the reef crest — west-southwest. This far down on the flank, the A2 carbonate dips do not show the proper direction to the reef crest. Dips in the B salt, A2 evaporite, and A1 evaporite are random.

Figure 12-12 shows the computed dipmeter for a flank well. The flank

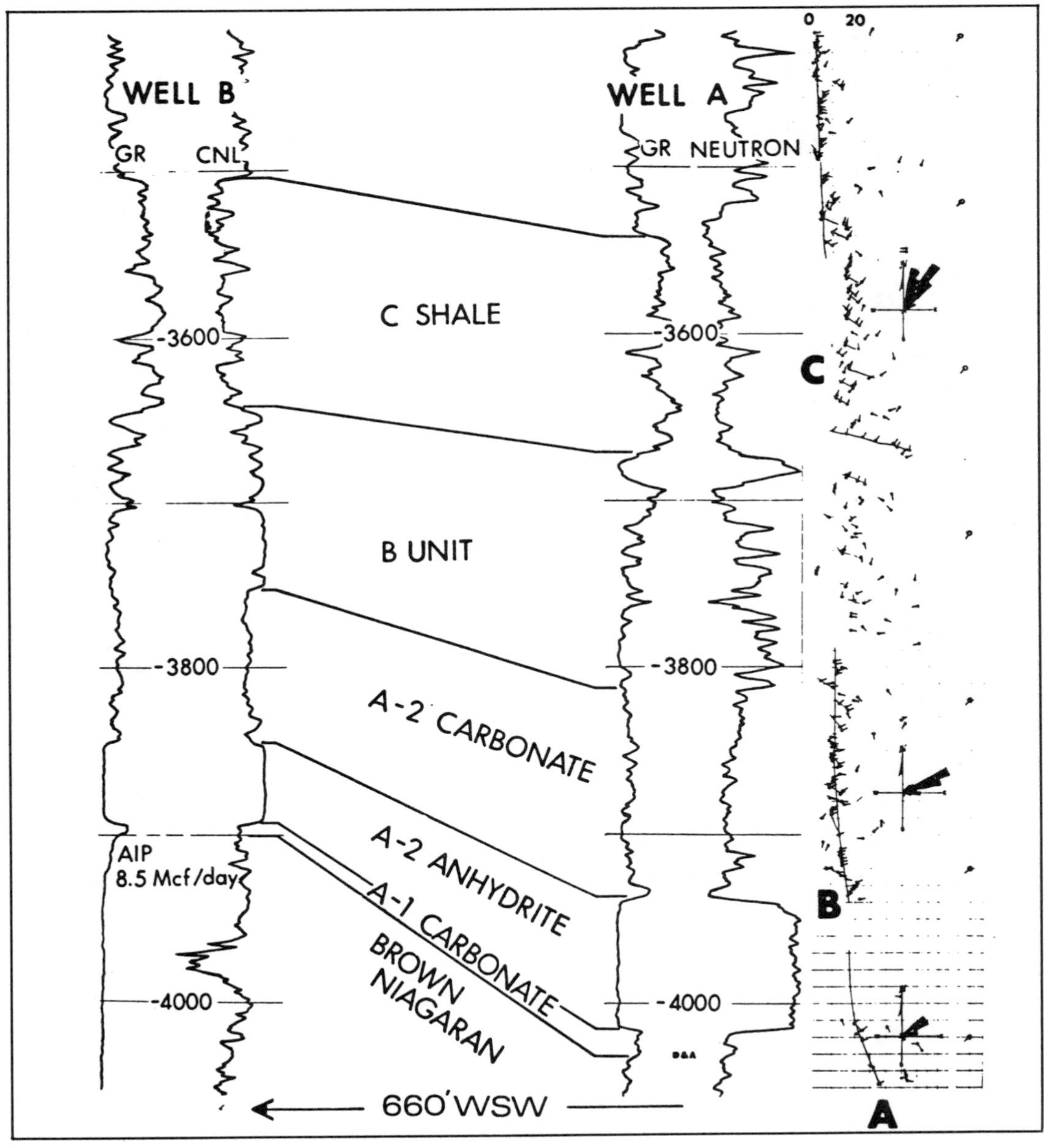

FIG. 12-10. The prime dip indicators for selecting a more favorable location on the reef are the A1 and A2 Carbonate formations (A and B). Dips in the salts and evaporites are random. The interreef infilling is nearly complete by C-shale time, and thus it exhibits predominately regional dip (C). (Bigelow, 1973, Courtesy Oil and Gas Journal.)

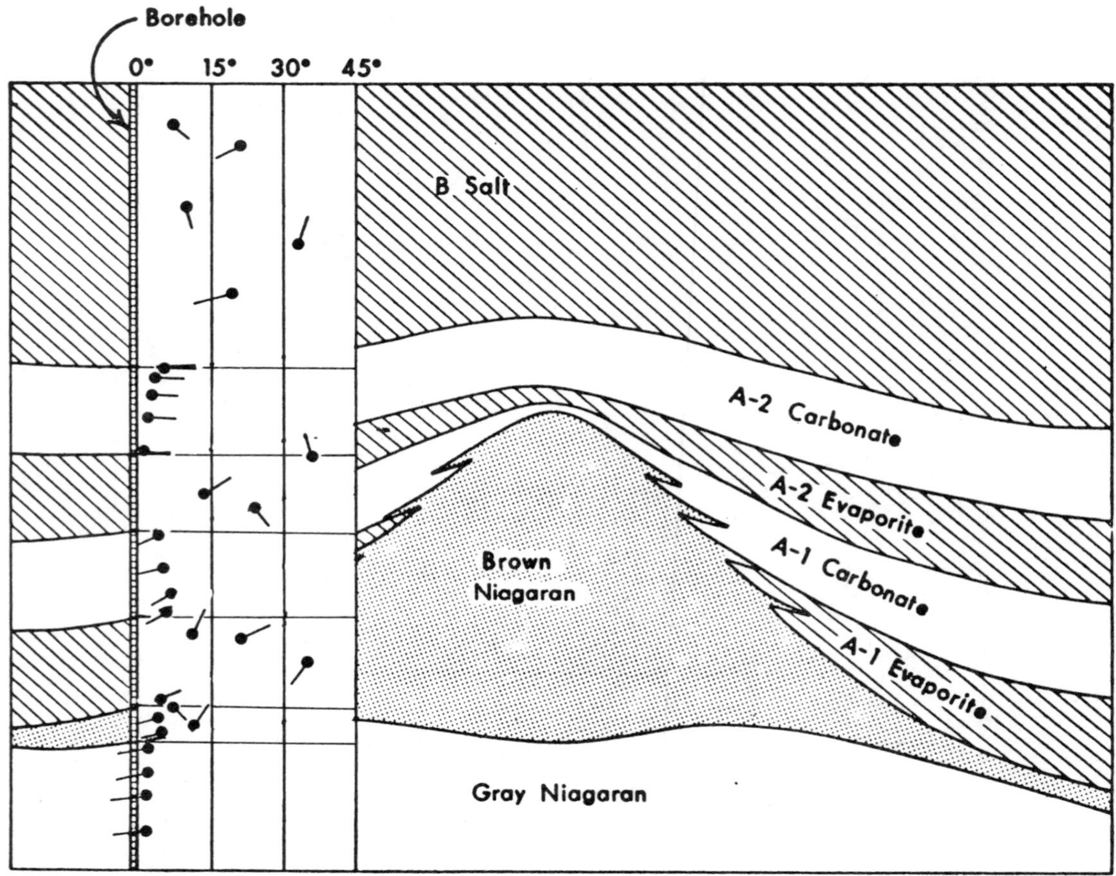

FIG. 12-11. Calculated dips from a near off-reef located borehole are ambiguous and can be misleading in locating the reef. This is shown by the opposite directions indicated at the A1 and A2 Carbonate levels, with the former indication correct. (Bigelow, 1973, Courtesy Oil & Gas Journal).

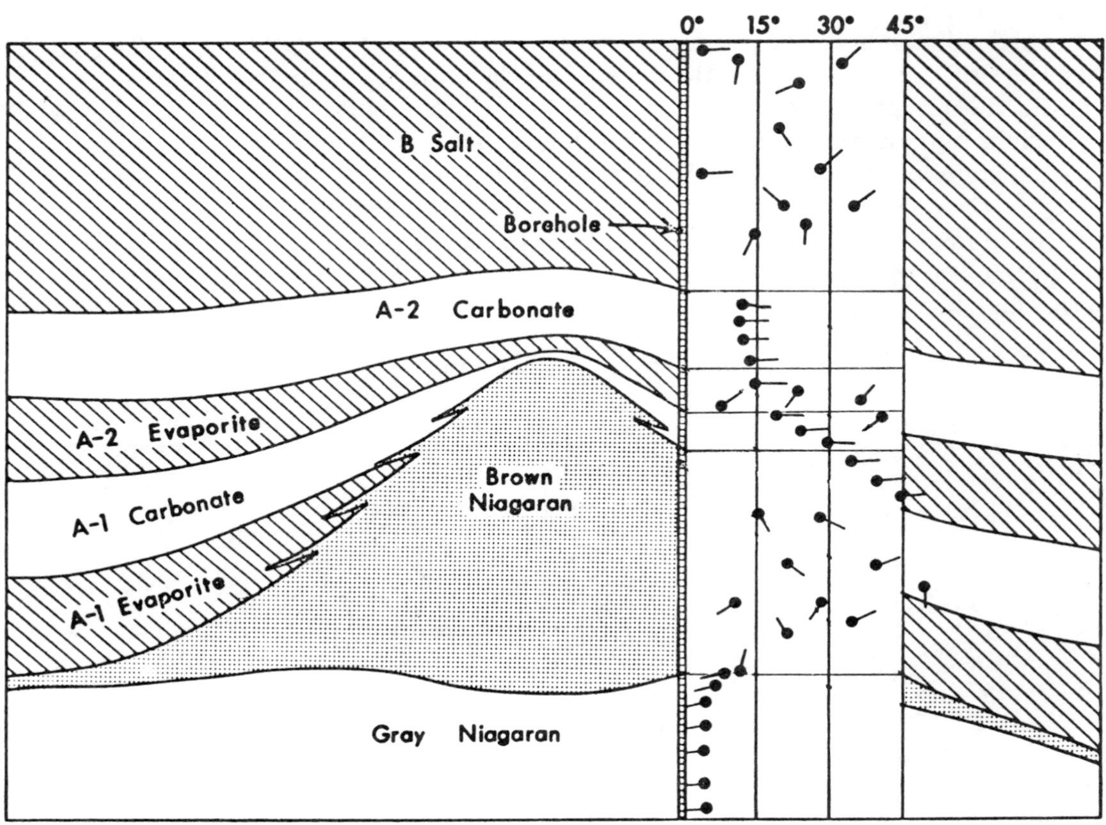

FIG. 12-12. Calculated dips from a borehole for a flank well show very good indications of the direction of the reef crest in both the A1 and A2 Carbonate formations. (Bigelow, 1973, Courtesy Oil & Gas Journal.)

position is indicated geologically by the absence of A1 evaporite, and a considerably greater than regional top of Clinton to top of Niagaran thickness. Again there are random calculated dips in the salt-evaporite sequences and in the body of the reef. The direction of the drape is strongly shown in the A2 and A1 carbonate formations, indicating that the reef crest is to the west.

On a near crest well (Figure 12-13) the computed dips become ambiguous. Dips in the A2 carbonate indicate two preferred directions.

From these examples it is obvious that the dipmeter can be used to indicate the direction of the reef crest, provided a reasonable section of the reef is drilled. Determining the distance to the crest is a matter of applying the indicated dip angles combined with experience in the expected reef height in the area. Note that if the well is drilled too far down the flank, there is a false indication of direction to the reef crest.

Figure 12-14 shows the tadpole plot for a reef interval. The reef was penetrated up on the flank, but at 383 ft from the top of A1 carbonate to Clinton, not the expected crestal thickness of 460 ft. To aid analysis of the dipmeter, azimuth-frequency plots were made for three formation

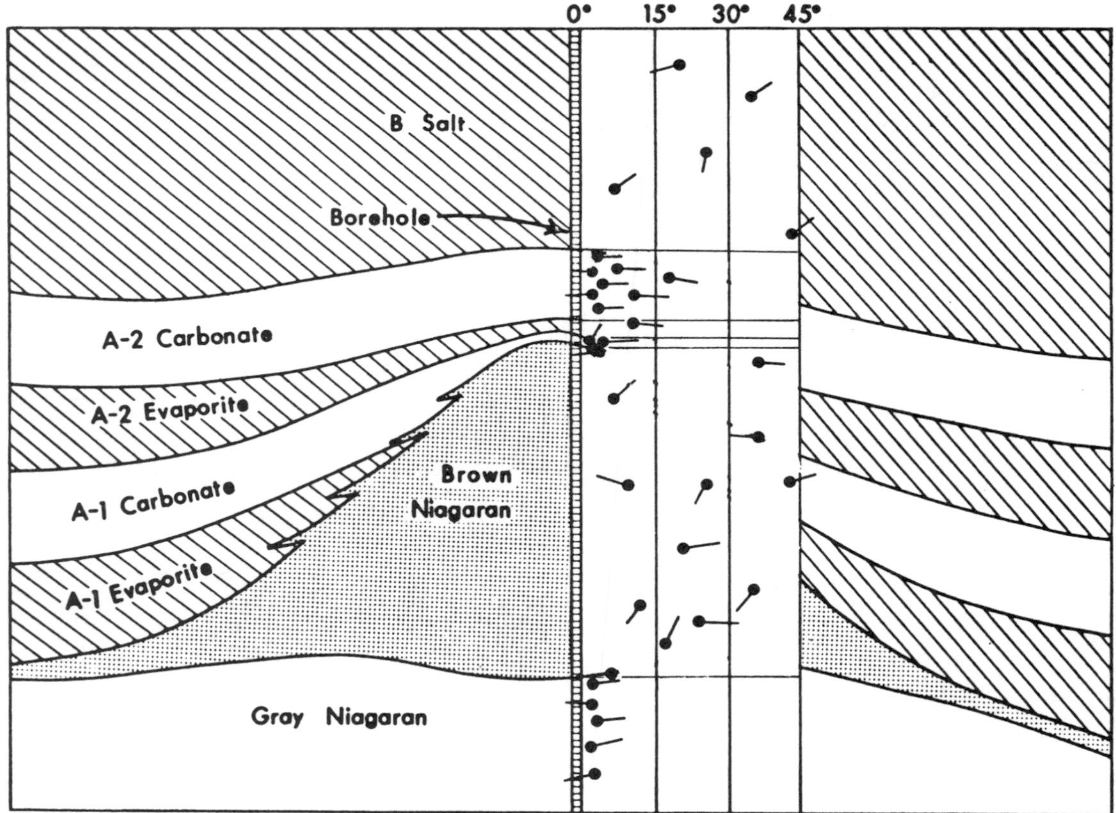

FIG. 12-13. Calculated dips from the borehole of a near crest well are very ambiguous, indicating a lack of formation draping near the reef crest. (Bigelow, 1973, Courtesy Oil & Gas Journal.)

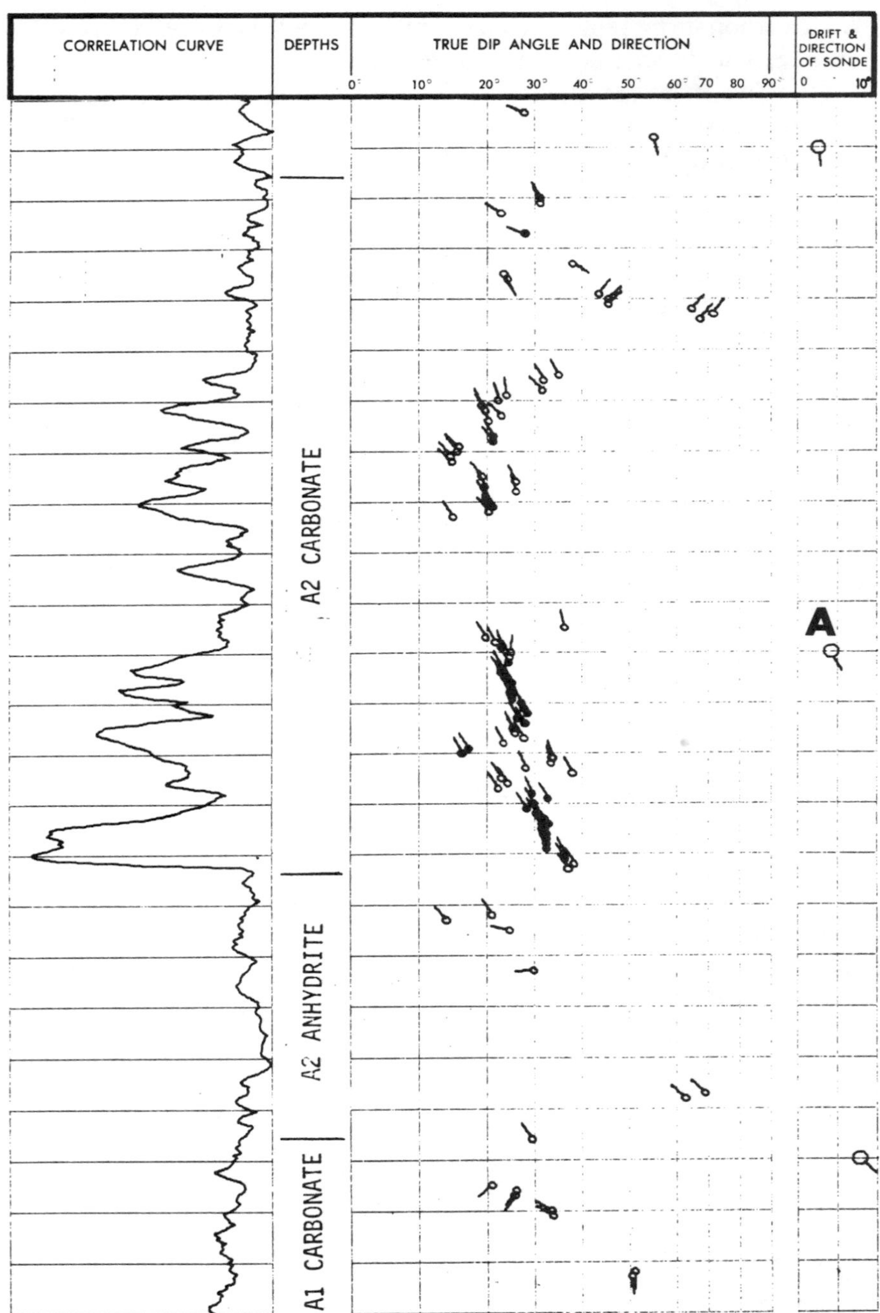

FIG. 12-14. Dipmeter reef example calculated dip tadpole plot. The good quality dips in the A2 Carbonate show increasing dip magnitude with depth, indicate the drape over the reef. Note that the borehole drift (A) indicates the drill bit's tendency to "walk up" the side of the reef, and thus is also a good near-reef direction indicator.

intervals (Figure 12-15). As expected, the A2 anhydrite does not show formation draping. The A1 carbonate and Niagaran dips are also random. The A2 carbonate pattern is well developed and indicates the crest is south-southeast. The dip angles are moderately strong, indicating a close, higher crest. The well was skidded 198 ft to the south-southeast, and the top of the reef was encountered 96 ft higher. This extra height contained hydrocarbons (Figure 12-16).

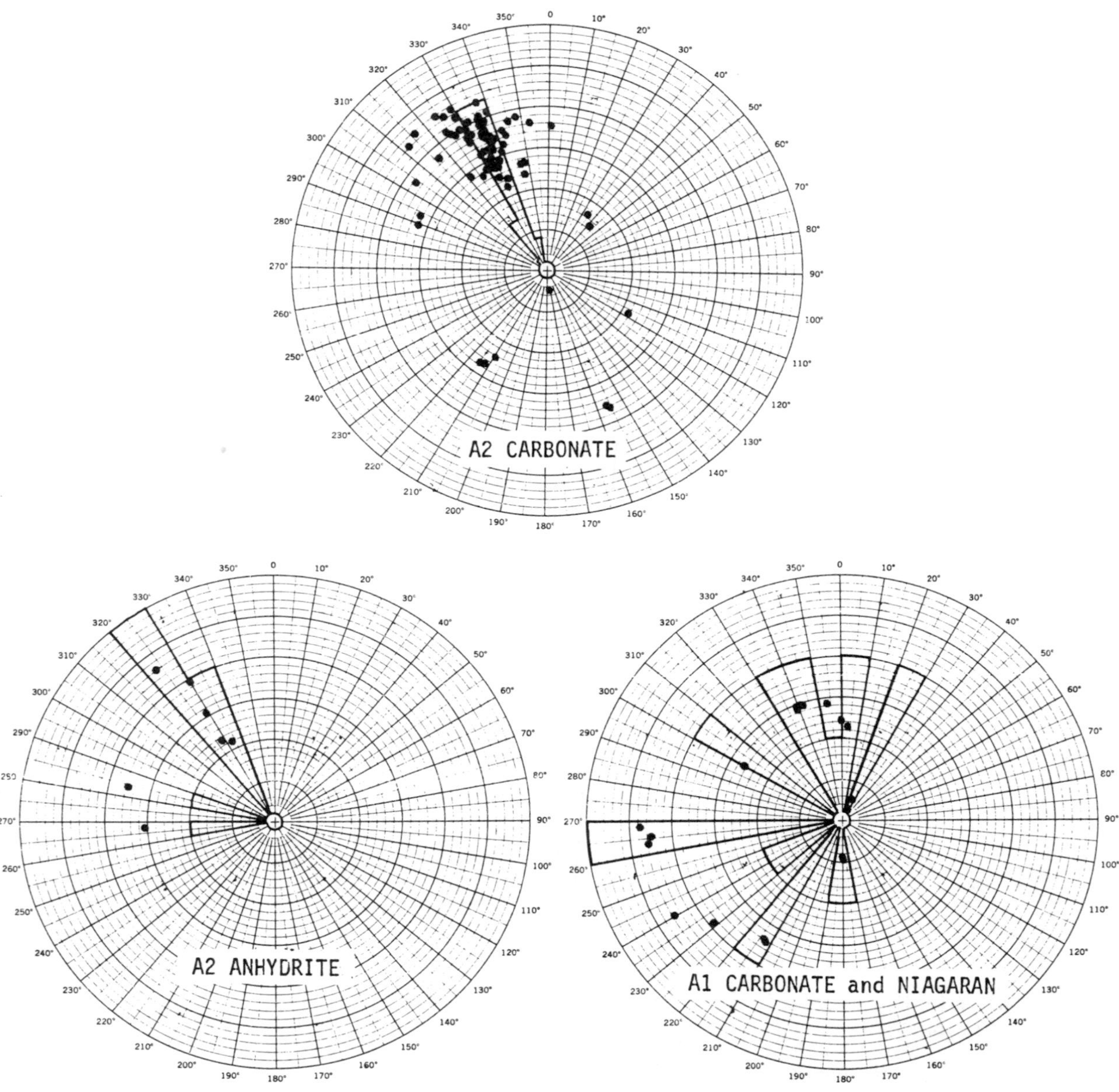

FIG. 12-15. Dipmeter reef example azimuth/frequency analysis of the calculated dips of Figure 12-14. The A2 Carbonate plot indicates the reef crest is south-southeast.

FRACTURE DETECTION

The dipmeter microresistivity measurements can also be used to detect the resistivity contrast of mud-filtrate filled vertical fractures to the surrounding matrix rock. As the sonde comes uphole it rotates with a corkscrew like path and one of the pads could cross horizontally a vertical fracture, while the adjacent pad does not. Thus a separation of adjacent pad measurements could indicate a vertical fracture (Figure 12-17). Because the sonde is rotating, a long vertical fracture should show up on different consecutive pad overlays. In practice this is confirmed by comparisons with core information as shown in Beck, et. al. (1977).

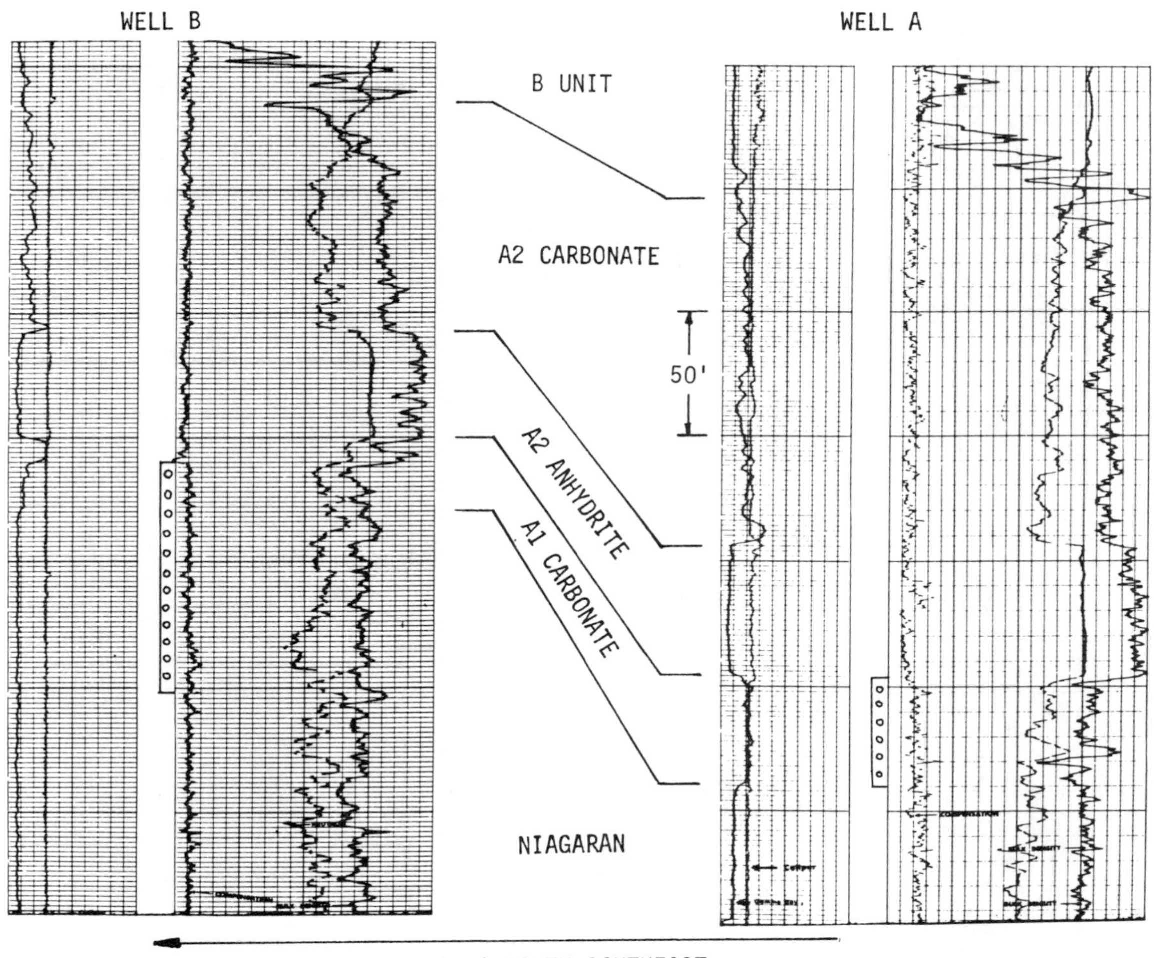

FIG. 12-16. Dipmeter reef example log cross-section. The hydrocarbon bearing reef drilled in Well B was located based on dipmeter information obtained in Well A (Figures 12-14 and 12-15).

BOREHOLE DIRECTIONAL SURVEY

The borehole directional measurements (Figure 12-9) needed to properly reduce the microresistivity measurements can also be used to determine the subsurface location of the borehole in relation to the base of casing and the true vertical depths of the borehole.

The dipmeter tool can be used without recording the microresistivity measurements to obtain just the borehole directional information. Input needed is borehole deviation and direction, plus distance along the borehole. The output is a listing and a plot (Figure 12-18) or a plot only.

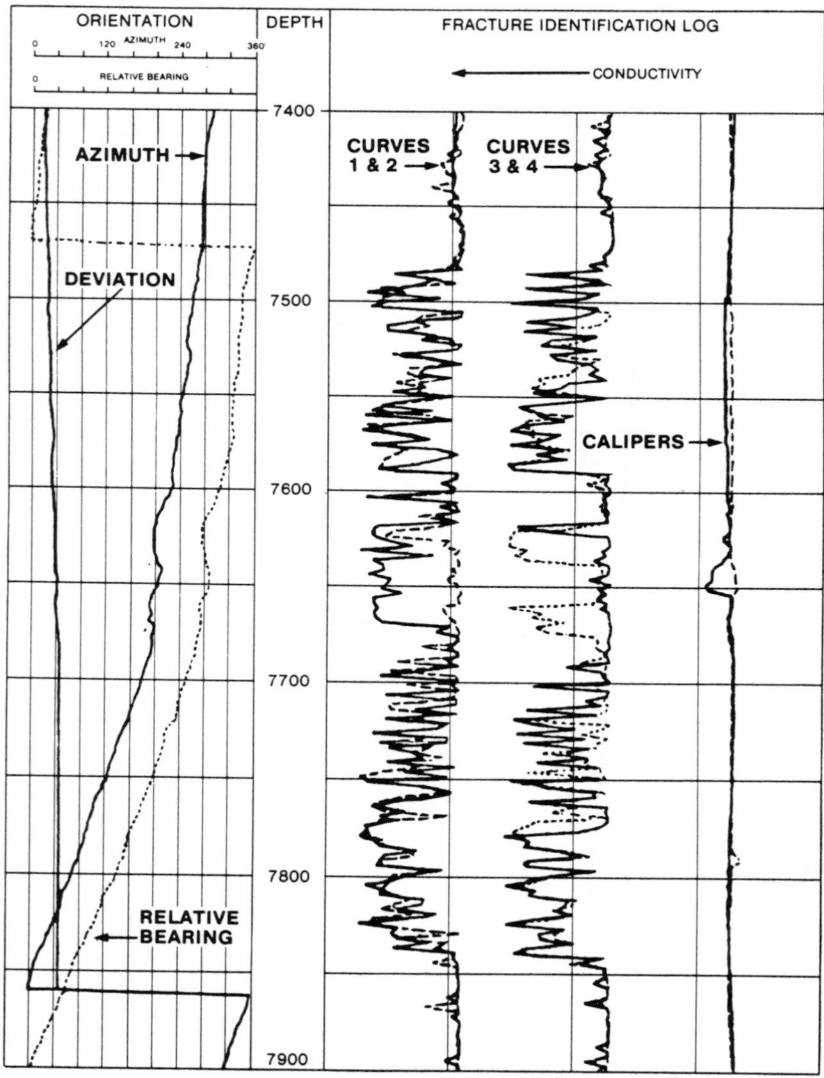

FIG. 12-17. Microresistivity curve separation can indicate vertical fractures in this dipmeter conductivity overlay presentation. (Schlumberger, 1981).

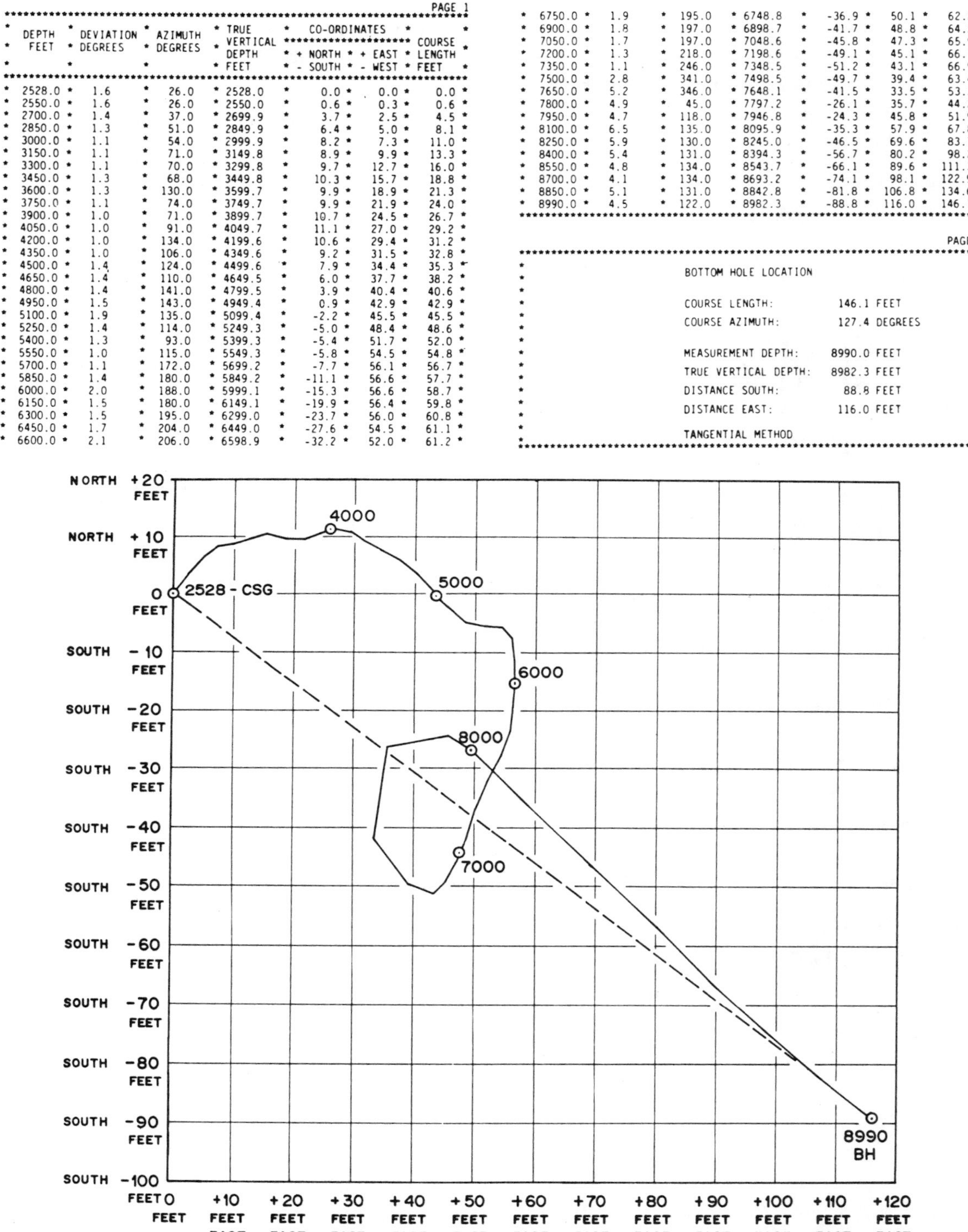

PAGE 1

DEPTH FEET	DEVIATION DEGREES	AZIMUTH DEGREES	TRUE VERTICAL DEPTH FEET	CO-ORDINATES + NORTH - SOUTH	CO-ORDINATES + EAST - WEST	COURSE LENGTH FEET
2528.0	1.6	26.0	2528.0	0.0	0.0	0.0
2550.0	1.6	26.0	2550.0	0.6	0.3	0.6
2700.0	1.4	37.0	2699.9	3.7	2.5	4.5
2850.0	1.3	51.0	2849.9	6.4	5.0	8.1
3000.0	1.1	54.0	2999.9	8.2	7.3	11.0
3150.0	1.1	71.0	3149.8	8.9	9.9	13.3
3300.0	1.1	70.0	3299.8	9.7	12.7	16.0
3450.0	1.3	68.0	3449.8	10.3	15.7	18.8
3600.0	1.3	130.0	3599.7	9.9	18.9	21.3
3750.0	1.1	74.0	3749.7	9.9	21.9	24.0
3900.0	1.0	71.0	3899.7	10.7	24.5	26.7
4050.0	1.0	91.0	4049.7	11.1	27.0	29.2
4200.0	1.0	134.0	4199.6	10.6	29.4	31.2
4350.0	1.0	106.0	4349.6	9.2	31.5	32.8
4500.0	1.4	124.0	4499.6	7.9	34.4	35.3
4650.0	1.4	110.0	4649.5	6.0	37.7	38.2
4800.0	1.4	141.0	4799.5	3.9	40.4	40.6
4950.0	1.5	143.0	4949.4	0.9	42.9	42.9
5100.0	1.9	135.0	5099.4	-2.2	45.5	45.5
5250.0	1.4	114.0	5249.3	-5.0	48.4	48.6
5400.0	1.3	93.0	5399.3	-5.4	51.7	52.0
5550.0	1.0	115.0	5549.3	-5.8	54.5	54.8
5700.0	1.1	172.0	5699.2	-7.7	56.1	56.7
5850.0	1.4	180.0	5849.2	-11.1	56.6	57.7
6000.0	2.0	188.0	5999.1	-15.3	56.6	58.7
6150.0	1.5	180.0	6149.1	-19.9	56.4	59.8
6300.0	1.5	195.0	6299.0	-23.7	56.0	60.8
6450.0	1.7	204.0	6449.0	-27.6	54.5	61.1
6600.0	2.1	206.0	6598.9	-32.2	52.0	61.2
6750.0	1.9	195.0	6748.8	-36.9	50.1	62.2
6900.0	1.8	197.0	6898.7	-41.7	48.8	64.2
7050.0	1.7	197.0	7048.6	-45.8	47.3	65.9
7200.0	1.3	218.0	7198.6	-49.1	45.1	66.7
7350.0	1.1	246.0	7348.5	-51.2	43.1	66.9
7500.0	2.8	341.0	7498.5	-49.7	39.4	63.4
7650.0	5.2	346.0	7648.1	-41.5	33.5	53.3
7800.0	4.9	45.0	7797.2	-26.1	35.7	44.2
7950.0	4.7	118.0	7946.8	-24.3	45.8	51.9
8100.0	6.5	135.0	8095.9	-35.3	57.9	67.8
8250.0	5.9	130.0	8245.0	-46.5	69.6	83.7
8400.0	5.4	131.0	8394.3	-56.7	80.2	98.2
8550.0	4.8	134.0	8543.7	-66.1	89.6	111.3
8700.0	4.1	134.0	8693.2	-74.1	98.1	122.9
8850.0	5.1	131.0	8842.8	-81.8	106.8	134.6
8990.0	4.5	122.0	8982.3	-88.8	116.0	146.1

PAGE 2

BOTTOM HOLE LOCATION

COURSE LENGTH:	146.1 FEET
COURSE AZIMUTH:	127.4 DEGREES
MEASUREMENT DEPTH:	8990.0 FEET
TRUE VERTICAL DEPTH:	8982.3 FEET
DISTANCE SOUTH:	88.8 FEET
DISTANCE EAST:	116.0 FEET

TANGENTIAL METHOD

FIG. 12-18. Borehole directional survey obtained from dipmeter borehole deviation, borehole direction, and depth measurements.

REFERENCES

Beck, J., Schulta, A., and Fitzgerald, D., 1977, Reservoir evaluation of fractured Cretaceous carbonates in South Texas: Trans. 18th Ann. Soc. Prof. Well Logging Analysts Logging Symposium, Houston.

Bigelow, E. L., 1973, High-resolution dipmeter uses in Michigan's Niagaran Reefs: Oil and Gas J., **71**, 36 (Sept. 3), 78-88.

Schlumberger Well Services, 1974, Calibration and quality standards.

Schlumberger Limited, 1981, Dipmeter interpretation, Volume I — Fundamentals.

REFERENCES FOR GENERAL READING

Allaud, L. A., and Ringot, J., 1969, The high resolution dipmeter tool: The Log Analyst, **10**, 3 (May-June 1969).

Allaud, L. A., and Martin, M. H., 1977, Schlumberger — The history of a technique: John Wiley and Sons, Inc.

Bigelow, E. L., 1985a, Making more intelligent use of log derived dip information, Part I: suggested guidelines: The Log Analyst, **26**, 1 (January-February), 41-53.

———, 1985b, Making more intelligent use of log derived dip information, Part II: wellsite data gathering considerations: The Log Analyst, **26**, 2 (March-April), 25-41.

———, 1985c, Making more intelligent use of log derived dip information, Part III: computer processing considerations: The Log Analyst, **26**, 3 (May-June), 18-31.

Bigelow, E. L. and Cox, J. W., 1975, High-resolution dipmeter uses in Michigan's Niagaran Reefs: 14th Ann. Ontario Petr. Inst. Conference, London.

Cox, J. W., 1970, The high-resolution dipmeter reveals dip-related borehole and formation characteristics: Trans. 11th Ann. Soc. Prof. Well Log Analysts Logging Symposium, Los Angeles.

Gilreath, J. A., and Maricelli, J. J., 1964, Detailed stratigraphic control through dip computations: Am. Assn. Petr. Geol. Bulletin, **48**, 1904-1910.

Holt, O. R., 1974, Relating diplogs to practical geology: Dresser Atlas.

Schlumberger, 1980, SOS-GC dipmeter workbook.

Appendix A

LOGGING TOOL ABBREVIATIONS

Tool/ Service Abbreviation	Service Company Code[1]	Tool type, tool trade name
A	P	acoustic log
A-BHC	S	acoustic amplitude
ABC	B	borehole compensated acoustic log
AC	D	acoustic log, conventional spaced
ACL	D	long spaced acoustic log
ALC	D	borehole compensated acoustic log
AMP	D	acoustic amplitude
APL	P	acoustic amplitude, Acoustic Parameter Log
AVL	W	acoustic log, Acoustic Velocity Log
A/L	W	acoustic log
BCN	N	dual detector neutron, Borehole Compensated Neutron
BCS	G	borehole compensated acoustic log
BGT	S	four arm caliper + orientation, Borehole Geometry log
BHC	D S	borehole compensated acoustic log
BHC-AL	D	borehole compensated acoustic log
BHGM		borehole gravimeter
BHTV		borehole televiewer
CAD	Shell	horizontal micro-acoustic log
CAL	W	wellsite computer interpretation, Computer Analyzed Log

–CAL		caliper log
CAVL	W	borehole compensated acoustic log
CA/L	W	borehole compensated acoustic log
CBL		cased hole acoustic amplitude, Cement Bond Log
CC	W	microresistivity, Contact Caliper
–CCL		casing collar locator
CDL	D G W	compensated density log
CDR	S	borehole directional survey
CET	S	Cement Evaluation Tool
CIS	S	customer supplied instrument service
CMT	S	horizontal microacoustic log
CNL	D S	dual detector neutron, Compensated Neutron Log
CNP	B	dual detector neutron, Compensated Neutron Porosity
CNS	G	dual detector neutron, Compensated Neutron Survey
CONC	W	microresistivity, Contact Caliper Log
–CL		caliper
CSNG	W	spectral natural gamma ray, Compensated Spectral Natural Gamma ray
CST	S	sidewall sampler gun, Core Sampler gun
CSVL	G	borehole compensated acoustic log
CVL		acoustic log, Continuous Velocity Log
DBC	B	compensated density log
DCL	G W	low frequency Dielectric Constant Log
DD	B D S	depth determination
DEL	D	low frequency Dielectric Log
DFL	D	Dip Fracture Log
DGL	W	dual laterolog
DIFL	D	Dual Induction-Focused Log
DIG	W	Dual Induction-Guard Log
DIL	S	Dual Induction-Laterolog
DIP	D	four arm dipmeter
DIR	D	directional survey
DISF	S	Dual Induction-Spherically Focused Log
DIRFS	G	Dual Induction-Radially Focused System
DL	G	uncompensated density log
DLC	D	compensated density log
DLL	D G S	simultaneous dual laterolog
DLS	G	dual laterolog
DML	G	dipmeter log
DNL	S	dual porosity compensated neutron log
DPT	S	low frequency dielectric log, Deep Propagation Tool

DSL	S	digital acoustic log
DSN	W	dual detector neutron, Dual Spaced Neutron log
DST	S	dual laterolog-microfocused resistivity
DSW	S	digital acoustic log waveforms
EL	D G	resistivity, Electric Log
EMC	B	microresistivity, Micro-Contact Caliper
ENP	B	single detector neutron, Epithermal Neutron Porosity
EPT	S	high frequency dielectric, Electromagnetic Propagation log
ES	B G S	resistivity, Electrical Survey
FCL	W	micro-resistivity, FoRxo-Caliper Log
FDC	S	compensated density log
F DIP	D	focused dipmeter
FDL	B S	uncompensated density log
FED	G	four electrode dipmeter
FIL	S	dipmeter derived Fracture Identification Log
FIT	S	Formation Interval Tester
FF	W	acoustic amplitude, Fracture Finder
FL	L	focused resistivity
FoRxo	W	microresistivity
FRAC	D	acoustic amplitude and wavetrain
FRT	S	compensated density-microfocused resistivity
FWS	W	acoustic wavetrain
GDS	B	focused resistivity, Guard log
GEO	D	geophone survey
GG	W	focused resistivity, Gamma Ray-Guard log
GL	G	focused resistivity, Guard Log
–GR		Gamma Ray log
GRN	S	Gamma Ray-thermal Neutron
GS	G	geophone survey
GST	S	spectral pulsed neutron, Gamma Ray Spectroscopy Tool
HDT	S	four arm dipmeter, High Resolution Dipmeter Tool
HRDIP	D	four arm dipmeter, High Resolution Dipmeter
I		acoustic travel time integration
IEL	D G P W	induction electric log
IES	B L S	induction electric survey
IL	G	induction log

IS	B	induction log
ISF	S	induction-spherically focused log
L		focused resistivity, Laterolog
LDT	S	compensated density-photoelectron lithology log
LGT	S	sequential dual laterolog, LL9
LL	D P S	focused resistivity log, Laterolog
LSS	D G S	long spaced acoustic log, Long Spaced Sonic
LSV	W	long spaced acoustic log
MEL	G	microresistivity, Micro-Electric Log
MGL	G	microfocused resistivity, Micro-Guard Log
ML	D G S	microresistivity, Minilog, Microlog
MLL	D G P S	microfocused resistivity, Microlaterolog
MOP	S	movable oil plot
MPL	S	computed Mechanical-Properties Log
MRL	P	microresistivity log
MS	P	microresistivity, Micro Survey
MSFL	G S	microfocused resistivity
MSG	W	acoustic wavetrain, Micro-Seismogram
N		neutron log
NBC	B	dual detector neutron log
NGS	G S	spectrum natural gamma ray
NL	B S	neutron log
NLL	D	pulsed neutron capture log, Neutron Lifetime Log
NML	D S	nuclear magnetism log
PDC	S	gamma ray/neutron-collar log, Perforating Depth Control
PDT	S	three arm powered dipmeter
PFC	D	gamma ray/neutron-collar log
PML	S	microfocused-microresistivity, Proximity-Microlog
PROLOG	D	wellsite computer interpretation
PROX	D	microfocused resistivity, Proximity Log
RFT	S	repeat formation tester
RH DIP	D	four arm dipmeter
SAT	S	geophone survey, Seismic Acquisition Tool
–SC	G	acoustic wavetrain, Signature Curve
SDL	W	Spectral Density Log
SFAL	G	acoustic amplitude, Sonic Formation Amplitude Log

SFL	S	focused resistivity, Spherically Focused Log
SGE	B	geophone survey, Seismic well Geophone survey
SGR	G	spectral natural gamma ray
SHDT	S	four arm stratigraphic dipmeter
SIG	D	acoustic signature
SL	S	uncompensated acoustic log, Sonic Log
SNL	G	epithermal neutron log, Sidewall Neutron Log
SNP	S	epithermal neutron log, Sidewall Neutron Porosity
SNAP	G	epithermal neutron log, Sidewall Neutron Analyzed Porosity
SPL	D	spectral natural gamma ray
SRS		geophone survey, Seismic Reference Service
–SS	G	acoustic wavetrain, Seismic Spectrum (variable density)
SVL	G	uncompensated acoustic log
SWAN	D	vertical microacoustic and epithermal neutron log
SWN	D W	epithermal neutron log, Sidewall Neutron
SWC	G	sidewall core gun
TDI	W	acoustic traveltime integration
TDT	S	pulsed neutron capture log, Thermal Decay Time
TMD	W	pulsed neutron capture log, Thermal Multigate Decay
–TTI	D G S	acoustic traveltime integration
–VDL	D S	acoustic wavetrain, Variable Density Log
V3D	B	acoustic wavetrain, 3 Dimensional Log
WST	S	geophone survey, Well Seismic Tool
WEL	G	wellsite computer interpretation, Well Evaluation Log
WFT	D	acoustic waveform taping
2IL	B	dual induction focused log

[1]Service Companies:
- B Birdwell
- D Dresser Atlas
- G Gearhart — GO International
- L Lane-Wells
- N NL McCullogh
- P PGAC
- S Schlumberger
- W Welex

Appendix B

ACOUSTIC LOG BOREHOLE COMPENSATION METHODS

Borehole compensation is the removal of the effects of an irregular shaped borehole from the Δt measurement of the acoustic log.

With a one-transmitter two-receiver system the signal from the transmitter arrives at the two receivers at different times. The difference in arrival times is the measured Δt value. When the borehole diameter is the same through the measured interval between the two receivers, Δt corresponds to the time for the wave to pass through the formation interval between the two receivers (Figure B-1). The traveltimes for the two receivers are:

$$TR_2: \quad A + B + \Delta t + A$$

$$TR_1: \quad A + B \qquad\quad + A.$$

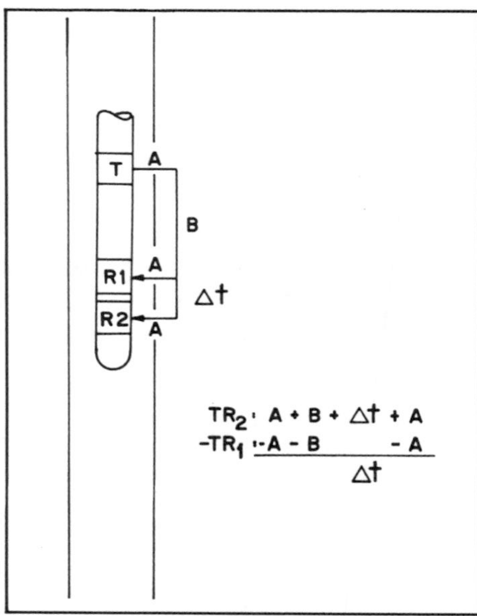

FIG. B-1. For a constant borehole diameter, subtracting the two-receiver arrival times leaves only formation traveltime (Δt).

When subtracted the result is a time difference, Δt, which is independent of borehole diameter as long as the diameter is the same for the two receivers.

If there is a variation in borehole diameter within the receiver span, the value recorded by this system is not exactly equal to the desired formation Δt (Figure B-2). For a diameter increase with increasing depth, the

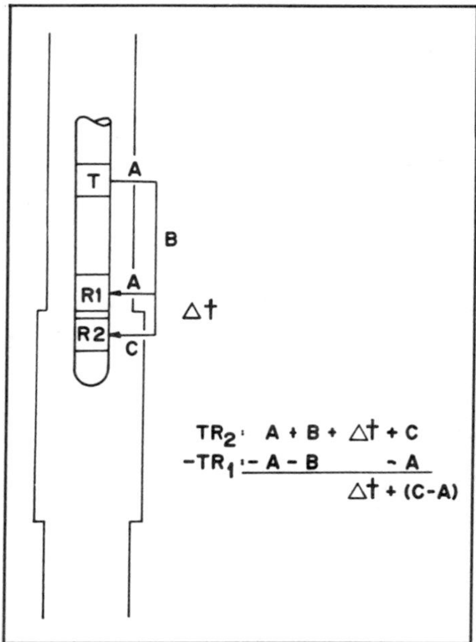

FIG. B-2. For varying borehole diameters, the apparent log traveltime is in error at the change of diameter.

traveltimes for the two receivers are:

$$TR_2: \quad A + B + \Delta t + C$$
$$TR_1: \quad A + B \qquad + A.$$

The time difference is:

$$TR_2 - TR_1 = \Delta t + (C - A).$$

In this case the difference between recorded and actual formation Δt is because of the difference in the mud transit times for paths A and C. Where the borehole enlarges, with increasing depth, there is an increase in Δt and where the borehole reduces in size, with increasing depth, the recorded Δt decreases (Figure B-3). The difference between the true formation and recorded Δt value is a function of the borehole diameter change and the receiver span distance (Figure B-4).

Various methods have been devised to provide borehole diameter change compensation. The first method uses a two-transmitter two-receiver system, which records differential times up and down across the receivers (Figure B-5). The resultant traveltimes are:

$$LR_1: \quad A \qquad + \Delta t + \; C + D$$
$$LR_2: \qquad\qquad\qquad 2C + D$$
$$UR_2: \quad A + B + \Delta t + \; C$$
$$UR_1: \quad 2A + B$$

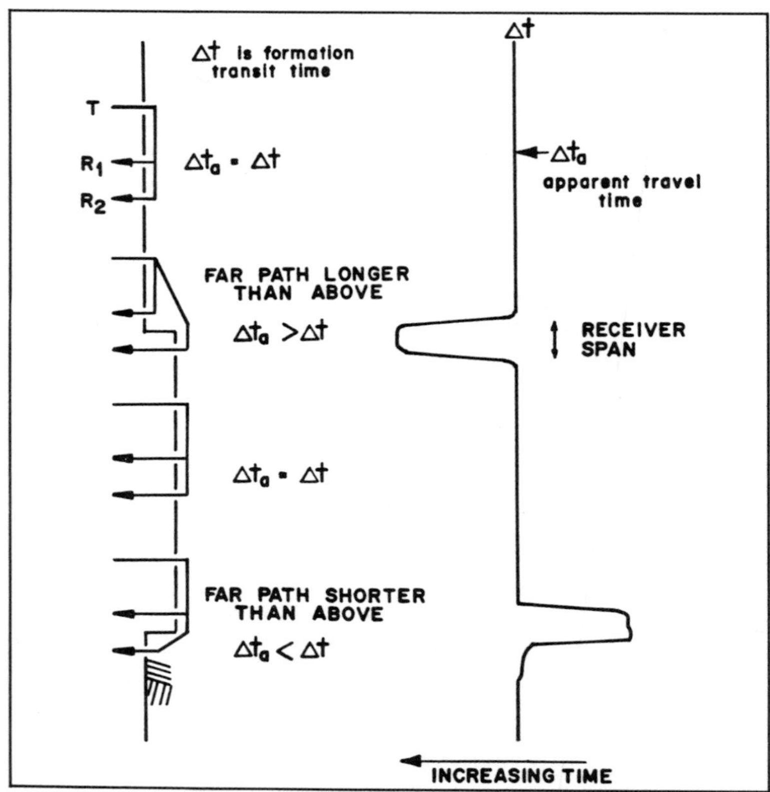

FIG. B-3. Varying borehole diameter produces spikes on the acoustic traveltime measurement (Δt) at the diameter change points. Their shape and magnitude are a function of the amount of diameter change, receiver span, difference between formation and mud velocity, and direction of measurement.

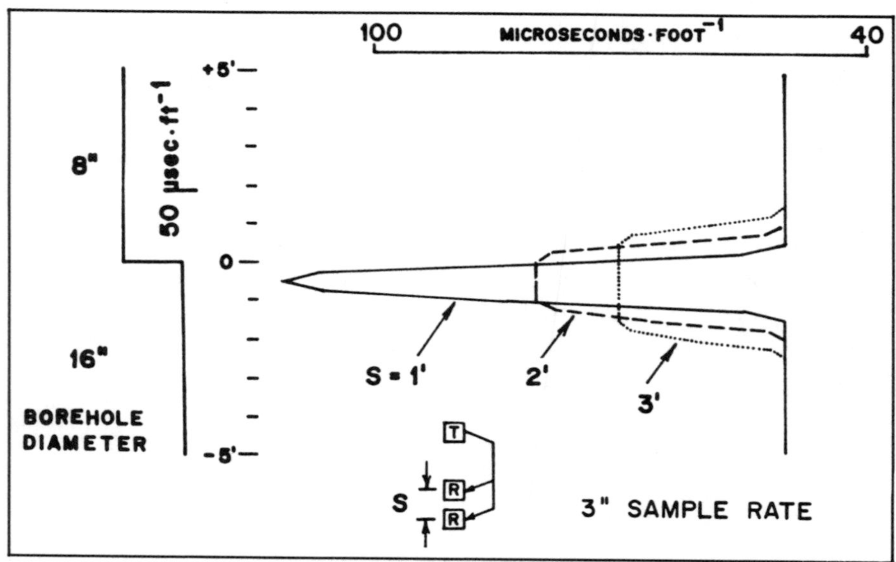

FIG. B-4. The one-transmitter two-receiver diameter change Δt spike's magnitude and width are a function of receiver span distance.

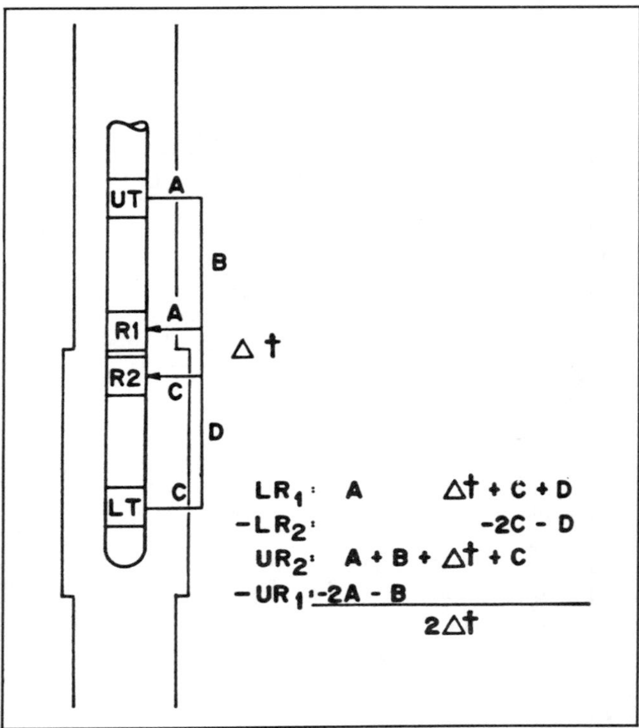

FIG. B-5. Eliminating the borehole diameter change problem by summing across the change in two directions.

The total time difference is:

$$LR_1 - LR_2 + UR_2 - UR_1 = 2\Delta t.$$

Thus, even though there is a change in the borehole diameter within the measured interval, the measurement is only a function of formation Δt.

A second method of borehole compensation involves measuring transit times at various borehole depths in relation to the interval of interest and then combining them in an appropriate manner (Figure B-6). This is accomplished at three depths with the single-transmitter two-receiver system. The traveltimes from the three depths are:

depth E $-TR_1$: $- C - 2D$

depth F $+TR_2$: $A + \Delta t + C + D$

depth G $+TR_2$: $A + B + \Delta t$

$-TR_1$: $-2A - B$

The total time difference is again $2\Delta t$. This method achieves the same result as the previous method but involves a complex storage and summing system.

A similar method is used for Schlumberger's long-spaced two-

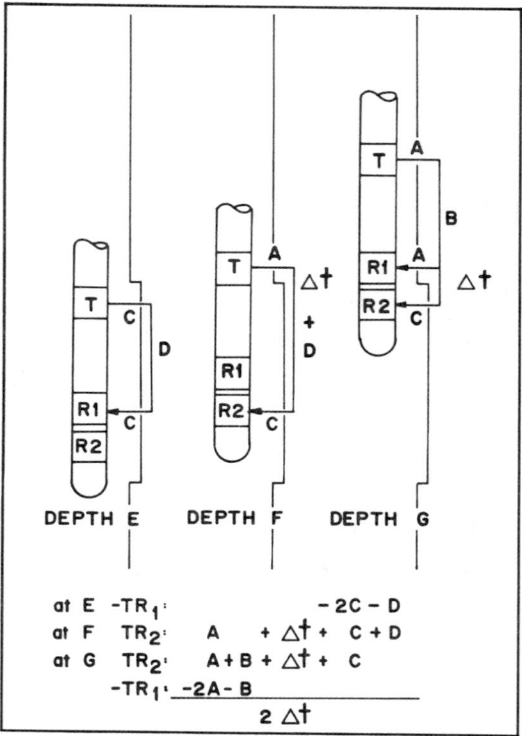

FIG. B-6. A single-transmitter two-receiver array can compensate for diameter changes by summing measurements at different depths across the diameter discontinuity. This method requires a complex storage and retrieval system.

transmitter two-receiver system (Figure B-7). The traveltimes for the shorter (8, 10 ft) spacing are:

$$
\begin{array}{llllll}
\text{depth } E & +UR_2: & A & + \Delta t + & C + D \\
 & -UR_1: & & & -\ 2C - D \\
\text{depth } G & +LR_1: & A + \mathrm{B} + \Delta t + & C & \\
 & -UR_1: & -\ 2A - \mathrm{B} & &
\end{array}
$$

The total time difference is $2\Delta t$. A second simultaneous Δt measurement with a longer spacing is also made (Figure B-7, right).

In reality there are two problems with these idealized explanations. First, the signals are refracted into the formation and back into the receivers, instead of arriving normally as presented in the above discussion. Thus the measurements for the two-transmitter two-receiver system don't quite measure the same formation interval (Figure B-8, left). Schlumberger compensates for ''average'' refraction paths by offsetting two pairs of receivers 5 inches to allow Δt measurements of the same formation interval (Figure B-8, right). (See Kokesh, 1965, page 283.)

Second, the Δt curve shape differs slightly depending on the direction the borehole discontinuity is measured (Figure B-9). The difference is somewhat distorted when a velocity difference is added to the borehole

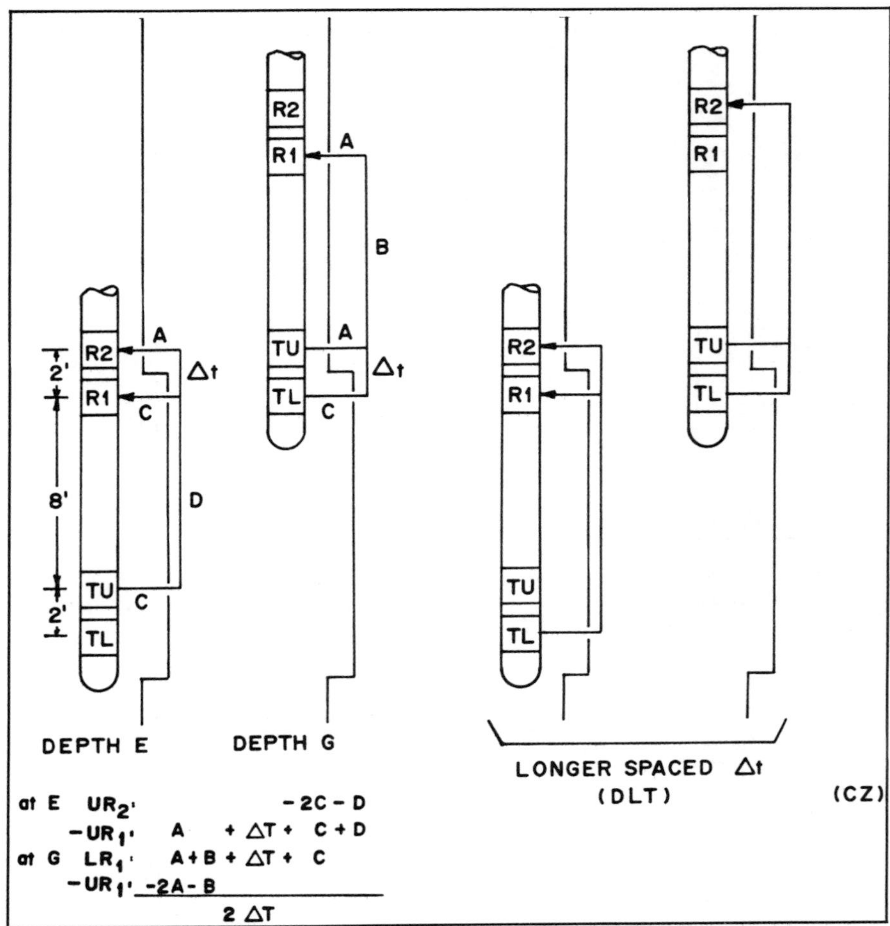

FIG. B-7. Schlumberger's long spaced two-transmitter two-receiver acoustic log allows measurement of Δt for two different spacings, using a complex storage and retrieval system.

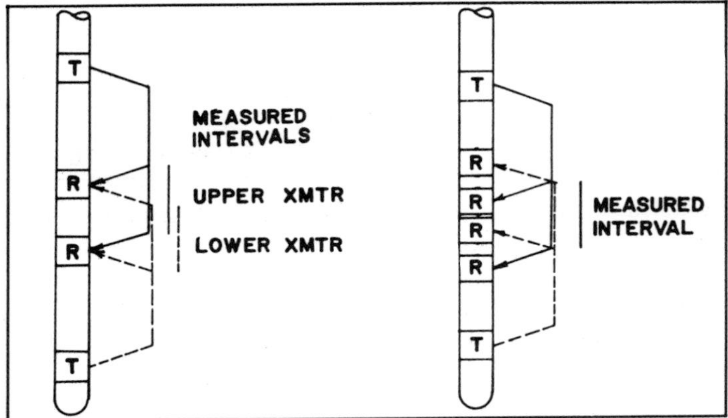

FIG. B-8. In reality the signal paths to the receivers are refractions and not perpendicular arrivals. Thus the two-transmitter two-receiver system (A) does not measure the same formation interval for each array. Schlumberger's four-receiver system (B) is designed to account for the refraction arrivals, and thus each array measures the same formation interval.

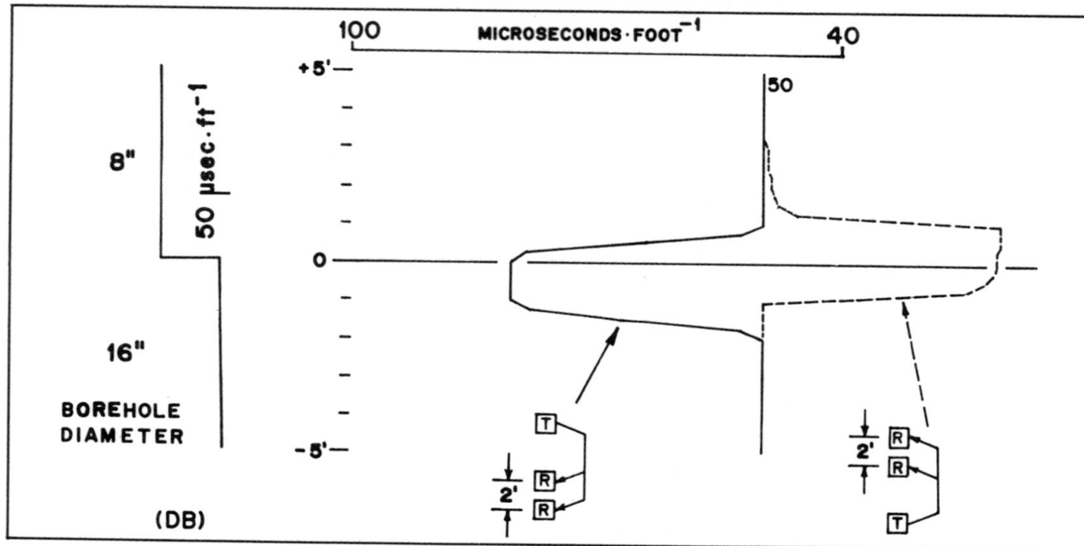

FIG. B-9. Modeled acoustic curve shapes for the same formation velocity across a borehole diameter change. Note the difference in curve shape and their depth offset due to refraction and direction of measurement.

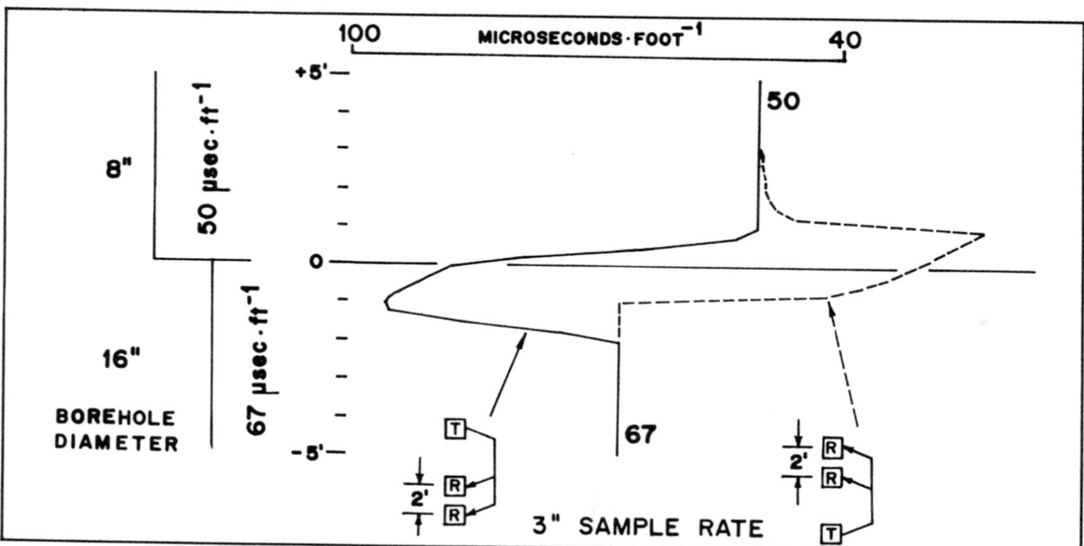

FIG. B-10. Modeled acoustic curve shapes for different formation velocities across a borehole diameter change.

diameter change (Figure B-10). Thus there is a slight residual spike for either the two-transmitter two-receiver or two-transmitter four-receiver borehole compensated systems.

REFERENCE

Kokesh, F. P., Schwartz, R. J., Wall, W. B., and Morris, R. L., 1965, A new approach to sonic logging and other acoustic measurements: J. Petr. Tech., **17**, 282-186.

Appendix C

ACOUSTIC TOOL SPECIFICATIONS

A list of acoustic tools, normal operating conditions (Figures C-1, C-2, and C-3), and a chart of tool specifications follow (Table C-1). Much other information in this appendix was obtained from the Reference Handbook of Selected Formation Evaluation Tools (out of print editions 1966 and 1972) by the Lafayette Chapter of the society of Professional Well Log Analysts.

Measurement-Company	Trade Name
BHC Traveltime	
Birdwell	Acoustic/Borehole Compensated Log
Dresser Atlas	BHC Acoustilog
Gearhart	Borehole Compensated Sonic
Lane-Wells	Dual Spaced Acoustilog
PGAC	Borehole Compensated Acoustic
Schlumberger	BHC Sonic
Welex	Compensated Acoustic Velocity
Amplitude	
Birdwell	3-D Velocity
Dresser Atlas	Signature, Acoustic Parameter, Fraclog
Gearhart	Sonic Formation Amplitude
Lane-Wells	Acoustic Amplitude
McCullough	Fracture Finder Log
PGAC	Acoustic Parameter
Schlumberger	Sonic Amplitude
Welex	Fracture Finder
Waveform	
Birdwell	3-D Velocity
Dresser Atlas	Variable Density
Gearhart	Seismic Spectrum, Signature Curve, X-Y Ploy
McCullough	Sonic Seismogram, Seismic Log
Schlumberger	Variable Density, Waveforms, Wave Train Display
Welex	Micro-Seismogram, Full Wave Acoustic Log

The results obtained when working in fresh water, salt water, oil base, or oil emulsion are good. When air or gas are encountered the acoustic measurements are not obtainable.

Measurements that can be made simultaneously are: gamma ray, spontaneous potential, caliper, compensated neutron (SWS), induction-electric, dual induction, traveltime integration, and spectral natural

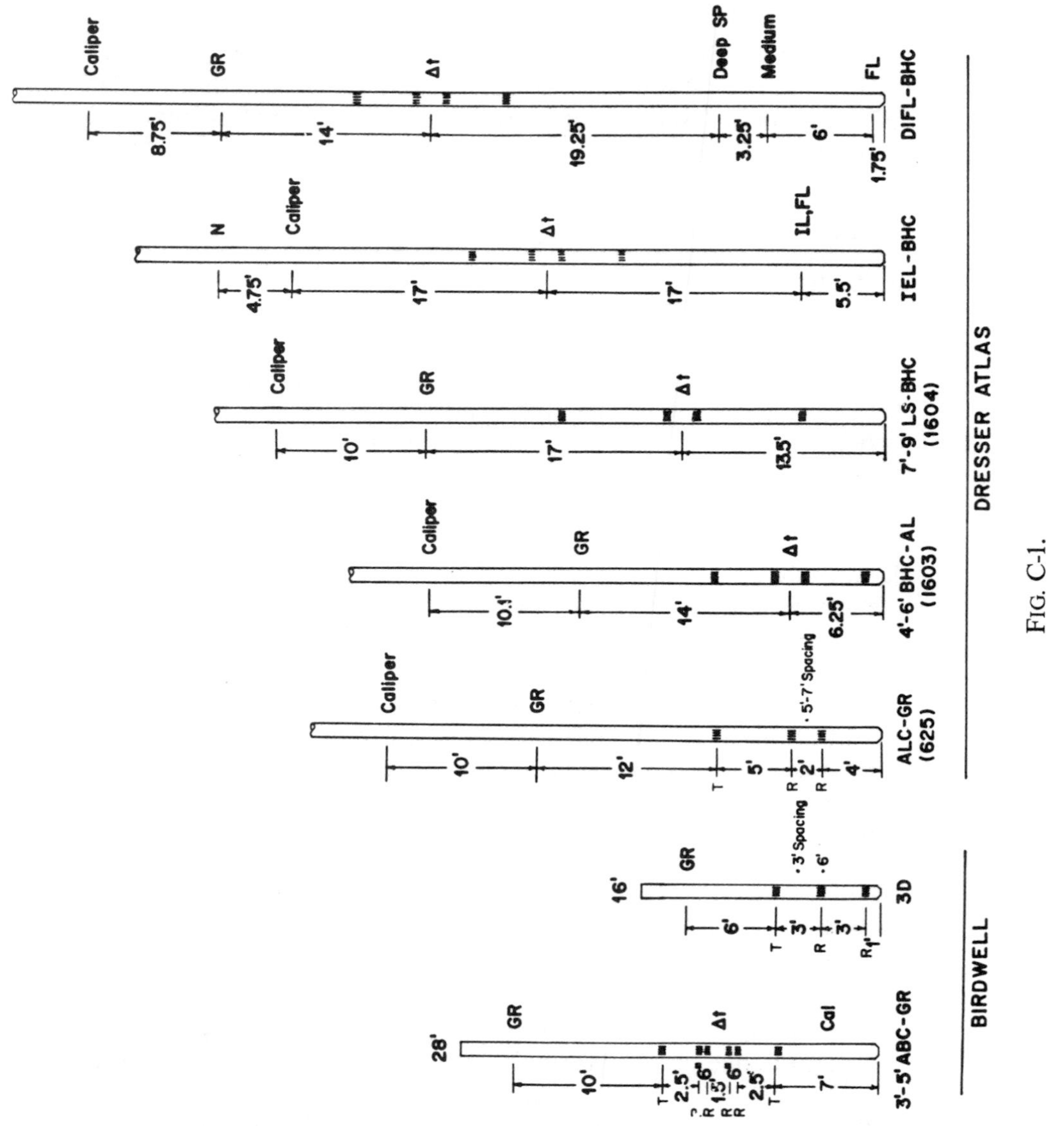

BIRDWELL
3'-5' ABC-GR
28'
GR
Δt
Cal
10'
2.5'
1.5'
6"
2.5'
7'
3D
16'
GR
3' Spacing
6'
DRESSER ATLAS
ALC-GR
(625)
Caliper
GR
5'-7' Spacing
10'
12'
5'
2'
4'
4'-6' BHC-AL
(1603)
Caliper
GR
Δt
10.1'
14'
6.25'
7'-9' LS-BHC
(1604)
Caliper
GR
Δt
10'
17'
13.5'
IEL-BHC
N
Caliper
Δt
IL,FL
4.75'
17'
17'
5.5'
DIFL-BHC
Caliper
GR
Δt
Deep SP
Medium
FL
8.75'
14'
19.25'
3.25'
6'
1.75'

FIG. C-1.

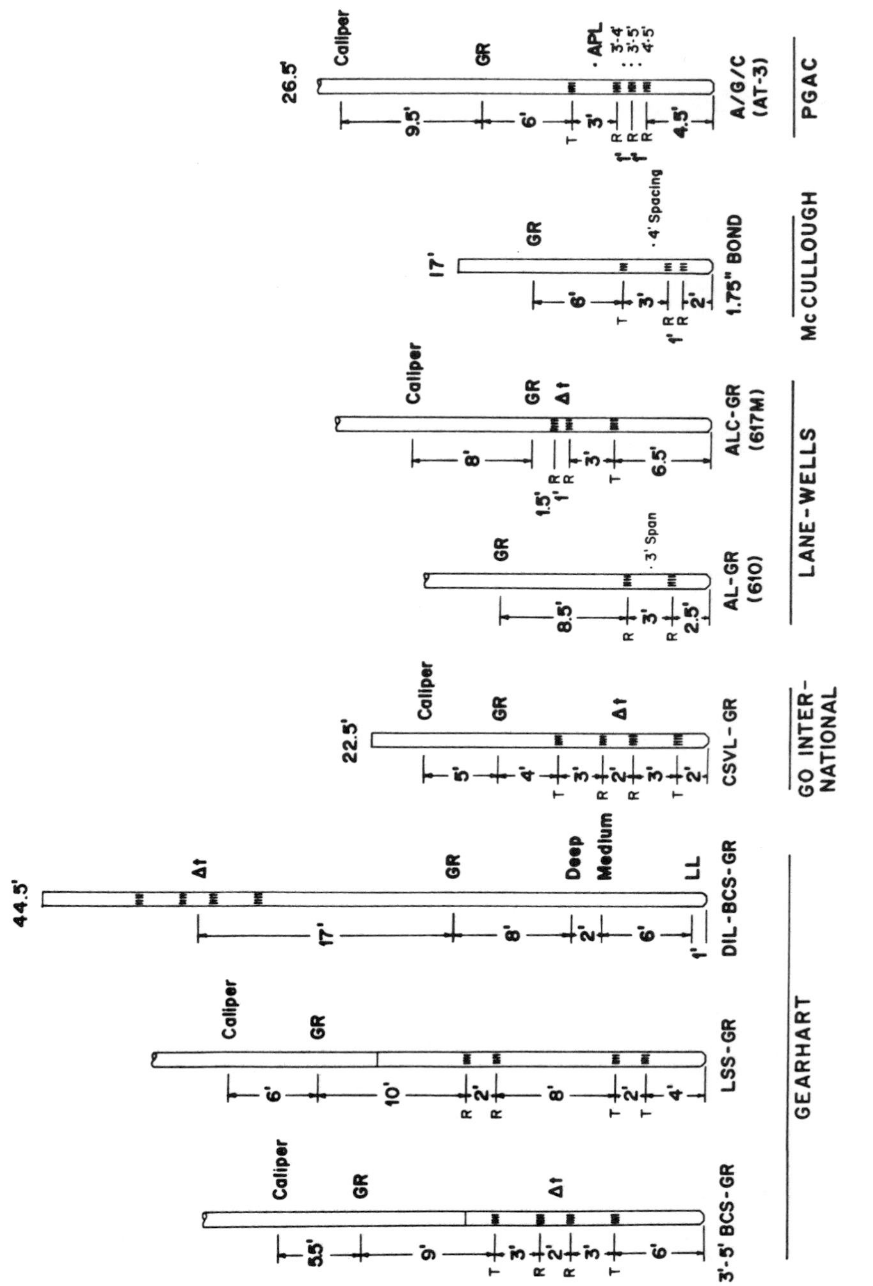
3'-5' BCS-GR
LSS-GR
DIL-BCS-GR
GEARHART
CSVL-GR
GO INTER-NATIONAL
AL-GR (610)
ALC-GR (617M)
LANE-WELLS
1.75" BOND
McCULLOUGH
A/G/C (AT-3)
PGAC
Caliper
GR
Δt
Deep
Medium
LL
44.5'
22.5'
17'
26.5'
· 3' Span
· 4' Spacing
· APL
3'-4'
3'-5'
4'-5'

Fig. C-2.

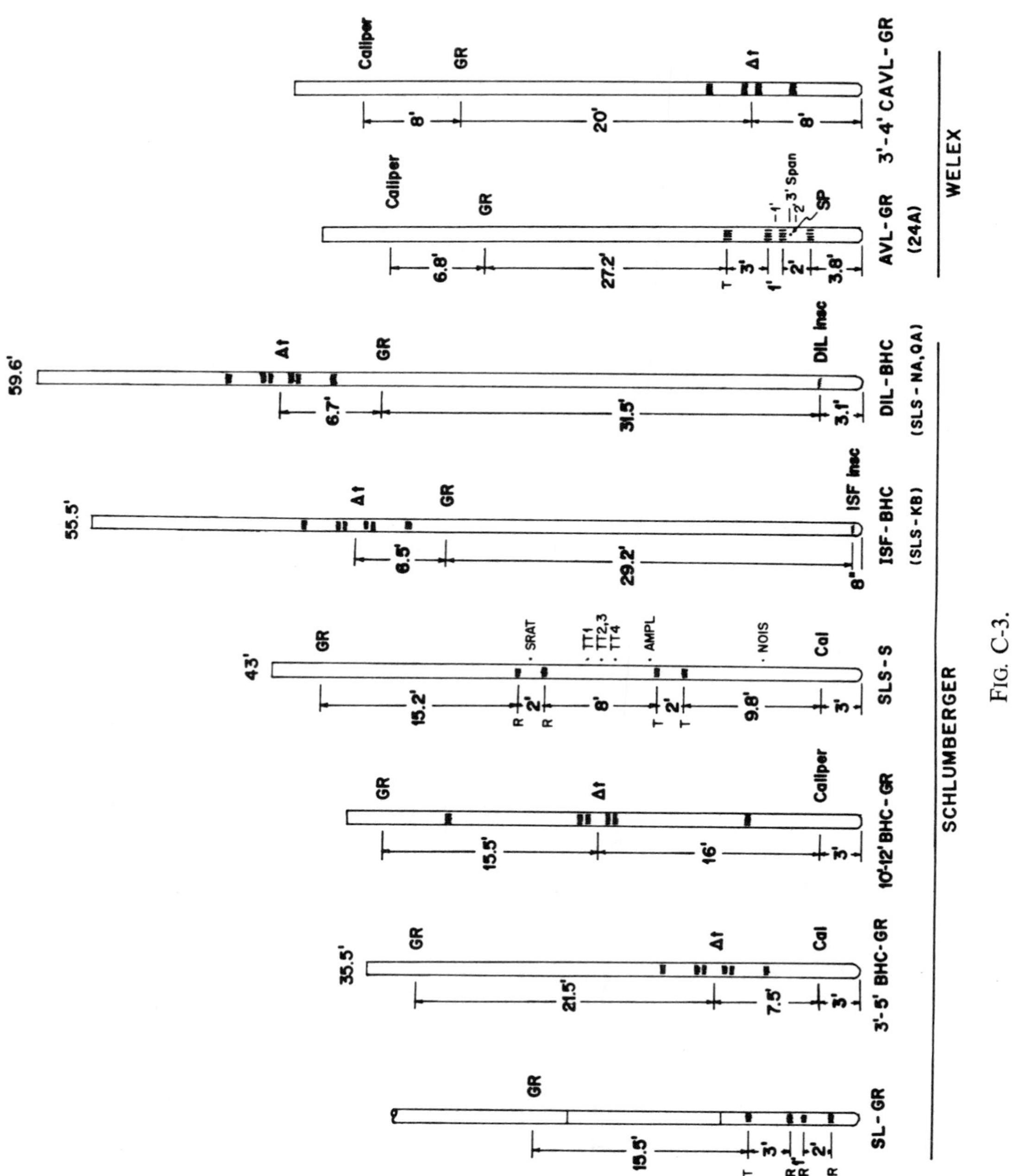

FIG. C-3.

Table C-1. Acoustic tool specifications.

Company	Model	Spacing (ft)	Span (ft)	Confg	Borehole Comp*	Tool Diameter (Inches)	Length (Ft/Inches)		Hole Diameter (Inches) Min	Max	Temp rating (°F)	Max pressure (PSI — in thousands)
Birdwell	ABC	3	2	2T-2R	2	3 3/4	16		6	16	250	10
	V3D	3,6	—	1T-2R	0	3 3/4	12		3	18	300	20
Dresser Atlas	625	3,5	1,2,4			3 5/8					400	20
	1412	4	—	1T-1R	0	1 11/16	11		2 7/8	5 1/2	400	17
	1415	3			0	3 3/8	12	3	4 1/2	12	500	20
	1601	4	2	2T-2R	2	4 1/8	19	5	4 3/4	16	350	20
	1603	4	2	2T-	2	3 3/8	19	6	4 1/2	16	350	20
	1604	7	2	2T-2R	2	3 7/8	25	2	4 1/2	16	350	20
	1607	3	2	2T-2R	2	3 7/8	17	4	4 1/2	12	400	20
	1610	4	2	2T-2R	2	4 1/4	23	4	6		300	20
	A0101E	4	2	2T	2	3						
Gearhart	BCS	3	2	2T-2R	2	3 1/2	15.96		5	12	350	18.5
	LSS	8,10	2	2T-2R	2m	4	22		6	24	400	20
GO-International	SVL			1T-2R	1	3 1/2	11	6	6		300	15
	CSVL			2T-2R	2	3 1/2	11	6	6		300	15
Lane-Wells	610	3	1,3		2							
	617	3	1	1T-2R	2	3 7/8	14		6	12 1/4	300	20
McCullouch	CBL	3,4	—	1T-2R	0	3 1/2	17		3.92		350	18.2
PGAC	AC	3,4	1,2	1T-3R	1	3 3/8	13		4 1/2	48	350	20
Schlumberger	SL	3	1,3	1T-3R	1	4	23	6	6	12 1/4	350	20
	BHC	3	2	2T-4R	2	3 5/8	27		6	12 1/4	400	20
	SLS-J	3,5	—	1T-2R	0	1 11/16	16		3	12	350	20
	LSS	7	2	2T-4R	2	3 5/8			6		350	20
	LSS	10	2	2T-4R	2	3 5/8			6		350	20
	SLS-S	8,10	2	2T-2R	2m	3 5/8	18	4	6	18	350	20
	SDT-A	3,5,8,10	2	2T-10R	2m	3 5/8	41	3	5	—	350	20
Welex	24A	3	1,2,3	1T-3R	1	3 5/8	21		4 3/5	12	325+	20
	CAVL	3	1	2T-2R	2						350	20
	307	4	2	2T-2R	2						350	20

*Comp: 0 - no borehole comp (T-R time), 1 - differential in one direction across receiver pair, 2 - differential effectively in two directions across receiver pair(s), m - done through memorization.

gamma ray (SWS). Proper borehole measurement position for an acoustic tool is centered in the borehole.

Vertical resolution depends on receiver-to-receiver span and is typically 24 inches for 2 ft span conventional BHC, and 6 inches for sidewall acoustic log. Horizontal resolution is 100 percent of the circumference of the borehole.

The approximate depth of investigation depends upon the transmitter-receiver spacing. For conventional BHC it is very shallow, only a few inches.

Appendix D

DENSITY TOOL SPECIFICATIONS

A list of density tools, normal operating conditions (Figures D-1 and D-2), and a chart of tool specifications (Table D-1) follow. Much older information presented was obtained from the Reference Handbook of Selected Formation Evaluation Tools (out of print editions 1966 and 1972) by the Lafayette Chapter of the Society of Professional Well Logging Analysts.

Company	Trade name
Birdwell	Density/Borehole Compensated Log
Dresser Atlas	Compensated Densilog
	Z-Densilog
Gearhart	Compensated Density Log
Lane-Wells	Densilog
McCullough	Compensated Density Log
PGAC	Compensated Gamma Gamma Density Log
Schlumberger	Compensated Formation Density Log
	Litho-Density Log
Welex	Compensated Density Log
	Spectral Density Log

Good results are obtained when using density tools in common borehole fluids (fresh water, salt water, oil base, oil emulsion, air, or gas). Density tools typically reach a 6-inch depth of investigation. The proper borehole measurement position is eccentric, against the borehole wall. Measurements which can be made simultaneously are: caliper, gamma ray, microresistivity (Dresser and SWS), compensated neutron, linear neutron (Welex), spontaneous potential, induction-electric, dual induction, spectral natural gamma ray (SWS), and photo-electron index (SWS, Dresser, and Welex).

Vertical resolution is typically 18 inches. Horizontal resolution is 12 percent of the circumference of the borehole, depending on the borehole diameter.

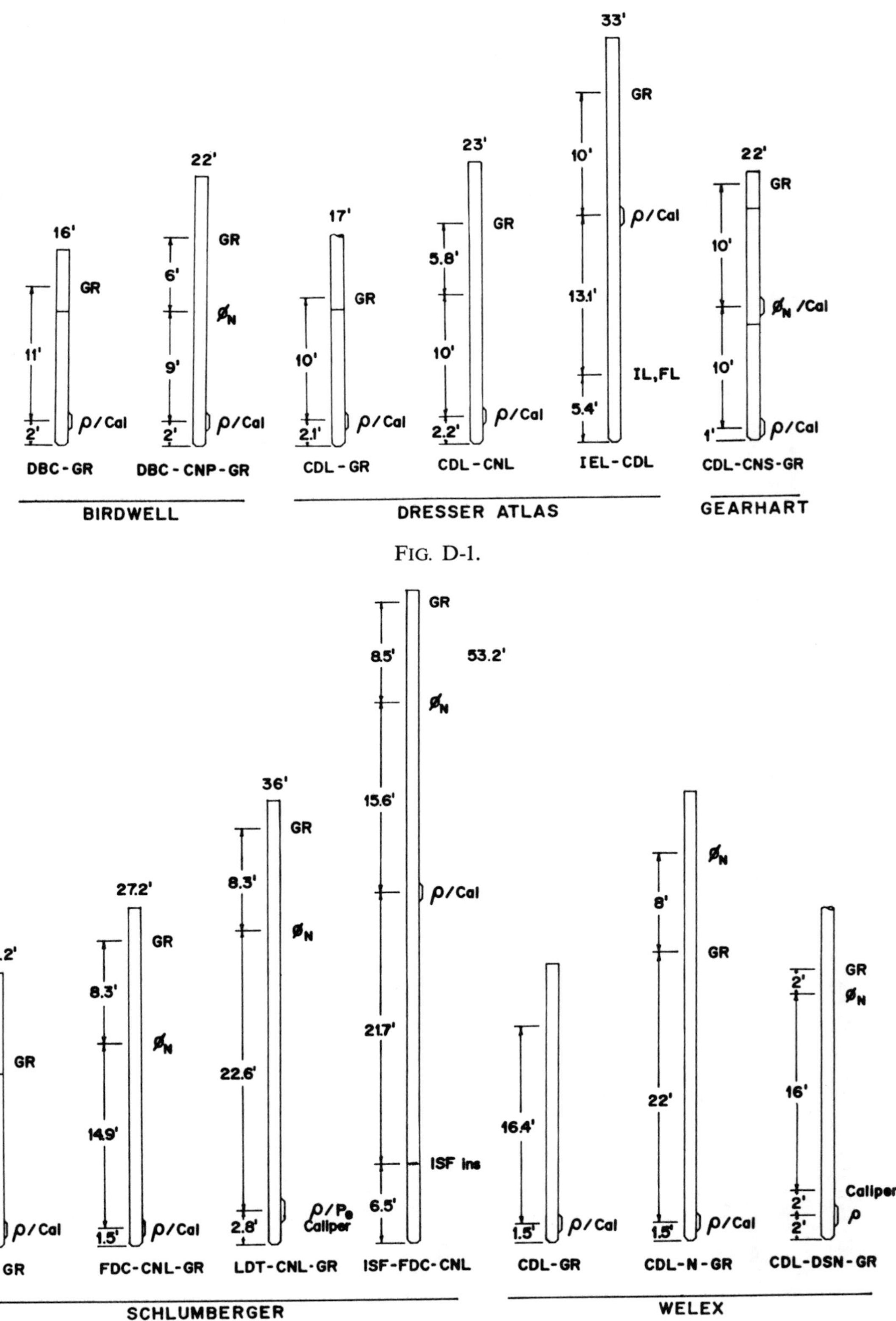
16'
GR
11'
2'
ρ/Cal
DBC-GR
22'
GR
6'
ϕ_N
9'
2'
ρ/Cal
DBC-CNP-GR
BIRDWELL
17'
GR
10'
2.1'
ρ/Cal
CDL-GR
23'
GR
5.8'
10'
2.2'
ρ/Cal
CDL-CNL
33'
GR
10'
ρ/Cal
13.1'
IL,FL
5.4'
IEL-CDL
DRESSER ATLAS
22'
GR
10'
ϕ_N /Cal
10'
1'
ρ/Cal
CDL-CNS-GR
GEARHART
22.2'
GR
13.5'
1.5'
ρ/Cal
FDC-GR
27.2'
GR
8.3'
ϕ_N
14.9'
1.5'
ρ/Cal
FDC-CNL-GR
36'
GR
8.3'
ϕ_N
22.6'
2.8'
ρ/P_e
Caliper
LDT-CNL-GR
GR
8.5'
53.2'
ϕ_N
15.6'
ρ/Cal
21.7'
ISF ins
6.5'
ISF-FDC-CNL
SCHLUMBERGER
16.4'
1.5'
ρ/Cal
CDL-GR
ϕ_N
8'
GR
22'
1.5'
ρ/Cal
CDL-N-GR
GR
2'
ϕ_N
16'
Caliper
2'
2'
ρ
CDL-DSN-GR
WELEX

FIG. D-1.

FIG. D-2.

Table D-1. Density Tool Specifications

Company	Model	Borehole Comp	Tool Diameter (Inches)	Length (Ft/Inches)		Hole diameter min (Inches)	Hole diameter max (Inches)	Temp rating (°F)	Max pressure (PSI) in thousands)
Birdwell	D/BC	yes	4 1/2	11		6	16	250	20
	8001	—	2 1/4	10	3	3	6	250	17
	6001	—	3 5/8	9		4	10	250	10
Dresser Atlas	2213	yes	3	13	9	4 1/2	16	400	20
		yes	3	12	6			400	20
	3411 (w/ML)	yes	4 3/4	11	5	6 1/2	16	400	20
	2212	yes	4 7/8	10	10	6	16	400	20
		yes	4 7/8	12				400	20
		yes	5 1/4	11	1			400	20
Gearhart	CDL-D	yes	4	8	9	5	22	350	20
	CDL-S	yes	2 3/4	10		4	10	350	15
GO International	DL	—	3 1/2	6	2	4 1/2		300	17
	CDL	yes	4	8	5	5		300	20
Lane-Wells	2203B		5 1/2	15	w/GR	6	16	300	20
McCullough		—	3 1/2	11		3.92		350	18.2
PGAC	E		4 7/8	9	6	6	16	400	20
Schlumberger	PGT-A	—	4 1/8	15	6	5	16	350	20
	PGT-E	yes	4 3/8	18		6	21 1/2	350	20
	LDT-C	yes	4 1/2	21		6	22	350	20
Welex	121	—	4	25		5 1/4	19	350	20
	125	yes	4	27		5 1/4	19	400	20
	126	yes	4					400	20
	127B	yes		21	w/GR	4 1/8	20	400	20

Appendix E

BOREHOLE GEOPHONE TOOL SPECIFICATIONS

A list of borehole geophone tools, their standard operating requirements, and a chart of tool specifications (Table E-1) follow.

Trade name	Tool name
Birdwell/SSL	Velocity Survey
Dresser Atlas	Seismic Logging Services
GO International	Velocity Survey
SRS	Velocity Well Geophone
Schlumberger	Well Seismic Tool
	Seismic Acquisition Tool

Good results are obtained in common borehole fluids (fresh water, salt water, oil base, oil emulsion) but where air or gas is encountered a special tool may be required. The proper borehole measurement position is against the borehole wall (eccentric).

The Schlumberger SAT simultaneously measures microresistivity and geophone orientation.

Table E-1. Borehole geophone tool specifications.

Company	Model	Geophone description	Measure components	Coupling arms	Tool Dia. (In)	Length (Ft)	Length (In)	Weight (lbs)	Hole Diameter min (In)	Hole Diameter max (In)	Temp rating (°F)	Max pressure (PSI in thousands)
Birdwell/SSC	Wellock	six 7.5 Hz HSI Model K	V	1	3 5/8	8	6	133	5	16	392	18
	HTH-1000	nine 8 Hz Geospace three per axis	3	1	3 5/8	12	8	275	4 5/8	16	430	25
CGG	Geolock H	Sensor SM4		2	4	5	6	225			356	17
Dresser Atlas		nine 14 Hz three per axis	3	1	2 5/8	12	0	220	4 5/8	16	400	20
GO International	GS				3 3/8	6	6		4 1/2		350	15
SIE (Geosource)	SWC-2C	three 14 Hz SM-6	V	1	3 7/8	4	5	120			392	20
	SWC-3C	three 14 Hz SMG-HT one per axis	3	1	4 1/2	4	11	130			392	20
SRS	L1-BU	three 20 Hz Mark Products	V	none	3 5/8	5	0	120	4		500	20
Schlumberger	WST	four 10 Hz Geospace HSI	V	1	3 5/8	16	3	275	5 1/2	20	350	20
	SAT	three 10 Hz Sensor SM-4 one per axis	3	1	4	11	0	155	5 1/2	19	400	20

Appendix F

CONVERSION TO SI UNITS

Well logs are commonly recorded in a mixture of english and metric units. Listed below are common well log units and their conversions to SI metric units.

Measurement	Well log unit	Conversion factor	SI unit
Acoustic traveltime	μs/ft	× 3.2808 =	μs/m
Acoustic velocity	ft/s	× 0.3048 =	m/s
Barrels (oil)	bbl	× 0.158 987 3 =	m^3
Borehole diameter		× 2.54 =	cm
(caliper)	inches	× 2.54 =	mm
Conductivity	mmhos/m	× 0.001 =	S/m
Density	g/cm^3	—	kg/m^3
Depth	ft	× 0.3048 =	m
Pressure	psi	× 6.894 757 =	Pa
Resistivity	Ω•m	—	Ω•m
Temperature	°F	273.15 + [(F-32)/1.8)]	K

INDEX